Applying Earth Observation Data for Urban Land-Use Change Mapping

Applying Earth Observation Data for Urban Land-Use Change Mapping

Guest Editor

Jūratė Sužiedelytė-Visockienė

MDPI

Basel • Beijing • Wuhan • Barcelona • Belgrade • Novi Sad • Cluj • Manchester

Guest Editor
Jūratė Sužiedelytė-Visockienė
Geodesy and Cadastre
Department, Environmental
Engineering Faculty
Vilnius Gediminas Technical
University
Vilnius
Lithuania

Editorial Office
MDPI AG
Grosspeteranlage 5
4052 Basel, Switzerland

This is a reprint of the Special Issue, published open access by the journal *Land* (ISSN 2073-445X), freely accessible at: https://www.mdpi.com/journal/land/special_issues/2FG28799PL.

For citation purposes, cite each article independently as indicated on the article page online and as indicated below:

Lastname, A.A.; Lastname, B.B. Article Title. *Journal Name* **Year**, *Volume Number*, Page Range.

ISBN 978-3-7258-3785-4 (Hbk)
ISBN 978-3-7258-3786-1 (PDF)
https://doi.org/10.3390/books978-3-7258-3786-1

Contents

land

MDPI

Article

Modelling Impact of Urban Expansion on Ecosystem Services: A Scenario-Based Approach in a Mixed Natural/Urbanised Landscape

Fatemeh Mohammadyari [1,*], Ardavan Zarandian [2], Mir Mehrdad Mirsanjari [1], Jurate Suziedelyte Visockiene [3] and Egle Tumeliene [3]

1 Department of Environmental Sciences, Malayer University, Malayer 65719-95863, Iran
2 Research Center for Environment and Sustainable Development (RCESD), Department of Environment, Tehran 19395-4697, Iran
3 Department of Geodesy and Cadaster, Vilnius Gediminas Technical University, Sauletekio av. 11, LT-10223 Vilnius, Lithuania
* Correspondence: m.fatima.1364@gmail.com

Abstract: The present study aims at predicting future land use/land cover (LULC) and quantifying and mapping the ecosystem services (ESs) of water yield, outdoor recreation opportunity and food production in current (here, 2017) and future landscapes in Northern Iran, using the InVEST, Recreation Opportunity Spectrum (ROS) and yield models. To that end, two LULC scenarios known as business as usual (BAU) and protection-based (PB) plan were applied for 2028, using the Markov Artificial Neural Network and Multi-objective land allocation (MOLA) models. The results show that rapid urbanisation, caused by the expansion of human settlements and industrial areas, has led to a decline in the ESs in the region. Compared to the ESs in 2017, the service of water yield increases as urban expansion increases, whereas food production and recreation services decrease as urban expansion increases, under the BAU scenario. On the other hand, in the PB scenario, relatively better conditions can be observed for all three ESs. Considering that the ecological structures of this region have been severely affected by rapid urban expansion, the results of this research will be useful for maintaining the existing ESs and can greatly affect planning and decision-making regarding future development towards urban sustainability.

Keywords: land use scenario; ecosystem services; Markov Artificial Neural Network; MOLA model; urban expansion; Karaj landscape

Citation: Mohammadyari, F.; Zarandian, A.; Mirsanjari, M.M.; Suziedelyte Visockiene, J.; Tumeliene, E. Modelling Impact of Urban Expansion on Ecosystem Services: A Scenario-Based Approach in a Mixed Natural/Urbanised Landscape. *Land* **2023**, *12*, 291. https://doi.org/10.3390/land12020291

Academic Editor: Jianjun Zhang

Received: 12 December 2022
Revised: 6 January 2023
Accepted: 17 January 2023
Published: 19 January 2023

1. Introduction

As a functional and dynamic unit of the biosphere, an ecosystem consists of living organisms and the physical environment in which interaction and material exchanges are witnessed [1]. Ecosystems, with the goods and services they provide, underlie all aspects of human, cultural, social and economic well-being [2] and provide an important material basis for development [3]. However, they are under threat from anthropogenic drivers including climate change, urbanisation and agricultural intensification and expansion [4], which have serious effects on nature (which provides such goods and services [5]). Meanwhile, since changes in ecological parameters are not tangible to most humans and development decision-makers, ecosystem services (ESs) are used to link ecological or biophysical changes to economic and social consequences [6].

ESs refer to delivering services to human beings through special ecosystems under appropriate ecological situations and functional and structural integrity [7]. ESs provide a large number of satisfaction-related advantages to human beings, such as well-being, selection, social relationships, personal security, social sustainability and human needs. Thus, ESs are indispensable to human health and sustainable development [8] and comprise four categories: provisioning services, regulating services, supporting services and cultural services [5].

Urban expansion, resulting from the development of human activities such as industrialization and urbanization, reduces the construction of ecological and natural infrastructure of a region and, thus, causes a decrease in the provision of ESs [9,10]. In this regard, urban expansion refers to a dynamic process, growing towards the surrounding natural regions [11], that leads to land use/land cover (LULC) change and the alteration of ecosystems and their services [12]. Due to human activities, especially in terms of land use or land management change, ESs undergo significant variation [13]. The changes in ESs reflect the impacts of anthropic activities on the ecological environment and directly or indirectly affect patterns, processes and ecosystem functions [14].

Therefore, investigating the effects of urban development on ESs has become an urgent and significant task in the better realisation of urban ecology and achieving urban stability and sustainable urban development [15,16]. In addition, evaluating and predicting the current and future condition of ES supply based on different plausible scenarios can assist decision-makers [17] in taking efficient measures to deal with environmental issues in a more informed way [18].

In a wide variety of studies, possible scenarios have been programmed based on local LULC policies or only focused on improving one particular ES [19]. In this regard, scenario-based modelling [20] and mapping [18] of multiple ESs has aroused widespread attention, leading to specific research aimed at implementing the concept of ESs [21]. In this research, due to the proximity of natural and urban areas where the supply and demand of ESs are exchanged, this approach was performed in a mixed urban/natural landscape.

In recent years, the evaluation and modelling of urban ecosystem services have also been taken into consideration. In most studies, regulating services (carbon sequestration and air pollution removal services) were the most common subjects [22–26]. In the meantime, a small number of studies were associated with food provision, recreation services and water yield. Considering the importance and priority of these services for city residents, this research focused on their modelling. In this sense, water yield is critical in water resource management [27] and the health and well-being of the urban population [28]. Outdoor recreational opportunity is also an important factor in human well-being [29]. Food production is vital to ensuring food safety and urban stability [30]. Thus, quantifying and modelling these services are essential in urban planning [31].

However, studies that have been performed on the effects of urban expansion on these services (food provision, outdoor recreational opportunity and water yield) are still insufficient. In this regard, Reference [16] investigated the effects of urban expansion on several ESs in a metropolitan area. It was found that food supply services, habitat quality, carbon storage and soil conservation decrease with increased urbanisation and water service increases as urbanisation increases. Similarly, Reference [32] modelled urban expansion and its relationship with multiple ESs (recreation opportunity, carbon storage, biodiversity conservation) in the Wuhan metropolitan area. They found that, to a large extent, global urban expansion causes the destruction of ESs and changes in relation to exchanges and synergies. Reference [33] found that the services of food production, water supply, raw materials, air quality and climate regulation are greatly affected by urbanisation, while recreation and habitat quality are less affected by urbanisation. By reviewing various studies, it was found that, in different parts of the world, especially the areas that face unplanned and unstable urban growth, natural and ecological structures and assets have been severely damaged. Consequently, people's livelihoods and well-being have been threatened due to the lack of supply of ecosystem goods and services.

A wide variety of methods has been used to model important ESs in urban ecosystems. For example, the Integrated Valuation of Ecosystem Services and Trade-offs (InVEST) water yield model has been extensively used by water resource managers to model the hydrological balance [34,35]. Moreover, the most common methods used in modelling recreation services are the InVEST model [36,37] and Recreation Opportunity Spectrum

(ROS) [12,38–40]. Eventually, various factors are involved when investigating food production, yield quality, nutritional value and the amount of land [41]. Therefore, food supply quantification has been performed in different ways, including the use of statistical data of the agricultural yield [42], the comprehensive evaluation of agricultural suitability by indicators [43], harvested energy [44] and a yield model [19].

Although ES studies in Iran have recently focused on urban ecosystems, these studies are limited and in their early stages. Meanwhile, in more advanced countries, the quantification of urban ESs has received much attention.

The dominant composition of the recent Karaj landscape has been green cover and it is known as a 'garden city'. The proximity of the Karaj metropolis to the capital city and communication highways to the west and south of the country has caused increased growth of the city's area and population density (beyond its capacity). The natural resources available have been dramatically degraded because of the rapidly growing population that migrated to this city due to rapid industrialisation. This means that ecological structures have been threatened, causing damage to the sustainability of the urban environment, so that the continuity of citizens' lives and their social well-being has decreased. In fact, the city of Karaj, near the capital of Iran, suffers from development instability, especially in terms of damage to its infrastructure and ecological assets. In this sense, urban expansion, LULC change, destroyed urban green spaces, recent droughts, the characteristics of climate seasonality, lack of rainfall, increased evapotranspiration, reduced water supply and increased water stress have all greatly reduced the capacity of ESs in this area. Accordingly, this city was chosen as a case study. Herein, we believe that the application of the ESs approach, as an integrated economic-social and ecological strategy, can greatly affect planning and decision-making regarding future developments towards urban sustainability. This investigation is the first attempt at quantifying and assessing ESs in the Karaj landscape, which has been experiencing dramatic urban change and intense environmental change.

In this research, the LULC map was initially generated using satellite imagery data of the Karaj landscape in 2006, 2011 and 2017. Afterwards, considering two different scenarios, an LULC simulation map was prepared for 2028. In this sense, the present study aims to predict future LULC by quantifying and mapping the ESs of water yield, outdoor recreation opportunity and food production in current and future landscapes. Comparing the results can contribute to proposing the most effective policy related to landscape-cover changes and sustainable urban development policies, achieving a sustainable city in the study area.

2. Materials and Methods

2.1. Study Landscape

The Karaj landscape (35°46′–36°09′ N, 50°46′–51°21′ E) covers an area of 117,520 ha in the east of Alborz Province, Iran. Karaj city is the fourth-largest city in Iran and the first populated city in Alborz Province, with a population of 1,592,492 people, according to the 2017 census. Moreover, about 96.2% of the population live in urban regions, with 3.8% living in rural areas (https://amar.org.ir/english, 20 January 2018). This region is mountainous, with an average altitude of 1300 m above sea level. The average annual rainfall and temperature are 247 mm and 14.4 °C, respectively. According to the Köppen–Geiger classification system, the Karaj metropolis is categorised as having a cold semi-arid climate. To better examine ESs and the influencing mechanism of urbanisation, the case study area was divided into two districts: upstream (ES supply locations) and downstream (ES demand locations) landscapes. In this regard, the upstream landscape is mainly covered by mountains, forest, grassland and gardens, which have a high potential for ecotourism. Regarding Alborz Province, there are two important water sources: the Karaj River and the Amir Kabir Dam; the Central Alborz Protected area is also in this district. The Karaj River is the most important waterway in the Alborz province and Karaj metropolis. Furthermore, most of the water in this river is transferred to Tehran city after being restrained in the Karaj dam. In addition, despite massive water production in the Amir Kabir Dam, a small part of it is consumed in the Karaj landscape but most of it is allocated to agriculture and drinking purposes in Tehran province. In addition, the popular Chalous Road—one of the busiest recreational roads—connects the district to the north of Iran. It is worth mentioning that the Karaj metropolis and agricultural land are located in the downstream landscape (Figure 1).

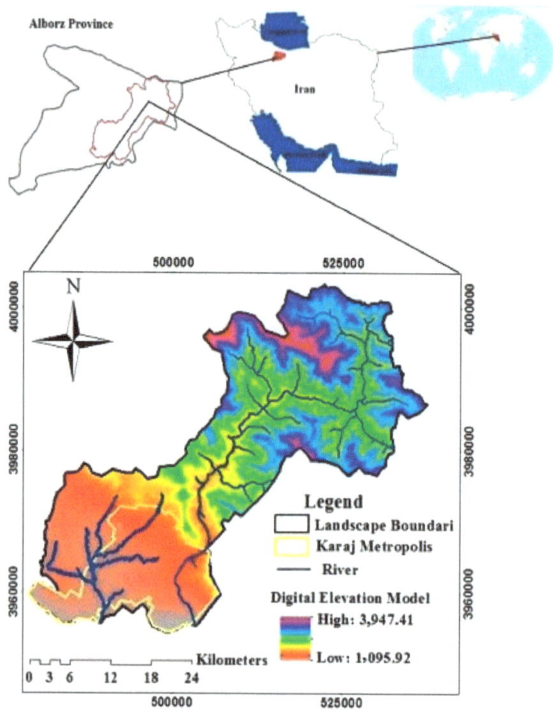

Figure 1. Location of the study area.

2.2. Data Sources and Methods to Quantify the ESs

In this study, three ESs were selected: water yield, food production and outdoor recreation opportunity. The selection of ESs was based on their importance in the region, data availability, interviews with local stakeholders, the knowledge of scientists from different disciplines and research centres and literature reviews for the ESs in an urban context [23,45]. The InVEST software tool was applied to model water yield [16,33], food production was evaluated with the Sun and Li yield model [19], and the ROS model was used to quantify outdoor recreation opportunity [37].

In addition, the Markov Artificial Neural Network (ANN)—Multi-objective land allocation (MOLA) model was used to model the future LULC (year 2028) in the first and second scenarios. The first scenario was defined on the basis that the process of LULC change will continue, compared to previous years, without any restrictions. On the other hand, the definition of the second scenario was based on the intervention of the government to prevent the destruction of agricultural land to stop the current process of LULC change. The second scenario, based on the protection of natural assets, was defined by assuming the existence of limits for the unsustainable development of human and industrial infrastructure and, if implemented, it is expected to change the current unsustainable development process towards gradual sustainability.

The sources and input data for ES assessment and LULC modelling are presented in Table 1. Moreover, the general structure of the study is indicated in Figure 2.

Table 1. The input data and sources for ES assessment and land use modelling.

Data Types	Model	Input Data	Description
LULC	Support Vector Machines (SVM) Markov Chain MOLA	LULC data	Landsat 5 and Landsat 8 satellite images were downloaded for 2006, 2011 and 2017 from the United States Geological Survey (www.usgs.gov, 1 June 2006, 2011 and 2017).
		Digital elevation model	Aster satellite
Provisioning Service Water yield	InVEST model [46]	LULC map	A GIS raster dataset with an LULC code for each cell
		Precipitation (mm)	A GIS raster dataset with a non-zero value for average annual precipitation for each cell.
		Average annual reference evapotranspiration (mm)	A GIS raster dataset with an annual average evapotranspiration value for each cell.
		Root restricting layer depth (mm)	A GIS raster dataset with an average root restricting layer depth value for each cell.
		Plant available water (PAWC)	A GIS raster dataset with a plant available water content value for each cell.
		Watersheds	One polygon per watershed (shape file).
		Sub-watersheds	A shape file with one polygon per sub-watershed within the main watersheds specified in the Watersheds shape file.
		Biophysical Table	Tables of LULC classes, including data on biophysical coefficients used in this tool.
		Demand Table	A table of LULC classes showing consumption water use for each LULC type.
Food production	Yield model [19]	LULC map	A GIS raster dataset with an LULC code for each cell.
Cultural Service Outdoor recreation opportunity	ROS model [37,38,47]	Cultivated land Number of fruit-producing tress	Statistical information on agricultural and garden products obtained from the Statistics Centre of Iran.
		Natural area	Degree of naturalness: Hemeroby index
		Water component	Water bodies (extracted from LULC) River (extracted from LULC)
		proximity	Road network Urban areas (extracted from LULC)

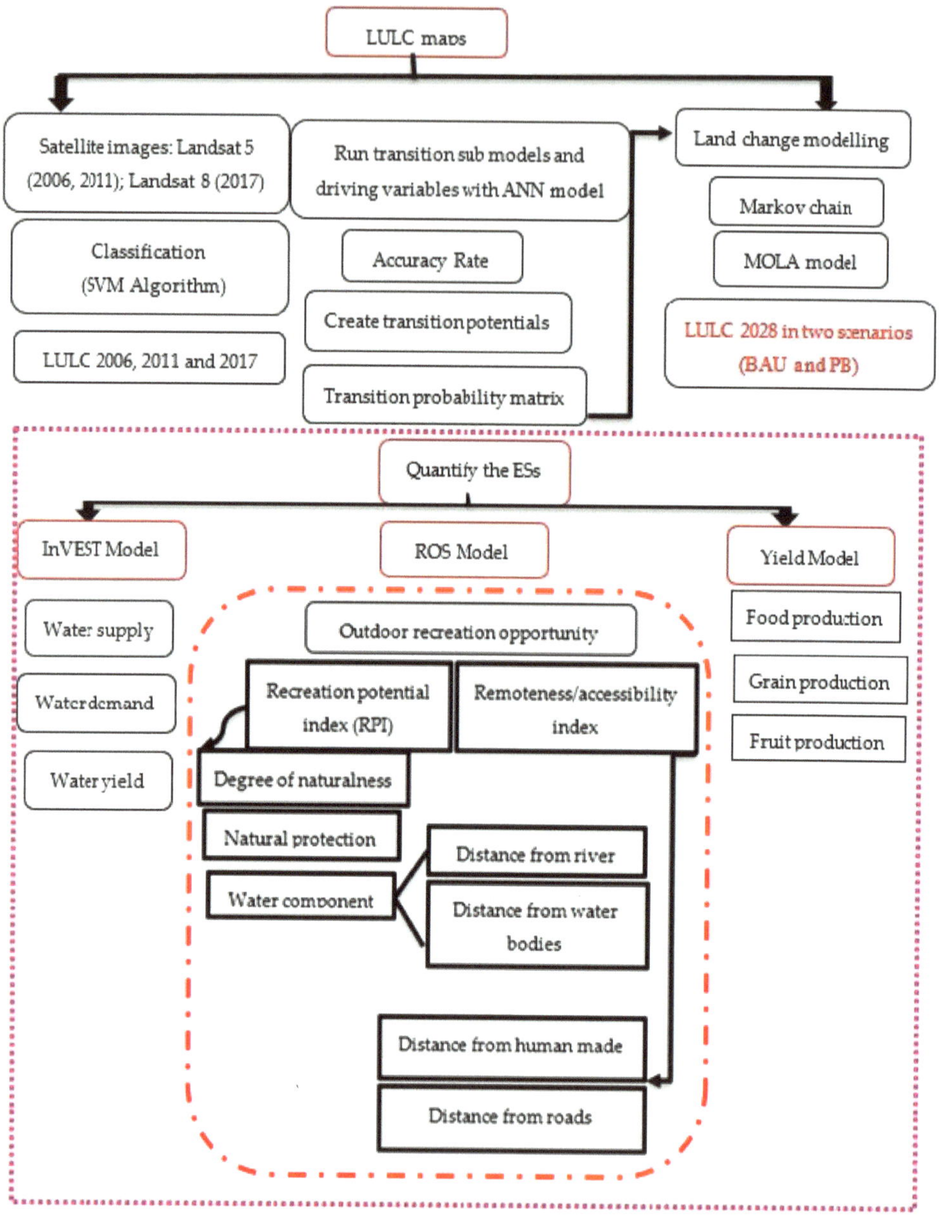

Figure 2. The general structure of the models used for urban expansion simulation and quantifying ESs.

2.3. *LULC Changes and Future Scenarios in the Karaj Landscape*

The Landsat images were classified using the Support Vector Machines (SVM) algorithm [11,48,49] in the ENVI 5.3 software. The generated LULC were categorised into ten classes: Human-made (including rural settlements, urban areas, industrial land and mining land); Agriculture (all areas used for crop production); Garden; Water bodies (the deep and shallow waters of the Amir Kabir Lake dam); Low dense grassland (including sparse vegetation surface); Dense grassland (including moderate to good vegetation and shrubs);

Barren (including uncovered, unutilised and barren land); Rocky outcrop; Greenspace (parklands); and River (Karaj River).

In order to confirm the accuracy of the classification, the kappa standard, kappa location, kappa no. and FoM in the IDRISI VALIDATE module were calculated. FoM is a number between 0 and 1, indicating complete overlap (1) and no overlap (0) between the simulated and real maps, respectively. FoM was obtained using Equation (1) [50].

$$\text{FoM} = \frac{\text{Hits}}{\text{Misses} + \text{Hits} + \text{False Alarms}} \tag{1}$$

where Hits denotes the correct pixels where land use change has occurred in the observed and simulated data; Misses means the pixels that were fixed in the simulated data, although in the observed data they have changed; and False Alarms are errors that the model predicts changed but did not do so in observation.

2.3.1. Future Scenarios in the Karaj Landscape

Business as Usual Scenario (BAU)

This scenario was created based on a recent investigation [51] that considered the continuation of the present transformation trend of natural and green covers (agriculture, garden and grassland) to human-made areas. The comparison of the current status (2017) and base map (2006) indicated that agricultural lands have been dramatically reduced and rapidly replaced by human-made classes. In this sense, Markov and ANN models were applied to simulate LULC in 2028, using the LCM tool in IDRISI Terrset 16.3 software [52–56]. The transition potential of each LULC was modelled using ANN. The output of this step was used as the input for the Markov model. The Markov chain model is a stochastic process model that explains how likely one state is to change into another. In addition, the transition probability matrix is created from the Markov chain analysis in the LCM model (for more details, see [57]). The combination of ANN and Markov models is a robust approach that can be used to successfully simulate future urban expansion [58].

In the present research, the LULC map in 2017 was modelled using LULC maps from 2006 and 2011 and applying Markov and ANN models. The accuracy of this model was 90% compared to the real LULC map in 2017. Afterwards, the LULC map in 2028 was modelled based on the maps from 2006 and 2017, utilising the integrated method.

Protection-Based Scenario (PB)

This scenario modelled the interference of governmental conservation policies towards preserving agricultural and garden lands that are at risk of being converted to human-made areas. Applying this scenario will prevent the conversion of 1316.97 ha of agricultural lands and 161.42 ha of garden lands to built-up areas. To implement this scenario, MOLA was used to simulate land use in 2028. This method was conducted to dedicate new land use transfer and predict variations [59]. It is worth mentioning that MOLA authorises the utilisation of suitability maps, according to Multi-Layer ANN, to help divide the amount of variation predicted by Multicriteria decision analysis (MCDA) in various LULC classes [60]. Moreover, the main purpose of using the MCDA method is to provide a basis for evaluating a number of alternative electoral possibilities, based on multiple criteria [59]. Furthermore, an Analytical Hierarchy Process (AHP) method was used to determine the weight of the criteria [59] and the Weighted Linear Combination (WLC) method was used to overlay maps [61,62].

2.4. *Assessing ESs*

2.4.1. Water Yield

The model was developed using InVEST software, based on precipitation, storage and evapotranspiration data [38]. In fact, this model shows how variation in LULC patterns influences the annual water yield [12]. Regarding the InVEST model, evapotranspiration is the main parameter in computing the water yield depth under the assumption that precipitation is constant [34]. Evapotranspiration is interceded by transpiration through plants [63] and the shading effect of vegetation cover also changes heat fluxes in the soil consequently leading to a reduction in evaporation [34,64]. The amount of rainwater permeating into the watershed's subsoil and groundwater is calculated through soil depth the available water content for plants and root depth [29]. The calculation of annual water yield is based on Equation (2).

$$\mathrm{Yx} = (1 - \frac{\mathrm{AETx}}{\mathrm{Px}}).\ \mathrm{Px} \tag{2}$$

where Y(x), AET(x) and P(x) refer to water yield, the annual actual evapotranspiration of grid unit x and the annual precipitation amount of grid unit x, respectively. Moreover, AET(x)/P(x) represents vegetation evapotranspiration [35]. In order to calculate the mean annual evapotranspiration, the 'Modified Hargreaves' equation [17] was applied, using:

- Daily average, maximum and minimum temperatures.
- Mean daily maximum and minimum differences.
- Extra-terrestrial radiation.

All information was acquired from the Karaj synoptic meteorological station. The data were collected from the Iranian Meteorology Organization's daily database (http://irimo.ir/far/, 17 June 2017). The factors related to the soil, such as plant available water and soil depth [46], were not accessible and so, accordingly, we used data from the Harmonized World Soil Database (HWSD) provided by FAO [17,65]. Moreover, plant available water was obtained from HWSD and then divided by soil depth in order to determine the plant available water content (PAWC) fraction throughout the landscape. PAWC is the fraction of water content in the soil profile that is available for plants [46]. Notably, the model assumptions are based on the processes at watershed and sub-watershed scales. The whole landscape was considered to be one large watershed containing five sub-watersheds. Regarding each LULC class, plant evapotranspiration coefficients (Kc) were computed based on the existing coefficients in the relevant literature: Human-made = 0.10; Agriculture = 0.65; Garden = 0.70; Low dense grassland = 0.80; Dense grassland = 0.90; Barren = 0.50; Water bodies, Rocky outcrop, Greenspace and River = 1.00 [38].

The total water demand was calculated as the quantity of water consumption in agricultural land and human-made areas. Consumptive water in farmland was assumed to be the water used by agricultural activities that are not returned to the watershed. Considering the human-made area, water-use was calculated according to the water consumption per person, multiplied by the relative population density per square km and generalised to each pixel of the raster map with a resolution of 30 m [1]. Water consumption was also calculated according to the water demands for each type of crop in the irrigation plan of the Karaj metropolis. Various biophysical variables for model implementation are represented in Table 1 and further information related to the water yield model can be found in the InVEST user guide [46].

2.4.2. Food Production

Food production was examined for the current situation (CU), the business as usual scenario (BAU) and the protection-based scenario (PB) in two steps, as described below.

Calculation of Food Production Using the Yield Model

In this step, the production of fruits and grains was considered in the area. Food production was calculated using the yield model (Equations (3) and (4)) [19].

$$\mathrm{PRO_G} = \sum_{i=1}^{i} \mathrm{A_i} \times \mathrm{R_{Gi}} \times \mathrm{P_{Gi}} \tag{3}$$

$$\mathrm{PRO_F} = \sum_{i=1}^{i} \mathrm{A_i} \times \mathrm{R_{Fi}} \times \mathrm{P_{Fi}} \tag{4}$$

where PRO_G and PRO_F indicate the production of fruits and grains, respectively. A_i is the area of district i in the Karaj landscape; R_{Fi} and R_{Gi} indicate the area proportions of fruits and grains in the range of i, respectively, and P_{Fi} and P_{Gi} indicate the yields of fruits and grains per area unit for each district, respectively.

Calculation of the Relevant Capacity of LULC Classes for Food Production

To calculate the present level of ES delivery produced by each LULC class, the viewpoints of thirty scientists were collected from various sectors of governmental institutions (five of these scientists were members of the Iranian Association for Environmental Assessment (http://www.iraneia.ir/en, 24 August 2018)). In this regard, a questionnaire was designed based on the opinions of Burkhard et al. [66,67]. The questionnaire included a matrix specifying the relationship or disaffiliation between each LULC class and food production service. Afterwards, each participant was asked to specify the disaffiliation or relationship between each LULC class and food production by signing 'Yes' or 'No' in a matrix, to identify the resources producing this ecosystem service. Subsequently, they gave a score for the effective resourcing of food supply in the range of 0 to 5: 0 = no capacity to supply the selected ESs; 1 = very low capacity; 2 = low capacity; 3 = medium capacity; 4 = high capacity and 5 = very high capacity. The scores were averaged and then the values of average capacity were transferred to Arc GIS software to prepare a food production map for current, BAU and PB scenarios for the study landscape.

2.4.3. Outdoor Recreation Opportunity

Outdoor recreation opportunities refer to an ecosystem's capacity [37] to provide outdoor recreation activities such as walking, running, outdoor sports and enjoyment in watching plants and animals. Therefore, the ROS model was conducted to determine the outdoor recreation service. The model simulates recreation opportunities provided by nature at a local level, as categorised in the ESs cascade [68]. This method was undertaken to assess outdoor recreation in the European Union [69]. Recreation potential covers three essential aspects of people's behaviour and preferences for outdoor activities [47]. This model is estimated according to two indicators: the recreation potential index (RPI) and the remoteness/accessibility index (RAI).

The first component of the RPI relates to people's preferences for more natural regions and concerns the degree of naturalness [12,19,38]. The second component of the RPI refers to the protected regions, as they indicate a high natural value [70], and the third one is the attractiveness of water bodies [71]. The degree of naturalness is simulated according to the Hemeroby index [38], which determines the human effect on vegetation and landscape, ranging from 1 (natural) to 7 (artificial). Moreover, the degree of naturalness was obtained for each LULC class [37,72,73]. The protected natural regions are scored based on the management classes of the International Union for Conservation of Nature (IUCN) [74]. The protected areas were classified by focusing on their significance for recreation aims in the range between 1 (with the highest natural value) and 0 (the lowest). In this regard, the protected areas were mapped using information from the database of the Department of Environment of Iran. Based on the database, there are two protected areas (the Central Alborz Area and Karaj River) in the study area.

The water landmark is considered a natural key factor for leisure and recreation activities [75]. The attractiveness of water bodies was determined by measuring the distance to all surface water bodies [38]. Two layers of distance from the river and from water sources were prepared with a buffer function. The layers were then standardised using the decreasing linear fuzzy method (Table 2).

Table 2. The shape of membership functions and control points of different factors.

Criteria	Shape of Membership Functions	Control Points			
		a	b	c	d
Distance from River	Decreasing linear	-	-	500	1000
Distance from Water bodies	Decreasing linear	-	-	30	2000
Distance from Human-Made	Decreasing linear	-	-	2000	5000
Distance from Roads	Decreasing linear	-	-	500	5000

Moreover, in this method the remoteness/accessibility index (RAI) or access to the recreation sites was examined [19]. The RAI was determined by applying the proximity analysis in the ArcGIS toolbox to compute the straight distance from roads and human-made areas with a buffer function [12,19]. Table 2 shows the control points and the selected membership function.

The final RAI map was prepared according to the parameters presented in Table A1 (Appendix A) [47]. Notably, this method was slightly modified to augment the precision of spatial distributions of recreation potential. The final ROS was obtained by combining the RPI and RAI, based on the parameters in Table A2 (Appendix B).

3. Results

3.1. LULC in CU Situation and Two Scenarios

Kappa indexes (K standard = 0.92; k location = 0.94 and k no. = 0.95) and FoM coefficient (0.89) indicate satisfactory results of the Markov chain and MOLA models in LULC modelling. The results confirm that these models are suitable for modelling urban expansion. From 2017 to 2028, the proportion of different classes was as follows (from the highest to the lowest, respectively): dense grasslands, rocky outcrop land, low dense grassland, human-made, garden, agriculture, rivers, and green space. It is worth mentioning that water bodies and barren land had the lowest proportion of LULC class in the Karaj landscape. As shown in Figure 2, the BAU scenario, agricultural, garden and low dense grassland increased in the downstream landscape and, in some parts, they were converted to human-made land. On the contrary, in the upstream landscape, dense grassland decreased, which can be related to overgrazing. According to the BAU scenario from 2017 to 2018, agricultural lands, gardens, dispersed grasslands and dense grassland decreased, while agricultural land and gardens experienced the highest reduction. Based on this scenario, it was expected that the area of agricultural lands would experience a reduction from 6511.86 ha to 5199.85 ha (i.e., a 20% reduction of agricultural lands in the next 11 years). Furthermore, the garden areas will be reduced by 202.38 hectares. According to this scenario, the development of human-made areas will continue to increase in the future. Thus, the area of this class will increase by 29%, from 14,478.66 ha to 18,729.64, compared to that of 2017 (CU situation).

In the PB scenario, applying efficient management practices will prevent the transformation of gardens and farmland to the human-made class (conserving 1273.31 ha of agricultural lands and 161.42 ha of gardens). In this scenario, human-made areas are still increasing because of high demand for settlements due to a high population, but the intensity of the increase is less than that of the BAU scenario. In general, regarding the current scenario, 12, 13 and 14% of the land area are under the human-made, agricultural and water body classes, respectively.

The modelling results indicate that, in the PB and BAU scenarios, 10% and 12% are agricultural and garden lands, and the remaining 76% and 73% are covered by natural land cover and water ecosystems, respectively. However, the spatial distribution of the landscape elements is different for each scenario (Figure 3). Moreover, the findings showed that the most significant LULC change was manifested by urban growth and garden, grassland and agricultural land reduction in the Karaj landscape during 2017–2028 (Table 3).

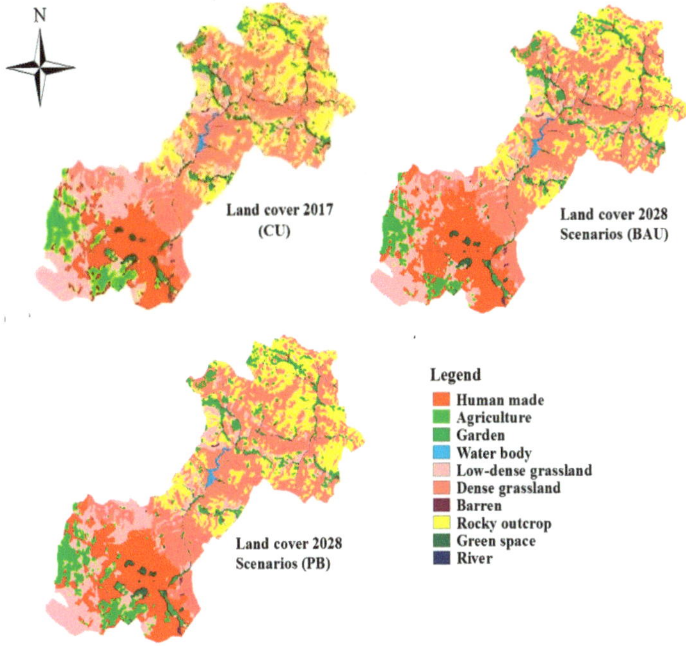

Figure 3. Spatial distributions of LULC in the Karaj landscape under CU and different scenarios.

Table 3. Changes in LULC in the Karaj landscape under each scenario.

Land Use Type	Area (ha)		
	CU	**BAU**	**PB**
Human-made	14,478.66	18,729.64	17,630.64
Agriculture	6511.86	5199.85	6473.16
Garden	7036.67	6834.29	6995.71
Water body	326.97	326.97	325.5
Low dense grassland	22,584.96	20,197.93	19,899.67
Dense grassland	40,167.09	39,810.59	39,774.65
Barren	227.43	227.43	226.63
Rocky outcrop	24,393.78	24,401.78	24,402.09
Green space	802.42	801.36	801.79
River	990.36	990.36	990.36

3.2. Changes to ES Flows

3.2.1. Water Yield

The outputs of the InVEST water yield model were the volume of water yield in the sub-watershed scale and the estimated water yield per pixel. One of the outputs is in the form of a shapefile and a table containing biophysical output values per sub-watershed and the other is in the form of a raster map. In CU and the two scenarios, the ESs were assessed at the sub-watershed scale and the results of the water balance are presented in Table 4. The results show that the average water balance and the amount of available water will be enhanced in all scenarios, since green areas and natural cover transformation (low dense grassland, dense grassland) lead to decreasing evapotranspiration. The increase in water yield due to the reduction in evapotranspiration is not a positive result for water but it signifies the sudden release of water due to the destruction of grassland and gardens, instead of being stored in plant tissue and then being gradually released throughout the year.

Table 4. The effects of changes in precipitation and evapotranspiration on landscape water yield in CU and two scenarios.

	Average Precipitation (mm)	Average Evapotranspiration (mm)	Water-Related (Million m³/Year)		
			Water Yield	Water Supply	Water Demand
CU	523.21	288.08	235.10	−105.62	340.72
BAU	523.19	282.58	240.61	−106.98	347.59
PB	523.19	283.75	238.41	−108.26	346.67

According to the model predictions, although the water yield would increase, the water demand outweighs the water resource due to the excessive water consumption in the region. This may result in increased access to groundwater aquifers, subsequently leading to a decrease in groundwater storage. The PB scenario indicates a minimum decrease in water supply potential, signifying more efficiency in terms of water consumption compared to the BAU scenario. Moreover, regarding the BAU scenario, the highest enhancement in water demand can follow. Due to the arid and semi-arid climatic conditions in the area, on the one hand, and the rapid development of man-made areas, on the other, a water supply deficit is witnessed in the current situation and, also, under both future scenarios. However, under the PB scenario, despite maintaining the existing farms, the demand for irrigation of the protected farms is reduced compared to the BAU scenario; this indicates the greater efficiency of our conservation (PB) scenario for optimal water management in the area. On the other hand, the amount of food production in farms also increases, which is an advantage for all stakeholders.

The volume of water production and other related variables in the five water basins in the Karaj landscape are shown in Table 5.

Table 5. The volume of water production and other related variables in the water basins in the Karaj landscape.

Sub-Basin Code	1	2	3	4	5
Area (ha)	76,114.00	963.18	3131.216	34,540.50	2771.277
Average Precipitation (mm/year)	573.62	407.49	407.44	431.46	454.05
Evapotranspiration (mm/pixel)					
Potential	550.91	127.31	273.22	450.4	597.59
Actual	338.77	92.26	139.98	198.6	245.32
Water yield volume (million m³/year)					
CU	234.7	315.2	267.5	232.9	208.7
BAU scenario	233.3	331.9	277.3	253.1	208.7
PB scenario	234.6	314.4	269.4	247.4	208.7
Water supply (million m³/year)					
CU	121.1	−868.8	−813.9	−539.0	134.4
BAU scenario	119.8	−666.1	−698.1	−556.7	131.5
PB scenario	120.0	−871.6	−830.5	−610.3	134.3
Water demand (million m³/year)					
CU	113.6	1184.0	1081.4	771.9	74.3
BAU scenario	113.5	998.0	975.4	809.8	77.2
PB scenario	114.6	1186.0	1099.9	857.7	75.4

In CU and the two scenarios, sub-basins 1 and 5 (possessing more than 120 million m^3/year) and sub-basins 2 and 3 (with 1000 million m^3/year) indicate the highest water supply and demand, respectively. In the BAU scenario, the water demand in watersheds 4 and 5 will increase by up to 38 million m^3/year and 3 million m^3/year, respectively. In watershed 1, the conditions will be the same as those of the CU situation, indicating the least effects on water-related ESs due to LULC transformation. Moreover, other watersheds show a decrease in water demand. In the PB scenario, the water supply potential is the same as the CU situation; however, the water demand will increase in all watersheds, especially in 3 (from 1081 to 1099 million m^3/year) and 4 (from 771 to 857 million m^3/year), because of the increasing agricultural activities.

Figure 4 shows the pixel-based (30 × 30 m) map of the water balance of the ecosystem in the study area. Water balance is higher in all three scenarios in the upstream landscape because the rainfall in these areas is greater than the evapotranspiration. In the northwest part of the upstream area, the water yield increases between 2017 and 2028. Significantly, in the central part of the study area, the water yield is low and the majority of change occurs in the central and downstream parts of the study area. The water yield decreases downstream (from 2017 to 2028) because of the transformation of vegetation cover to other LULC classes with higher water demand and, also, due to the fact that evapotranspiration is higher and precipitation is lower than the upstream areas. The results indicate that the volume of water produced in each of the three scenarios is 1259, 1304 and 1274 million m^3/year, respectively. Accordingly, in 2028, the volume of water production in the BAU and PB scenarios will increase by 45 and 15 million m^3/year, respectively.

3.2.2. Food Production

Food production showed various distribution patterns. In this sense, grain planting was only provided by the downstream landscape in the rather flat and fertile areas. In the west and southwest of the region, there is the least human habitation. Fruit production was mainly provided by the upstream landscape (Figure 4). Accordingly, from 2017 to 2028, grain production decreased in both the BAU and PB scenarios by 19% and 0.6%, respectively. Total grain production in the Karaj landscape will decrease in both BAU and PB scenarios. In this regard, grain production is 110,024.95, 88,170.88 and 109,271.36 tons per hectare for the CU situation and BAU and PB scenarios, respectively. During the 2017–2028 period, the cultivated lands decrease in the BAU and PB scenarios by 1312.01 and 38.7 ha, respectively. The expansion of urban and industrial areas of the downstream landscape is the reason for this decrease. The crop yield also decreases, mainly due to the reduction in cultivated land. Considering the fruit production, an increasing trend is seen in the upstream landscape (from 56,602.34 to 57,588.55 ton/ha), while in the downstream landscape—from the CU situation to the BAU scenario—a significant reduction is witnessed (from 1259.22 to 1015.15 ton/ha) and, in the PB scenario, it reaches the highest value (1361.53 ton/ha). In general, the total fruit production in the Karaj landscape decreases in the BAU scenario and increases in the PB scenario. Fruit production in the CU situation is 57,861.56 ton/ha and for the BAU and PB scenarios, it could be 57,774.38 and 58,950.91 ton/ha, respectively. During the 2017–2028 period in the upstream region, the area of garden land reduces in the BAU and PB scenarios by up to 74.15 and 70.23 ha, respectively. On the contrary, regarding the downstream area, the area of garden land experiences a reduction in the BAU scenario (from 930.7 to 802.47 ha) but a specific increase in the PB scenario (from 930.7 to 959.97 ha) (Table 6). The reason for this decrease is rooted in the massive villa construction and road expansion in the upstream area and the urbanisation of the downstream area. In general, the results indicate that the amount of food production will only improve under the PB scenario.

Table 6. Changes in food production under different scenarios.

Food	Landscape District	Area Agriculture and Garden Land (ha)			Production (Ton/ha)		
		CU	BAU	PB	CU	BAU	PB
Grains (Agriculture land)	upstream	0	0	0	0	0	0
	downstream	6511.8	5199.8	6473.1	110,024.9	88,170.88	109,271.3
Fruits (Garden land)	upstream	6105.9	6031.8	6035.7	56,602.3	56,759.23	57,588.5
	downstream	930.7	802.4	959.9	1259.2	1015.15	1361.5

3.2.3. Outdoor Recreation Opportunity

Figure 5 shows the percentage of land within each ROS category. It can be seen that categories 9 (high provision–not easily accessible) and categories 3 (low provision–not easily accessible) are the largest and smallest areas, respectively. From 2017 to 2028, the value of the outdoor recreation opportunity index in the BAU scenario decreased, but in the PB scenario it increased.

Despite no considerable variation in recreation potential, this service represented clear spatial heterogeneity. Additionally, recreation services decreased from the upstream landscape to the downstream landscape because in the upstream landscape there are natural and semi-natural habitats such as water bodies, grassland, lakes, etc. It is worth mentioning that most of this area is mountainous with forest cover. Moreover, regions with a low population and high distance from the roads and urban areas always presented higher values of outdoor recreation opportunity. However, this service became worse in the downstream landscape. The rapid growth of the human-made class decreased recreational opportunity. The parts of the downstream landscape with green spaces showed a high recreation opportunity (Figure 4). According to the results, the PB scenario would be more favourable than the BAU scenario.

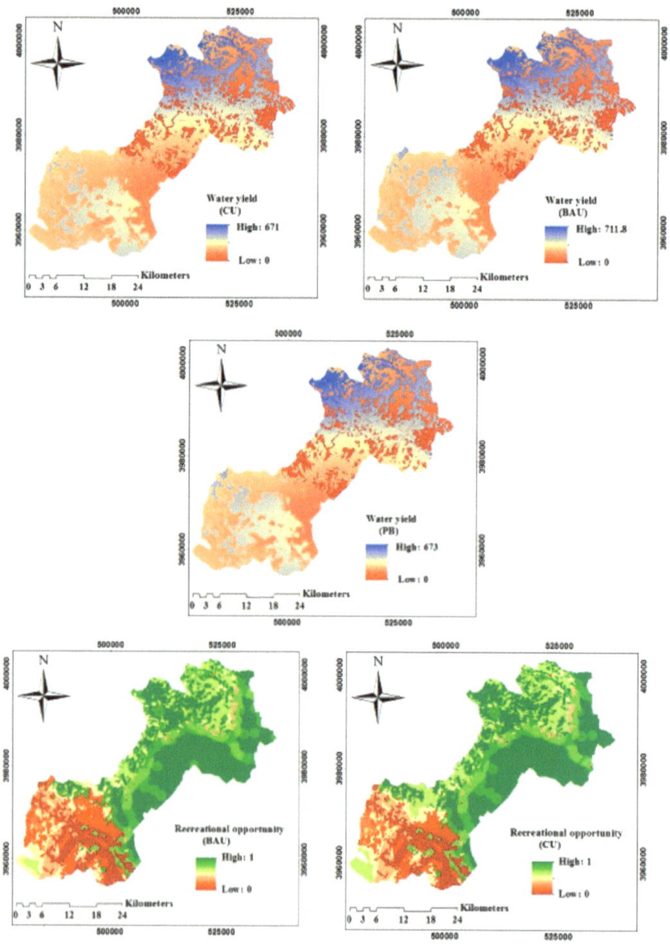

Figure 4. *Cont.*

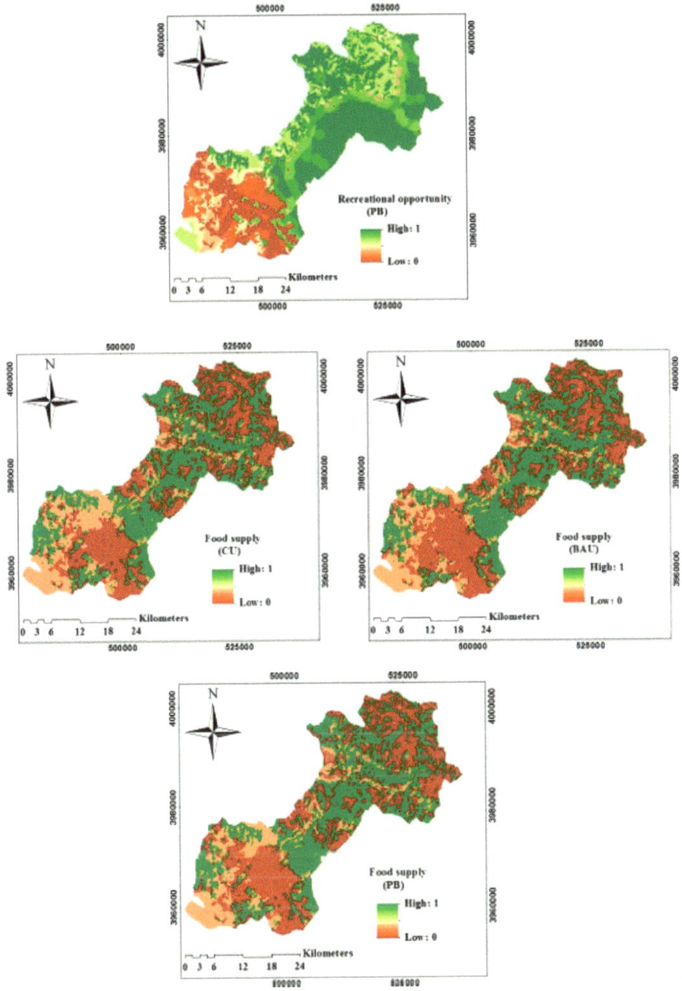

Figure 4. Spatial patterns of the ESs in the Karaj landscape.

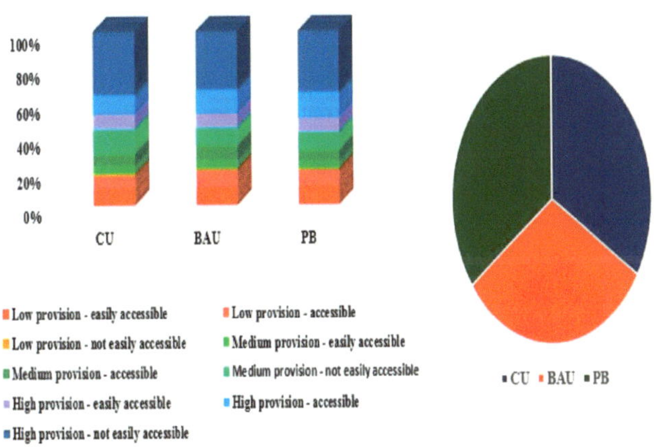

Figure 5. Percentage of land within each ROS category and the indicator value of outdoor recreational opportunity index in the Karaj landscape.

4. Discussion

4.1. Changes to LULC under Scenarios in the Future

Scenario simulation is one of the essential methods for predicting ES delivery [76], which can provide recommendations for LULC planning scenarios [77]. Therefore, in this paper, the effects of LULC changes were investigated on multiple ES delivery and supply under the current situation and two other scenarios in the future (BAU and PB scenarios).

There are some limitations in previous studies. Some researchers, such as [78,79], did not perform scenario analysis, which usually leads to different results. Additionally, in the studies that considered scenario analysis, such as Gao et al. [80], they did not study LULC effects on ESs in future scenarios.

The development of human-made settlements and infrastructure, due to urbanisation, will be the main LULC change in the Karaj landscape under both scenarios in 2028. Another important change is the reduction in agricultural and garden areas in the BAU scenario and their increase in the PB scenario. This augmentation can be attributed to lower anthropogenic pressure on the agricultural, garden and grassland areas as a result of better management practices followed by all stakeholders in the PB scenario. Furthermore, rapid economic development and population growth in the Karaj metropolis significantly changes the LULC.

4.2. Spatial Distribution of ESs in the Karaj Landscape

4.2.1. Spatial Distribution of Water Yield

In the InVEST model, the raster output is used to interpret and identify areas with water balance, as compared to areas facing water stress. Considering the problems of drought and lack of water output, our model indicates that, in the current situation, a major part of the downstream area suffers from water stress. By implementing the protection scenario and preserving the existing natural assets, it can be hoped that the water stress situation in the studied area will not worsen. Based on the yield, supply and demand of water in the sub-basins (Table 5), the water balance status of the Karaj landscape can be classified into three categories: balanced, water stressed and critical. In the CU situation and two scenarios, sub-basins 1 and 5 are in the balanced class, sub-basin 4 is in the water-stress class, and sub-basins 2 and 3 are in the critical class. Sub-basin 1 is dominated by green cover, mountainous and blue areas located in the upstream landscape, and sub-basins 2, 3 and 4 are dominated by agricultural lands and urban areas in the downstream area, which demand the most water. Overall, due to the topographic conditions [26,81,82] and LULC coverage [83–87], the water balance in the upstream landscape is higher than in the downstream area.

The water yield in the central parts of the downstream area is higher than other parts. This includes Karaj city and urban areas. The expansion of human-made areas increases the impervious region and, accordingly, a large amount of precipitation is collected and stored in the form of runoff, which decreases land surface evaporation and results in a high water yield depth in the urban area [88]. Therefore, in this section, urban growth was conducive to the augmentation of water yield [34]. In general, regarding the Karaj landscape, the destruction and conversion of green cover (such as agriculture and gardens) can be seen but the rate of this conversion is higher in the downstream region. The result of these LULC changes leads to evapotranspiration reduction, consequently increasing water yield [33,84] and floods [89] in the arid region of Karaj. In total, LULC change has a little positive impact on water yield but a powerful negative impact on water demand, water supply and, consequently, water stress. The positive effect of LULC change on water yield is mainly via decreased evapotranspiration and vegetation. This result is also compatible with the results of studies performed by [17,89–91]. On the other hand, rapid urban expansion and population growth have led to an increased water demand. In fact, population change has caused a change in water supply and demand, its increase leading to an increase in water demand. Therefore, in the Karaj landscape, despite the large supply of water, there is no balance between supply and demand. The mismatch between water supply and demand limits sustainable development and the economy of the region [79]. In this regard, given that most water resources in the region have been used more than the existing capacity, the need to pay attention to water resources and their management is very important for informing decision- and policy-makers regarding regional water security.

4.2.2. Spatial Distribution of Food Production

In terms of food supply, large amounts of fruit and grain production were dispensed in the upstream and downstream. According to the results (Table 6), although the cultivation level of garden products in the upstream decreases by 74.15 and 70.23 ha in 2028 compared to 2017, production in both scenarios is higher than the base year. Increasing garden products, despite reducing the area under cultivation of crops, indicates an increase in production per unit area and a productivity improvement. Generally, compared to the ESs in 2017, food production will decrease under the BAU scenario but will increase under the PB scenario in 2028. According to the different results of food production in the two scenarios, it can be argued that food production is dependent on intact natural LULC [92,93]. However, food production was also affected by other factors, such as the amount of land, yield, food quality, soil properties, climate, management practices and agricultural technology [19]. Herein, the model shows that grain production will decrease in the Karaj landscape by 2028. Due to the fact that there is no possibility of agricultural activities in the upstream areas, downstream agricultural activities must be developed and the preservation of these lands must be accompanied by strict management to ensure food security in the region. [87] argued that grain production decreases due to the degradation of agricultural land, particularly in developing countries. Consequently, managing the relationship between food security, urban construction and ecological conservation is vital in all countries [94].

4.2.3. Spatial Distribution of Outdoor Recreation Opportunity

The classification of LULC percentage in each ROS class (Figure 4) is based on the two factors of provision and accessibility. According to this classification, the easier the access and the more the provision, the higher the value of the recreational potential of the area would be. The total area percentage of these two categories in the region is 21%, 19% and 23% in the CU situation and BAU and PB scenarios, respectively. Based on the ROS map (Figure 5), the highest outdoor recreation opportunities were observed in the northeast, east and southeast parts and a few parts of the northwest upstream, in both future scenarios. The dominant coverage of these areas is natural habitats, water bodies and river areas, gardens and dense grassland. On the other hand, considering the downstream area of the region in both future scenarios, areas with high recreational value are distributed in parts of the north, southeast, southwest and a few parts of the centre that are urban parks. Although the spatial distribution of this service is the same in both future scenarios, the area of high recreational value is much more distributed in the PB scenario than in the BAU scenario. Overall, the identified recreation area is almost fine for both scenarios, despite obvious spatial heterogeneity. However, in the BAU scenario, compared to the PB scenario, the Karaj landscape would have lower recreational opportunities.

Outdoor recreation opportunities were highly influenced by topography. Due to its mountainous nature and other natural attractions in the upstream landscape, there are more recreation opportunities than downstream, which is located at lower elevations and is predominantly covered by the human-made class. Therefore, in confirmation of previous studies [12,19,40], outdoor recreation opportunities are affected by LULC changes.

4.3. *The Impact of Urban Expansion on ESs*

LULC is a determinative variable causing changes in ESs which are affected by the intensity of variations in the composition, patterns and structure of LULC [95]. When compared to the ESs in 2017 (CU situation), the water yield increases as urban expansion increases, whereas food production and recreational services decline as urban expansion increases, under the BAU scenario in 2028. On the other hand, in the PB scenario, water yield, food production and recreational opportunities increase.

Regarding the continuation of unsustainable urban development through extensive conversion of natural land cover to man-made uses, the ecological infrastructure of the studied metropolis has been threatened. This is because the management of natural assets, such as grassland, gardens, etc., is not in a favourable condition. Under these conditions, even with the implementation of protection scenarios, it is a difficult task to control LULC changes in the areas that are facing increasing demand for the creation of human and industrial infrastructures. However, in this study, it was shown that by adopting the conservation scenario, even if it is not possible to completely continue the current trend towards more sustainable LULC changes, it can be hoped that by preserving existing natural assets, the rate of deterioration of ESs and natural functions will be reduced as much as possible. Our modelling shows the PB scenario with fewer adverse consequences for ESs, indicating that the protection scenario is an approach that can establish a more sustainable form of urban development, compared to the scenario of the continuation of the existing situation in the future time perspective (year 2028).

Previous studies have demonstrated that the increase in the water yield due to LULC change, high precipitation and urban expansion [16,19,26,34,84,87] decreases outdoor recreation opportunity due to urban expansion and LULC change [39,40,96], as well as food production due to urban expansion [1,30,43,76,97].

In the current research, it was found that the rapid and wide variation of LULC types, urban expansion and the absence or weak implementation of sustainable urban LULC planning methods cause some specific changes to the spatial distribution of ESs. Undoubtedly, these changes will cause the Karaj landscape to move from having an ecological function to a sociological one and reduce the sustainability of ecological networks [98]. Therefore, by adopting effective management policies and creating a balance between ecological and sociological functions, it is possible to achieve the sustainability of the urban ecosystem and, consequently, reduce the effects of urban expansion on ESs. In this sense, it is worth mentioning that urban expansion is one of the most significant reasons for changes in ES functions (increasing water yield and reducing outdoor recreation opportunity and food production) in the Karaj landscape. These results validate the results of previous investigations [7,12,16,17,24,26,99,100]. Significantly, the increased water yield, which can increase nutrient transport and flood risk [16], will have negative effects on urban sustainability and social well-being. On the other hand, water balance and the regional water cycle (which are vital in urban areas [101]) have been disrupted in the Karaj landscape due to excessive water consumption. Therefore, planners and managers of water resources should implement appropriate strategies and policies to deal with water supply deficiencies in the Karaj landscape. In addition, recreation potential is important for citizens' well-being [102]. Thus, due to the reduction of this service in the Karaj landscape, and in order to improve the quality of the recreational and mental comfort in society, urban expansion strategies should increasingly aim to strengthen the natural and artificial network of green spaces by planting more trees, increasing the diversity of plant combinations and properly managing green spaces. Finally, it is worth mentioning that food can be regarded as a basic material for a good life [15]. The lack of food production will further deteriorate food safety in the Karaj landscape. Considering that there is no possibility for agriculture in the upstream areas and the reduction of the area of these lands, along with the garden lands in the downstream areas (especially the eastern and western parts), a significant decrease will be experienced in the capacity of the land to provide food for the city's residents. Thus, urban planners must provide appropriate protection and sustainable plans for these uses.

Urbanisation leads to the spatial heterogeneity of ESs as a result of urban growth and population settlement [30]. Therefore, regarding the urban growth process in the Karaj landscape, the loss of ESs that were affected by LULC changes can negatively influence urban sustainability and, consequently, human well-being. However, this negative effect in the PB scenario was significantly reduced. It is due to the fact that, in this scenario, future management policies and land use planning were implemented. In this regard, to achieve favourable welfare, the planners and managers of Karaj should end the continuation of the current process by defining more informed protection and management scenarios, such as our protection scenario.

4.4. Limitations

In this study, there are restrictions in the InVEST model and the ROS model that result in less accurate parameters. These tools were applied due to the limited available data and information for the case study area. Moreover, InVEST models have the potential to beget a basic underpinning for decision- and policy-making [103]. Although the InVEST annual water supply does not contain the precipitation patterns all over the year, it overrates runoff and neglects groundwater recharge [37]. Moreover, the limitation of the stout testing restricts their credit [103]. The ROS method focuses on the recreational activities of natural infrastructure (usually the water component) but overlooks the recreational value of artificial substructures in the human-made areas [19]. Therefore, these limitations can lead to wrong approximations of ES supply [38]. Furthermore, we were able to support the accuracy of our estimations with the results provided in the literature and the experimental data, arranged to downgrade the inaccuracies. In this study, we did not consider the demand for recreation and food due to a lack of data. We suggest that the demand for these ESs be considered in future studies.

5. Conclusions

The research presented proposes a framework to evaluate spatial-temporal variations of multiple ESs in the Karaj landscape from 2017 to 2028. Models which are used to quantify ESs in Iran, although previously used for natural areas, have not yet been used in the urban landscape. Therefore, this study can form a good model to highlight key research priorities in urban ESs and can be considered as a basic study in urban landscapes for other researchers in Iran.

According to the results, fast urban growth will lead to a kind of decline in provisioning and support services in the Karaj landscape in the future—since urban expansion will destroy the grassland, farmlands and gardens. In contrast, water yield increases and water balance is disrupted in the region because of the process of urban expansion. In this regard, changes in water yield increase in most of the Karaj landscape, while outdoor recreation opportunity and food production mainly occur in the downstream areas and have spatial consistency with urban expansion.

Herein, recommendations were provided for urban planners in order to manage ESs. Generally, urban planners should take measures to improve food production in the downstream, especially in the western parts; outdoor recreation opportunities in the downstream areas, especially in the central region; and water supply in the central region and the western and northwestern parts of the downstream. Hence, regarding the process of fast urbanisation, the government must efficiently protect the sustainable capacity of ESs and logically plan for LULC, especially in places where the urban ecosystem is adjacent to natural lands and changes the natural structures in the appearance of Karaj's landscape.

Scenario analysis shows that the PB scenario has a better performance in providing ESs. Consequently, urban ESs are spatially heterogeneous within the region, vary in relation to different scenarios, and highlight important issues in ESs, amounting to sustainable urban expansion policies and schematisation. Overall, the results presented in this study justify further research to ensure multiple ESs in the Karaj landscape. In addition, the present study gives a comparative result to areas having a similar scale. Finally, it is hoped that the results presented in this study will help managers and urban planners to make sensible decisions in ecosystem management and achieve urban sustainability. Regarding future research, we will present the effects of urban expansion on ESs in the Karaj landscape in several conservation and management scenarios, to provide urban planners with guidelines for urban sustainability and balanced development.

Author Contributions: All authors (F.M., A.Z., M.M.M., J.S.V. and E.T.) contributed equally to the study, conception and design of this paper. Material preparation, data collection and analysis were performed by all authors. F.M. wrote the manuscript, and F.M., A.Z. and J.S.V. performed revisions All authors (F.M., A.Z., M.M.M., J.S.V. and E.T.) contributed ideas to the study. All authors have read and agreed to the published version of the manuscript.

Funding: This research received no external funding.

Data Availability Statement: The datasets used or analysed during the current study are available from the corresponding author on reasonable request.

Acknowledgments: The authors of this article would like to thank the Department of Environmental Sciences at Malayer University, Research Centre for Environment and Sustainable Development (RCESD), Tehran and the Department of Geodesy and Cadaster, Vilnius Gediminas Technical University, for their support.

Conflicts of Interest: The authors declare no conflict of interest.

Appendix A

Table A1. Remoteness/accessibility (proximity) parameters.

		Distance From Roads (KM)			
		<1	1–5	5–10	>10
Distance from human-made areas (KM)	<5	1	2	2	4
	5–10	2	2	2	4
	10–25	3	3	3	4
	25–50	3	4	4	4
	>50	4	4	4	5
1	Neighbourhood				
2	Proximity				
3	Almost far				
4	Remote				
5	Very remote				

Appendix B

Table A2. Parameters for the recreational opportunity spectrum.

			Recreation Potential Index (RPI)		
			1 <0.19	2 0.19–0.25	3 >0.25
Remoteness/accessibility (proximity)	1	Neighbourhood	1	4	7
	2	Proximity	1	4	7
	3	Almost far	2	5	8
	4	Remote	3	6	9
	5	Very remote	3	6	9
1 Low provision—easily accessible 2 Low provision—accessible 3 Low provision—not easily accessible 4 Medium provision—easily accessible 5 Medium provision—accessible 6 Medium provision—not easily accessible 7 High provision—easily accessible 8 High provision—accessible 9 High provision—not easily accessible					

References

1. Das, M.; Das, A. Estimation of ecosystem services (EESs) loss due to transformation of Local climatic zones (LCZs) in Sriniketan-santiniketan planning area (SSPA) West Bengal, India. *Sustain. Cities Soc.* **2019**, *47*, 101474. [CrossRef]
2. Wood, S.L.; Jones, S.K.; Johnson, J.A.; Brauman, K.A.; Chaplin-Kramer, R.; Fremier, A.; Girvetzg, E.; Gordon, L.; Kappel, C.; Mandle, L.; et al. Distilling the role of ecosystem services in the Sustainable Development Goals. *Ecosyst. Serv.* **2018**, *29*, 70–82. [CrossRef]
3. Zhai, T.; Zhang, D.; Zhao, C. How to optimize ecological compensation to alleviate environmental injustice in different cities in the Yellow River Basin? A case of integrating ecosystem service supply, demand and flow. *Sustain. Cities Soc.* **2021**, *75*, 103341. [CrossRef]
4. Connor, J.D.; Summers, D.; Regan, C.; Abbott, H.; Van Der Linden, L.; Frizenschaf, J. Sensitivity analysis in economic evaluation of payments for water and carbon ecosystem services. *Ecosyst. Serv.* **2022**, *54*, 101416. [CrossRef]
5. Millennium Ecosystem Assessment (MEA). *Ecosystems and Human Well-Being: Current State and Trends*; Island Press: Washington, DC, USA, 2015.
6. Sharon, O.; Fishman, S.N.; Ruhl, J.B.; Olander, L.; Roady, S.E. Ecosystem services and judge-made law: A review of legal cases in common law countries. *Ecosyst. Serv.* **2018**, *32*, 9–21. [CrossRef]
7. Xie, Z.; Li, X.; Chi, Y.; Jiang, D.; Zhang, Y.; Ma, Y.; Chen, S. Ecosystem service value decreases more rapidly under the dual pressures of land use change and ecological vulnerability: A case study in Zhujiajian Island. *Ocean Coast Manag.* **2021**, *201*, 105493. [CrossRef]
8. Balasubramanian, M. Economic value of regulating ecosystem services: A comprehensive global level review. *Environ. Monit. Assess.* **2019**, *191*, 616. [CrossRef]
9. Blanco, E.; Raskin, K.; Clergeau, P. Towards regenerative neighbourhoods: An international survey on urban strategies promoting the production of ecosystem services. *Sustain. Cities Soc.* **2022**, *80*, 103784. [CrossRef]
10. Athukorala, D.; Estoque, R.C.; Murayama, Y.; Matsushita, B. Ecosystem Services Monitoring in the Muthurajawela Marsh and Negombo Lagoon, Sri Lanka, for Sustainable Landscape Planning. *Sustainability* **2021**, *13*, 11463. [CrossRef]
11. Karimi, F.; Sultana, S.; Shirzadi Babakan, A.; Suthaharan, S. An enhanced support vector machine model for urban expansion prediction. *Comput. Environ. Urban Syst.* **2019**, *75*, 61–75. [CrossRef]
12. Sun, X.; Crittenden, J.C.; Li, F.; Lu, Z.; Dou, X. Urban expansion simulation and the spatio-temporal changes of ecosystem services, a case study in Atlanta Metropolitan area, USA. *Sci. Total Environ.* **2018**, *622–623*, 974–987. [CrossRef] [PubMed]
13. Gao, J.; Tang, X.; Lin, S.; Bian, H. The influence of land use change on key ecosystem services and their relationships in a mountain region from past to future (1995–2050). *Forests* **2021**, *12*, 616. [CrossRef]
14. Li, S.; He, Y.; Xu, H.; Zhu, C.; Dong, B.; Lin, Y.; Wang, K. Impacts of urban expansion forms on ecosystem services in urban agglomerations: A case study of Shanghai-Hangzhou Bay urban agglomeration. *Remote Sens.* **2021**, *13*, 1908. [CrossRef]
15. Zhang, D.; Huang, Q.; He, C.H.; Wu, J. Impacts of urban expansion on ecosystem services in the Beijing-Tianjin- Hebei urban agglomeration, China: A scenario analysis based on the Shared Socioeconomic Pathways. *Resour. Conserv. Recycl.* **2017**, *125*, 115–130. [CrossRef]
16. Wang, S.; Hu, M.; Wang, Y.; Xia, B. Dynamics of ecosystem services in response to urbanisation across temporal and spatial scales in a mega metropolitan area. *Sustain. Cities Soc.* **2022**, *77*, 103561. [CrossRef]
17. Zarandian, A.; Baral, H.; Stork, N.E.; Ling, M.A.; Yavari, A.R.; Jafari, H.R.; Amirnejad, H. Modelling of ecosystem services informs spatial planning in lands adjacent to the Sarvelat and Javaherdasht protected area in northern Iran. *Land Use Policy* **2017**, *61*, 487–500. [CrossRef]
18. Jopke, C.; Kreyling, J.; Maes, J.; Koellner, T. Interactions among ecosystem services across Europe: Bag plots and cumulative correlation coefficients reveal synergies, trade-offs, and regional patterns. *Ecol. Indic.* **2014**, *49*, 46–52. [CrossRef]
19. Sun, X.; Li, F. Spatiotemporal assessment and trade-offs of multiple ecosystem services based on land use changes in Zengcheng, China. *Sci. Total Environ.* **2017**, *609*, 1569–1581. [CrossRef]
20. Robinson, C.J.; Maclean, K.; Hill, R.; Bock, E.; Rist, P. Participatory mapping to negotiate indigenous knowledge used to assess environmental risk. *Sustain. Sci.* **2016**, *11*, 115–126. [CrossRef]
21. Zhao, C.; Sander, H. Assessing the sensitivity of urban ecosystem service maps to input spatial data resolution and method choice. *Landsc. Urban Plan.* **2018**, *175*, 11–22. [CrossRef]
22. Shoyama, K.; Kamiyama, C.; Morimoto, J.; Ooba, M.; Okuro, T. A review of modelling approaches for ecosystem services assessment in the Asian region. *Ecosyst. Serv.* **2017**, *26*, 316–328. [CrossRef]
23. Xie, V.; Huang, Q.; He, C.H.; Zhao, X. Projecting the impacts of urban expansion on simultaneous losses of ecosystem services: A case study in Beijing, China. *Ecol. Indic.* **2018**, *84*, 183–193. [CrossRef]
24. Mirsanjari, M.M.; Zarandian, A.; Mohammadyari, F.; Suziedelyte-visockiene, J. Investigation of the impacts of urban vegetation loss on the ecosystem service of air pollution mitigation in Karaj metropolis, Iran. *Environ. Monit. Assess.* **2020**, *192*, 501. [CrossRef] [PubMed]
25. Nie, X.; Lu, B.; Chen, Z.; Yang, Y.; Chen, S.; Chen, Z.; Wang, H. Increase or decrease? Integrating the CLUMondo and InVEST models to assess the impact of the implementation of the Major Function Oriented Zone planning on carbon storage. *Ecol. Indic.* **2021**, *118*, 106708. [CrossRef]

26. Ma, S.H.; Wang, L.J.; Zhu, D.; Zhang, J. Spatiotemporal changes in ecosystem services in the conservation priorities of the southern hill and mountain belt, China. *Ecol. Indic.* **2021**, *122*, 107225. [CrossRef]
27. Pessacg, N.; Flaherty, S.; Brandizi, L.; Solman, C.; Pascual, M. Getting water right: A case study in water yield modelling based on precipitation data. *Sci. Total Environ.* **2015**, *537*, 225–234. [CrossRef]
28. Gunnarsdottir, M.J.; Persson, K.M.; Andradottir, H.O.; Gardarsson, S.M. Status of small water supplies in the Nordic countries: Characteristics, water quality and challenges. *Int. J. Hyg. Environ. Health* **2017**, *220*, 1309–1317. [CrossRef]
29. Levrel, H.; Cabral, P.; Feger, C.; Chambolle, M.; Basque, D. How to overcome the implementation gap in ecosystem services? A user-friendly and inclusive tool for improved urban management. *Land Use Policy* **2017**, *68*, 574–584. [CrossRef]
30. Li, T.; Cui, Y.; Liu, A. Spatiotemporal dynamic analysis of forest ecosystem services using Big data: A case study of Anhui province, central-eastern China. *J. Clean. Prod.* **2017**, *142*, 589–599. [CrossRef]
31. Peng, J.; Tian, L.; Liu, Y.; Zhao, M.; Hu, Y.; Wu, J. Ecosystem services response to urbanisation in metropolitan areas: Thresholds identification. *Sci. Total Environ.* **2017**, *607–608*, 706–714. [CrossRef]
32. Sun, Y.; Liu, D.; Wang, P. Urban simulation incorporating coordination relationships of multiple ecosystem services. *Sustain. Cities Soc.* **2022**, *76*, 103432. [CrossRef]
33. Li, R.; Shi, Y.; Feng, C.C.; Guo, L. The spatial relationship between ecosystem service scarcity value and urbanisation from the perspective of heterogeneity in typical arid and semi-arid regions of China. *Ecol. Indic.* **2021**, *132*, 108299. [CrossRef]
34. Li, G.; Jiang, C.; Zhang, Y.; Jiang, G. Whether land greening in different geomorphic units are beneficial to water yield in the Yellow River Basin? *Ecol. Indic.* **2021**, *120*, 106926. [CrossRef]
35. Hu, W.; Li, G.; Gao, Z.; Jia, G.; Wang, Z.; Li, Y. Assessment of the impact of the Poplar Ecological Retreat Project on water conservation in the Dongting Lake wetland region using the InVEST model. *Sci. Total Environ.* **2020**, *733*, 139423. [CrossRef] [PubMed]
36. Cunha, J.; Elliott, M.; Ramos, S. Linking modelling and empirical data to assess recreation services provided by coastal habitats: The case of NW Portugal. *Ocean Coast. Manag.* **2018**, *162*, 60–70. [CrossRef]
37. González-García, A.; Palomo, I.; González, J.; López, G.; Montes, G. Quantifying spatial supply-demand mismatches in ecosystem services provides insights for land-use planning. *Land Use Policy* **2020**, *94*, 104493. [CrossRef]
38. Früh- Müller, A.; Hotes, S.; Breuer, L.; Wolters, V.; Koellner, T. Regional Patterns of Ecosystem Services in Cultural Landscapes. *Land* **2016**, *5*, 17. [CrossRef]
39. Morse, W.C.; Stern, M.; Blahna, D.; Stein, T. Recreation as a transformative experience: Synthesizing the literature on outdoor recreation and recreation ecosystem services into a systems framework. *J. Outdoor Recreat. Tour.* **2022**, *38*, 100492. [CrossRef]
40. Lavorel, S.; Rey, P.L.; Grigulis, K.; Zawada, M.; Byczek, G. Interactions between outdoor recreation and iconic terrestrial vertebrates in two French alpine national parks. *Ecosyst. Serv.* **2020**, *45*, 101155. [CrossRef]
41. Lopes, R.; Videira, N. Modelling feedback processes underpinning management of ecosystem services: The role of participatory systems mapping. *Ecosyst. Serv.* **2017**, *28*, 28–42. [CrossRef]
42. Baró, F.; Gómez-Baggethun, E.; Haase, D. Ecosystem service bundles along the urban-rural gradient: Insights for landscape planning and management. *Ecosyst. Serv.* **2017**, *24*, 147–159. [CrossRef]
43. Gong, J.; Liu, D.; Zhang, J.; Xie, Y.; Cao, E.; Li, H. Trade-offs/synergies of multiple ecosystem services based on land use simulation in a mountain-basin area, Western China. *Ecol. Indic.* **2019**, *99*, 283–293. [CrossRef]
44. Kain, J.H.; Larondelle, N.; Haase, D.; Kaczorowska, A. Exploring local consequences of two land-use alternatives for the supply of urban ecosystem services in Stockholm year 2050. *Ecol. Indic.* **2016**, *70*, 615–629. [CrossRef]
45. Mexia, T.; Vieira, J.; Príncipe, A.; Anjos, A.; Silva, P.; Lopes, N.; Freitas, C.; Reis, M.S.; Correia, O.; Branquinho, C.; et al. Ecosystem services: Urban parks under a magnifying glass. *Environ. Res.* **2018**, *160*, 469–478. [CrossRef]
46. Sharp, R.; Tallis, H.T.; Ricketts, T.; Guerry, A.D.; Wood, S.A.; Chaplin-Kramer, R.; Nelson, E.; Ennaanay, D.; Wolny, S.; Olwero, N.; et al. *InVEST 3.7.0 User's Guide*; The Natural Capital Project; Stanford University, University of Minnesota, The Nature Conservancy, and World Wildlife Fund: Stanford, CA, USA, 2020; Available online: https://invest-userguide.readthedocs.io/en/3.5.0/ (accessed on 1 December 2022).
47. Zulian, G.; Paracchini, M.L.; Maes, J.; Liquete, C. *ESTIMAP: Ecosystem Services Mapping at European Scale*; EUR 26474; Publications Office of the European Union: Luxembourg, 2013; p. JRC87585. [CrossRef]
48. Heydari, S.H.; Mountrakis, G. Meta-analysis of deep neural networks in remote sensing: A comparative study of mono-temporal classification to support vector machines. *ISPRS J. Photogramm. Remote Sens.* **2019**, *152*, 192–210. [CrossRef]
49. Alassery, F.; Alzahrani, A.; Khan, A.; Irshad, K.; Kshirsagar, S.R. An artificial intelligence-based solar radiation prophesy model for green energy utilization in energy management system. *Sustain. Energy Technol. Assess.* **2022**, *52*, 102060. [CrossRef]
50. Zhang, Y.; Chang, X.; Liu, Y.; Lu, Y.; Wang, Y.; Liu, Y. Urban expansion simulation under constraint of multiple ecosystem services (MESs) based on cellular automata (CA)-Markov model: Scenario analysis and policy implications. *Land Use Policy* **2021**, *108*, 105667. [CrossRef]
51. Mohammadyari, F.; Mirsanjari, M.M.; Suziedelyte Visockiene, J.; Zarandian, A. Evaluation of Change in Land-Usage and Land-Cover in IRAN, KARAJ City. In Proceedings of the Environmental Engineering 11th International Conference, Vilnius Gediminas Technical University, Vilnius, Lithuania, 21–22 May 2020; VGTU Press: Vilnius, Lithuania, 2020; pp. 1–8.
52. Mohammadyari, F.; Purkhabbaz, H.; Aghdar, H.; Tavakoli, M. Predicted trends in land use City Behbahan years 2014 to 2028 Using LCM model. *Geogr. Space* **2019**, *65*, 37–56.

53. Silva, L.P.; Xavier, A.; Silva, R.M.; Santos, G. Modelling land cover change based on an artificial neural network for a semi-arid river basin in north-eastern Brazil. *Glob. Ecol. Conserv.* **2019**, *21*, e00811. [CrossRef]
54. Mohammad, A.; Worku, H. Simulating urban land use and cover dynamics using cellular automata and Markov chain approach in Addis Ababa and the surrounding. *Urban Clim.* **2020**, *31*, 100545. [CrossRef]
55. Al Kafy, A.; Rahman, M.; Al- Faisal, A.; Hasan, M.M.; Islam, M. Modelling future land use land cover changes and their impacts on land surface temperatures in Rajshahi, Bangladesh. *Remote Sens. Appl. Soc. Environ.* **2020**, *18*, 100314. [CrossRef]
56. Islam, N.; Irshad, K. Artificial ecosystem optimization with Deep Learning Enabled Water Quality Prediction and Classification model. *Chemosphere* **2022**, *309*, 136615. [CrossRef] [PubMed]
57. Mirsanjari, M.M.; Visockienė, J.S.; Mohammadyari, F.; Zarandian, A. Modelling of expansion changes of Vilnius city area and impacts on landscape patterns using an Artificial Neural Network. *Ecol. Chem. Eng.* **2021**, *28*, 429–447. [CrossRef]
58. Simwanda, M.; Murayama, Y.; Phiri, D.; Nyirenda, V.R.; Ranagalage, M. Simulating Scenarios of Future Intra-Urban Land-Use Expansion Based on the Neural Network–Markov Model: A Case Study of Lusaka, Zambia. *Remote Sens.* **2021**, *13*, 942. [CrossRef]
59. Hajehforooshnia, S.; Soffianian, A.; Mahiny, A.; Fakheran, S. Multi objective land allocation (MOLA) for zoning Ghamishloo Wildlife Sanctuary in Iran. *J. Nat. Conserv.* **2011**, *19*, 254–262. [CrossRef]
60. Quesada-Ruiz, L.C.; Perez, L.; Rodriguez-Galiano, V. Spatiotemporal analysis of the housing bubble′s contribution to the proliferation of illegal landfills—The case of Gran Canaria. *Sci. Total Environ.* **2019**, *687*, 104–117. [CrossRef]
61. Rahimi, V.; Pourkhabbaz, H.R.; Aghdar, H.; Mohammadyari, F. Comparison of fuzzy AHP buckley and ANP models in forestry capability evaluation (Case Study: Behbahan City Fringe). *Iran. J. Appl. Ecol.* **2015**, *4*, 15–31. [CrossRef]
62. Ghosh, P.; Lepcha, K. Weighted linear combination method versus grid-based overlay operation method—A study for potential soil erosion susceptibility analysis of Malda district (West Bengal) in India. *Egypt. J. Remote Sens. Space Sci.* **2019**, *22*, 95–115. [CrossRef]
63. Seneviratne, S.I.; Corti, T.; Davin, E.L.; Hirschi, M.; Jaeger, B.; Lehner, I.; Orlowsky, B.; Teuling, A. Investigating soil moisture–climate interactions in a changing climate: A review. *Earth Sci. Rev.* **2010**, *99*, 125–161. [CrossRef]
64. Cascone, S.; Coma, J.; Gagliano, A.; Perez, G. The evapotranspiration process in green roofs: A review. *Build. Environ.* **2019**, *147*, 337–355. [CrossRef]
65. Nachtergaele, F.; Van Velthuizen, H.; Verelst, L.; Batjes, N.; Dijkshoorn, K.; VanEngelen, V.; Fischer, G.; Jones, A.; Montanarella, L.; Petri, M. *Harmonized World Soil Database*; Food and Agriculture Organization of the United Nations: Rome, Italy, 2008.
66. Burkhard, B.; Kroll, F.; Nedkov, S.; Müller, F. Mapping supply, demand and budgets of ecosystem services. *Ecol. Indic.* **2012**, *21*, 17–20. [CrossRef]
67. Burkhard, B.; Kroll, F.; Müller, F.; Windhorst, W. Landscapes′ capacities to provide ecosystem services-a concept for land-cover based assessments. *Landsc. Online* **2009**, *15*, 1–22. [CrossRef]
68. Haines-Young, R.H.; Potschin, M.P. The links between biodiversity, ecosystem services and human well-being. In *Ecosystem Ecology: A New Synthesis*; Raffaelli, D.G., Frid, C.L.J., Eds.; Cambridge University Press: Cambridge, UK, 2010; p. 162. [CrossRef]
69. Paracchini, M.L.; Zulian, G.; Kopperoinen, L.; Maes, J.; Schägner, J.P.; Termansen, M.; Zandersen, M.; Perez-Soba, M.; Scholefield, P.A.; Bidoglio, G. Mapping cultural ecosystem services: A framework to assess the potential for outdoor recreation across the EU. *Ecol. Indic.* **2014**, *45*, 371–385. [CrossRef]
70. Maes, J.; Paracchini, M.L.; Zulian, G.; Dunbar, M.B.; Alkemade, R. Synergies and trade-offs between ecosystem service supply, biodiversity, and habitat conservation status in Europe. *Biol. Conserv.* **2012**, *155*, 1–12. [CrossRef]
71. Kienast, F.; Degenhardt, B.; Weilenmann, B.; Waeger, Y.; Buchecker, M. GIS-assisted mapping of landscape suitability for nearby recreation. *Landsc. Urban Plan.* **2012**, *105*, 385–399. [CrossRef]
72. Paracchini, M.L.; Capitani, C. *Implementation of an EU Wide Indicator for the Rural-Agrarian Landscape: In Support of COM (2011) 508 Development of Agri-Environmental Indicators for Monitoring the Integration of Environmental Concerns into the Common Agricultural Policy*; Publications Office: Luxembourg, 2011; ISBN 978929223952. [CrossRef]
73. Tavakoli, M.; Mohammadyari, F. Modeling the spatial distribution of multiple ecosystem services in Ilam dam watershed, Western Iran: Identification of areas for spatial planning. *Urban Ecosyst.* **2022**, 1–20. [CrossRef]
74. Vallecillo, S.; La Notte, A.; Zulian, G.; Ferrini, S.; Maes, J. Ecosystem services accounts: Valuing the actual flow of nature-based recreation from ecosystems to people. *Ecol. Model.* **2019**, *392*, 196–211. [CrossRef] [PubMed]
75. Ghermandi, A. Benefits of coastal recreation in Europe: Identifying trade-offs and priority regions for sustainable management. *J. Environ. Manag.* **2015**, *152*, 218–229. [CrossRef]
76. Xiaonan, Y.; Zixiang, Z.; Jing, L.; Xin, F.; Xingmin, M.; Ting, L. Trade-offs between carbon sequestration, soil retention and water yield in the Guanzhong-Tianshui Economic Region of China. *J. Geogr. Sci.* **2016**, *26*, 1449–1462. [CrossRef]
77. Shi, M.; Wu, H.; Fan, X.; Jia, H.; Dong, T.; He, P.; Jiang, P. Trade-offs and synergies of multiple ecosystem services for different land use scenarios in the yili river valley, China. *Sustainability* **2021**, *13*, 1577. [CrossRef]
78. Wang, J.; Zhang, J.; Xiong, N.; Liang, B.; Wang, Z.; Cressey, E.L. Spatial and Temporal Variation, Simulation and Prediction of Land Use in Ecological Conservation Area of Western Beijing. *Remote Sens.* **2022**, *14*, 1452. [CrossRef]
79. Baig, M.F.; Mustafa, M.R.U.; Baig, I.; Takaijudin, H.B.; Zeshan, M.T. Assessment of land use land cover changes and future predictions using CA-ANN simulation for selangor, Malaysia. *Water* **2022**, *14*, 402. [CrossRef]
80. Gao, L.; Tao, F.; Liu, R.; Wang, Z.; Leng, H.; Zhou, T. Multi-scenario simulation and ecological risk analysis of land use based on the PLUS model: A case study of Nanjing. *Sci. Total Environ.* **2022**, *85*, 104055. [CrossRef]

81. Lang, Y.Q.; Song, W.; Zhang, Y. Responses of the water-yield ecosystem service to climate and land use change in Sancha River Basin, China. *Phys. Chem. Earth* **2017**, *101*, 102–111. [CrossRef]
82. Yang, D.; Liu, W.; Tang, L.; Chen, L.; Li, X.; Xu, X. Estimation of water provision service for monsoon catchments of South China Applicability of the InVEST model. *Landsc. Urban Plan.* **2019**, *182*, 133–143. [CrossRef]
83. Betru, T.; Tolera, M.; Sahle, K.; Kassa, H. Trends and drivers of land use/land cover change in Western Ethiopia. *Appl. Geogr.* **2019** *104*, 83–93. [CrossRef]
84. Chen, Y.L.; Wang, S.S.; Ren, Z.G.; Jingfeng, H.; Xiuzhen, W.; Shanshan, L.; Haijun, D.; Wenke, L. Increased evapotranspiration from land cover changes intensified water crisis in an arid river basin in northwest China. *J. Hydrol.* **2019**, *574*, 383–397. [CrossRef]
85. Gaertner, B.A.; Zegre, N.; Warner, T.; Fernandez, R.; He, Y.; Merriam, E.R. Climate, forest growing season, and evapotranspiration changes in the central Appalachian Mountains, USA. *Sci. Total Environ.* **2019**, *650*, 1371–1381. [CrossRef]
86. Aghsaei, H.; Dinan, N.M.; Moridi, A.; Asadolahi, Z.; Delavar, M.; Fohrer, N.; Wagner, P.D. Effects of dynamic land use/land cover change on water resources and sediment yield in the Anzali wetland catchment, Gilan, Iran. *Sci. Total Environ.* **2020**, *712*, 136449 [CrossRef]
87. Chemura, A.; Rwasoka, D.; Mutanga, O.; Dube, T.; Mushore, T. The impact of land-use/land cover changes on water balance of the heterogeneous Buzi sub-catchment, Zimbabwe. *Remote Sens. Appl. Soc. Environ.* **2020**, *18*, 100292. [CrossRef]
88. Zhan, C.S.; Xu, Z.X.; Ye, A.Z.; Su, H.B. LUCC and its impact on run-off yield in the Bai River catchment–upstream of the Miyun Reservoir basin. *J Plant Ecol.* **2011**, *4*, 61–66. [CrossRef]
89. Daneshi, A.; Brouwer, R.; Najafinejad, A.; Panahi, M.; Zarandian, A.; Maghsood, F.F. Modelling the impacts of climate and land use change on water security in a semi-arid forested watershed using InVEST. *J. Hydrol.* **2021**, *593*, 125621. [CrossRef]
90. Abera, W.; Tamene, L.; Abegaz, A.; Solomon, D. Understanding climate and land surface changes impact on water resources using Budyko framework and remote sensing data in Ethiopia. *J. Arid Environ.* **2019**, *167*, 56–64. [CrossRef]
91. Zhang, H.; Wang, B.; Li Liu, D.; Zhang, M.; Feng, P.; Cheng, L.; Yu, Q.; Eamus, D. Impacts of future climate change on water resource availability of eastern Australia: A case study of the Manning River basin. *J. Hydrol.* **2019**, *573*, 49–59. [CrossRef]
92. Zarandian, A.; Baral, H.; Yavari, A.R.; Jafari, H.R.; Stork, N.E.; Ling, M.A.; Amirnejad, H. Anthropogenic Decline of Ecosystem Services Threatens the Integrity of the Unique Hyrcanian (Caspian) Forests in Northern Iran. *Forests* **2016**, *7*, 51. [CrossRef]
93. D'Amour, C.B.; Reitsma, F.; Baiocchi, G.; Barthel, S.; Güneralp, B.; Erb, K.; Haberl, H.; Creutzig, F.; Seto, K. Future urban land expansion and implications for global croplands. *Proc. Natl. Acad. Sci. USA* **2017**, *114*, 8939–8944. [CrossRef] [PubMed]
94. Liu, J.; Zhang, G.; Zhuang, Z.; Cheng, Q.; Gao, Y.; Chen, T.; Huang, Q.; Xu, L.; Chen, D. A new perspective for urban development boundary delineation based on SLEUTH-InVEST model. *Habitat Int.* **2017**, *70*, 13–23. [CrossRef]
95. Locher-Krause, K.E.; Lautenbach, S.; Volk, M. Spatio-temporal change of ecosystem services as a key to understand natural resource utilisation in Southern Chile. *Reg. Environ. Chang.* **2017**, *17*, 2477–2493. [CrossRef]
96. Tolessa, T.; Senbeta, F.; Kidane, M. The impact of land use/land cover change on ecosystem services in the central highlands of Ethiopia. *Ecosyst. Serv.* **2017**, *23*, 47–54. [CrossRef]
97. Van Vliet, J.; Eitelberg, D.A.; Verburg, P.H. A global analysis of land take in cropland areas and production displacement from urbanisation. *Glob. Environ. Chang.* **2017**, *43*, 107–115. [CrossRef]
98. Mohammadyari, F.; Mirsanjari, M.M.; Zarandian, A. Evaluating Ecological Networks of Urban Landscape (Case Study: Karaj Metropolis). *Town Ctry. Plan.* **2018**, *10*, 225–247.
99. Song, W.; Deng, X. Land-use/land-cover change and ecosystem service provision in China. *Sci. Total Environ.* **2017**, *576*, 705–719. [CrossRef] [PubMed]
100. Yang, S.Q.; Zhao, W.W.; Liu, Y.X.; Wang, S.; Wang, J.; Zhai, R.J. Influence of land use change on the ecosystem service trade-offs in the ecological restoration area: Dynamics and scenarios in the Yanhe watershed, China. *Sci. Total Environ.* **2018**, *644*, 556–566. [CrossRef]
101. Jie, X.; Yu, X.; Na, L.; Hao, W. Spatial and temporal patterns of supply and demand balance of water supply services in the Dongjiang Lake Basin and its beneficiary areas. *J. Resour. Ecol.* **2015**, *6*, 386–396. [CrossRef]
102. Cortinovis, S.; Geneletti, D. Ecosystem services in urban plans: What is there, and what is still needed for better decisions. *Land Use Policy* **2018**, *70*, 298–312. [CrossRef]
103. Redhead, J.W.; Stratford, C.; Sharps, K.; Jones, L.; Ziv, G.; Clarke, D.; Oliver, T.H.; Bullock, N.M. Empirical validation of the InVEST water yield ecosystem service model at a national scale. *Sci. Total Environ.* **2016**, *569*, 1418–1426. [CrossRef] [PubMed]

land

MDPI

Article

Spatial Study of Enzymatic Activities from Bacterial Isolates in a Mediterranean Urban Park

Sergi Maicas [1,*] and Jaume Segura-Garcia [2,*]

1 Department de Microbiologia i Ecologia, Facultat de Biologia, Universitat de València, 46100 Burjassot, País Valencià, Spain
2 Departament d'Informàtica, Escola Tècnica Superior d'Enginyeria, Universitat de València, 46100 Burjassot, País Valencià, Spain
* Correspondence: sergi.maicas@valencia.edu (S.M.); jaume.segura@valencia.edu (J.S.-G.)

Abstract: Urban parks are a rich source of microbial diversity, as they are heavily used by city residents. In this study, we sampled a Mediterranean park and were able to isolate bacteria that have the ability to inhibit the growth of control microorganisms. Out of the 560 bacteria we tested, many displayed antibacterial activity, particularly against *Salmonella* sp. and *K. pneumoniae*. These results suggest that the microorganisms in the park are in close proximity to the human population. Additionally, the isolated bacteria demonstrated diverse enzymatic activities, possibly as a response to the environmental substances available to them, which could aid in the degradation of different compounds of interest. The study of the spatial distribution of soil parameters and the inhibition against relative-safe pathogens in an urban park in València (Spain) demonstrated a higher proportion of isolates in certain areas. These spatial data maps can help researchers understand the behaviors of bacterial populations on a regional level, which can assist in the creation of novel antimicrobial agents and promote advancements in public health.

Keywords: urban garden; protease; lipase; DNase; lipase; spatial statistics

Citation: Maicas, S.; Segura-García, J. Spatial Study of Enzymatic Activities from Bacterial Isolates in a Mediterranean Urban Park. *Land* **2023**, *12*, 655. https://doi.org/10.3390/land12030655

Academic Editor: Jūratė Žuziedelytė-Visockienė

Received: 11 February 2023
Revised: 8 March 2023
Accepted: 9 March 2023
Published: 10 March 2023

1. Introduction

Urban gardens play a role in maintaining biodiversity in anthropogenic areas. Their potential value may be considerable, as they can function as semi-natural habitats that offer a sanctuary for bacterial biodiversity [1]. Some of these bacteria include strains that are able to produce antibiotics and other interesting products for the industry, such as extracellular enzymes [2]. Additionally, anthropogenic activities frequently impact urban garden soils, potentially leading to bacterial community composition variations distinct from those observed in conventional agriculture. Such variations can manifest in antibiotic-producing soil strains [2]. Urban gardens often constitute a significant proportion of green spaces in urbanized cities, with some cities having 23 to 36% of their areas dedicated to these gardens [3–5]. These gardens provide support for numerous local, landscape, and sociopolitical features that aid in the preservation of biodiversity. By comprehending spatial connectivity, it is possible to anticipate bacterial resource diversity, abundance, distribution, composition, and species distribution. Opportunistic pathogens that are typically present in soil microbiota or those that colonize it (enteric pathogens) can contaminate the soil through the deposition of human or animal excreta. Animal-based excreta, such as manure, and the improper disposal of human excrement in gardens can introduce substantial quantities of enteric pathogens into the soil environment. The emergence of antimicrobial resistance in urban areas should not be ruled out either [6]. One of the major global health concerns is antibiotic-resistant bacteria with ubiquitous phenotypes and genotypes in parks and gardens [7]. Matthiessen et al. [8] advocate that the high prevalence of antibiotic-resistant bacteria in the environment is one of the most important threats to public health today due to their direct contact with humans. The overuse and misuse of antibiotics have inevitably increased the environmental concentration of antibiotic resistance among bacteria, especially among enterobacteria [9,10]. The problem of antimicrobial resistance in parks is only becoming worse [11]. John Snow's work on cholera and the acceptance of germ theory show that urban dwellers interact with microbes in markedly different ways than those living in rural areas [12]. Microbes in an urban environment are a potential source of contagion [13], and certain conditions, such as allergies, are associated with increasing urbanization [14]. It is more than evident that cities (and, thus, their parks and gardens) have an impact on human health. The mechanisms of this impact are variable and not well understood. The analysis and mapping of microbial dynamics in these urban environments outside pandemics have only just begun [15]. The use of maps for analyzing and illustrating the occurrence of bacterial pathogens at a national level has been employed on a European-wide basis since 1999 [16]. Their contribution to predicting urban environments will enable significant new research on the impact of urban microbiomes on human health [17].

The present investigation scrutinized bacterial resources within an urban garden situated in València (Eastern Spain) to establish the relationship between alterations in soil composition and bacterial abundance, as well as the spatial distribution of bacteria within the garden. Benicalap park is one of the most important green spaces in the city of València. It is an outdoor space that has large sports areas, swimming pools, a theater forum, and children's areas. The park was built in 1983 on the periphery of the city. It covers an approximate area of 80,000 m^2. Vegetation is present in most of the park in the form of Mediterranean wooded groves, e.g., olive trees, cypresses, pines, laurels, holm oaks, mulberry trees, and strawberry trees. The park can be accessed via Burjassot Avenue, Luis Braille Street, and Andreu Alfaro Street. In this study, the microbiological biodiversity was specifically tested, e.g., fungi and bacteria (Enterobacteria). In total, 28 samples were studied extensively in terms of soil bacteria composition, antibiotic effects against test strains, and the ability to produce extracellular hydrolytic enzymes. Bacteria and fungi are incapable of performing endocytosis processes, which prevents them from ingesting nourishing particles or taking advantage of macromolecules in suspension in the external environment as sources of nutrients. It is not uncommon for these microorganisms to produce and excrete medium enzymes with the ability to extracellularly hydrolyze various types of macromolecules, producing smaller molecules (monomers, dimers, oligomers) that will later be incorporated by active transport to the cytoplasm. Moreover, we examined the relationship between bacterial abundance, spatial distribution, and other soil characteristics, such as pH and color.

2. Materials and Methods

2.1. Sample Collection

During the 2018 and 2019 seasons, soil sampling was carried out in an urban community park located in Benicalap, València, Spain (Figure 1). The garden utilized municipal and rainwater for irrigation, which is a common practice in many urban community gardens throughout València. To isolate bacteria, a total of 28 soil samples were collected from a depth of 0–10 cm and transferred to screw tubes, following the procedures detailed in a prior publication [18]. The unused samples were kept at 4 °C for subsequent analysis.

Figure 1. Park in Benicalap. Geolocation of sampling points (Google Maps).

2.2. Soil Characterization

2.2.1. pH Measurement

To determine soil pH, 1 g of soil was dissolved in 5 mL of distilled water, agitated for 2 min, and allowed to settle for 30 min. The soil pH was determined in triplicate using a pH meter (Consort, Turnhout, Belgium) (Figure 2).

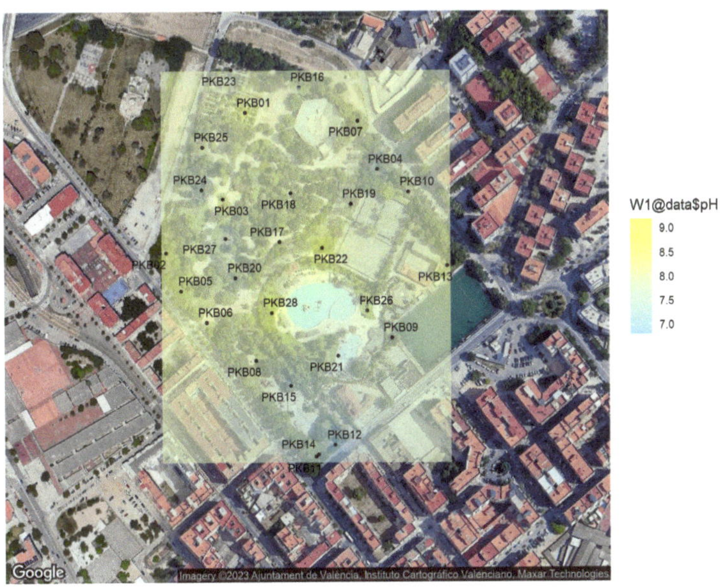

Figure 2. Kriging distribution of pH measurements (black dots) at Benicalap park.

2.2.2. Color and Texture Determination

The soil color was assessed through a visual examination of the samples and compared against a Munsell standard table [19]. The texture of the soil was determined to enable classification based on the sizes of the constituent particles, as described elsewhere [20–22]

2.3. *Isolation and Antibiosis Characterization of Soil Bacteria*

To prepare serial dilutions, 1 g of the sample was suspended in 10 mL of sterile water, and 0.1 mL of the resulting 1:10 to 1:100,000 dilutions were plated on 10% trypticase soy agar (TSA, Conda, Spain). After an incubation period of 48 h, 15–20 random colonies were chosen and transferred onto a fresh TSA "mother" plate using a grid. Calibrated suspensions of safe relatives (*Escherichia coli* CECT101, *Bacillus cereus* CECT495, *Salmonella sp.* CECT443, *Staphylococcus aureus* CECT4013, *Pseudomonas fluorescens* CECT378, *Klebsiella pneumoniae* CECT143, *Enterobacter cloacae* CECT194, or *Enterococcus faecalis* CECT184) were spread on TSA plates using sterile swabs. All grown microorganisms from the mother grid plate were individually replicated on these plates using sterile toothpicks to evaluate their antibiotic effects.

2.4. *Microbial Identification*

2.4.1. 16S rDNA Partial Sequence

Bacterial DNA was extracted and amplified following the method described by Arahal et al. [23]. The 16S rRNA gene was amplified using primers SWI-F (5′-AGAGTTTGATCCTG GCTCAG-3′) and SWI-R (5′-GGTTACCTTGTTACGACTT-3′) [24]. The amplification reaction was performed in a Primus 25 thermocycler (MWG, Ebersberg, Germany). Amplification products were separated by electrophoresis on a 1% (w/v) agarose gel. The PCR amplifications were purified and washed using a high-pure PCR product amplification kit (Boehringer, Mannheim, Germany). The direct sequencing of the PCR products was carried out using the ABIPrism BigDye Terminator Cycle Sequence Ready Reaction Kit (Applied Biosystems, Stafford, TX, USA) at the SCSIE service (Universitat de València, Paterna, Spain). The sequences were aligned using the BLAST program with complete 16S rDNA gene sequences obtained from the EMBL nucleotide sequence data libraries [25].

2.4.2. MALDI-TOF

The bacterial strain identification was conducted following the recommended protocol by Bruker Daltonics (http://www.bdal.de) utilizing the extended direct transfer method. Freshly cultured strains were analyzed using the matrix-assisted laser desorption/ionization-time of flight mass spectrometry (MALDI-TOF MS) technique. A Microflex L20 mass spectrometer (Bruker Daltonics) equipped with an N2 laser was utilized, and all spectra were acquired in a positive linear ion mode with an acceleration voltage of 20 kV [26]. The spectra were acquired as the sum of 240 shots per target within the mass range of 2000–20,000 Da. Three spectra were obtained per strain using the MALDI Biotyper Real-time Classification (RTC) method, and the resulting identification against the database MBT 7854 and MBT 7311_RUO (Bruker Daltonics) corresponded to the profile with the highest log score.

2.5. *Assays for Extracellular Hydrolytic Enzymes Production*

To demonstrate the production of extracellular enzymes (hydrolases), the microorganisms were grown in a standard culture media, provided with the basic nutrients necessary for good growth of the strain, and supplemented with the macromolecule under investigation (proteins, polysaccharides, lipids, or nucleic acids). After timely incubation, the presence of the macromolecule was investigated using an appropriate reagent to show its presence. Sometimes the presence of the macromolecule gives the environment a characteristic aspect (opacity, solidity). This disappearance will indicate that hydrolysis has occurred.

All of the strains isolated in this project were spread onto four replicated Petri plates using sterile toothpicks, each containing media with the following composition in g L^{-1}.

- Casein agar: 10 bacteriological peptone, 4 NaCl, 3 meat extract, 15 agar; 10% of skim milk was sterilized in a separate vessel. Once sterile, both parts were poured onto Petri plates. The plates were incubated at 28 °C for 3 days. The presence of casein gives the environment an opaque appearance that disappears when casein is hydrolyzed. A transparent halo must appear in the middle of the proteolytic microorganisms grown on casein. The non-proteolytic ones do not produce a change in the original aspect of the environment.
- TWEEN-80 agar: This medium consists of a synthetic lipid containing ester links between sorbitol and oleic acid (Tween-80) and calcium salts. When microorganisms possessing esterase activity (lipase) are present, they can hydrolyze the ester link, leading to the release of oleic acid from Tween-80. In the presence of an excess amount of Ca^{2+}, the released oleic acid forms small crystalline oleate crystals that create an opaque halo around the growth area.
- Starch Agar: To reveal the presence of starch, it was necessary to dye it with Lugol. We added 2 milliliters of Lugol to the plate and observed the development of the dark violet color in areas where there was starch. If transparent halos appeared around the growth of a microorganism, this indicated that there was polysaccharide hydrolysis.
- DNase agar: 20 tryptose, 2 deoxyribonucleic acid, 5 sodium chloride, 12 agar. The agar medium was inoculated and incubated with the organisms, then the surface growth was flooded with 1N hydrochloric acid. Polymerized DNA, which is present in the medium, will precipitate in the presence of 1N HCl and cause the medium to become opaque. If the organisms produce enough DNase enzymes to hydrolyze the DNA, clear zones will be observed around the colonies. As a control, *Staphylococcus aureus* CECT4013 was used.

2.6. Kriging Method for Biological Spatial Statistics

Spatial statistics is a collection of methods used to analyze and interpret geographical data by using various interpolation techniques such as inverse distance weighted (IDW), splines, and Kriging. IDW is a simple technique that predicts outcomes based on nearby sample values, but it can produce a bull's-eye pattern and an uneven surface. Splines also create a smooth surface by fitting a mathematical function to the input data, but they are deterministic methods. Kriging, on the other hand, is a statistical method that incorporates spatial autocorrelation and is preferred for spatial interpolation [27].

The soil samples yield valuable information that is recorded as a dataset linking specific locations to the GPS coordinates, longitude, and latitude. The data are represented by the number of colonies or isolates that inhibit the growth of control strains at a given location x, denoted as $Y(x)$. The dataset is defined as $Y(x),, x \in \mathcal{D}$, where $\mathcal{D}$ includes all of the locations being analyzed. Kriging techniques [28] are then utilized on this dataset to facilitate spatial analysis.

3. Results and Discussion

Urban park soil is capable of sustaining microbial life in different degrees and conditions. Depending on the degree of anthropogenization, areas with higher concentrations of microorganisms are observed (top-right and bottom-left extremes) Figure 3.

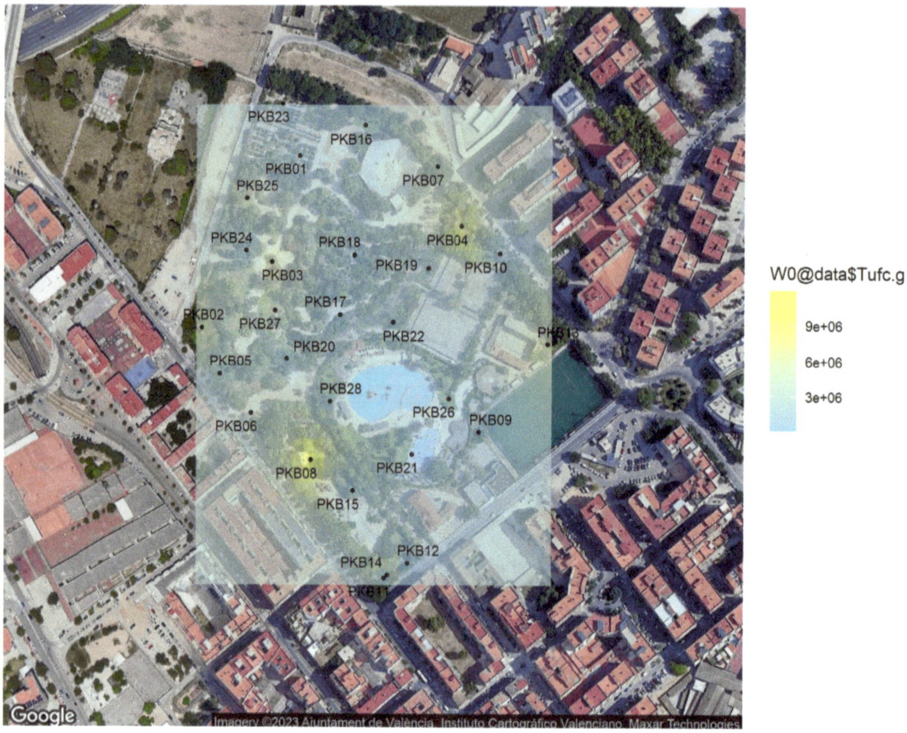

Figure 3. Kriging mapping of isolated microorganisms at the park in Benicalap.

Based on the possible different characteristics of these soils, different tests were carried out and they allowed us to characterize them minimally. On the one hand, one of the most significant properties of the soil, the color, was determined. It was generally conditioned by the existence and proportion of organic and mineral compounds.

Organic matter produces dark colors, usually blackish or brown. Most of the world's soils are dominated by two large lineages of bacteria, i.e., *Proteobacteria* and *Actinobacteria* [29]. They are very efficient competitors for organic carbon, which is a limiting resource in the soil. Therefore, their abundance increases the concentrations of organic carbon. The darker the soil, the greater the chance of finding a large number of microorganisms. Most of the analyzed soils showed a dark or brownish-dark coloration, which suggests a great number of organic compounds and, consequently, high numbers of microorganisms (Table 1).

Although the samples were taken at different depths, ranging from the surface to 10 cm, similar results were observed. The soil type is strongly associated with pH, which is determined by the constituents of the soil. The pH value influences the type of microorganisms that can be found in each soil sample. Bacteria prefer neutral or slightly alkaline conditions, while fungi are more abundant at acidic pH [29]. In the soils in Mediterranean regions, the calcium content is high and the pH is alkaline [30]. The pH values in these soils are linked to the presence or absence of $CaCO_3$ [31].

The soils analyzed exhibited a range of pH from 6.79 to 9.24. We decided to only isolate bacteria for further experiments, following previous strategies [32,33].

Table 1. Geo-location and basic soil characterization.

Sample Code	GPS Coord	Depth	pH	Color	Texture	Total ufc·g^{-1}
PKB01	39.4986, −0.3973	2–5	8.71	Dark	Sandy-loam	1.70×10^5
PKB02	39.4974, −0.3982	0–2	8.46	Brown	Sandy-loam	4.70×10^6
PKB03	39.4979, −0.3976	0–2	8.63	Brownish dark	Sandy-loam	6.00×10^6
PKB04	39.4981, −0.3958	2–5	7.60	Grey	Sandy-loam	9.00×10^6
PKB05	39.4971, −0.3981	0–2	8.57	Brown	Sandy-loam	9.40×10^5
PKB06	39.4968, −0.3978	0–2	8.53	Dark	Sandy-loam	3.20×10^6
PKB07	39.4985, −0.3960	5–10	8.48	Dark	Sandy-loam	4.00×10^6
PKB08	39.4965, −0.3972	2–5	8.19	Dark	Sandy-loam	1.20×10^7
PKB09	39.4967, −0.3957	2–5	7.80	Dark	Sandy-loam	7.40×10^5
PKB10	39.4979, −0.3955	0–2	8.13	Brown	Sandy-loam	7.00×10^5
PKB11	39.4957, −0.3965	2–5	7.50	Dark	Sandy-loam	6.60×10^5
PKB12	39.4958, −0.3963	2–5	6.79	Brownish dark	Sandy-loam	9.70×10^4
PKB13	39.4973, −0.3950	2–5	8.68	Brownish dark	Sandy-loam	6.40×10^6
PKB14	39.4957, −0.3965	0–2	7.21	Brownish dark	Sandy-loam	5.70×10^6
PKB15	39.4963, −0.3968	0–2	7.33	Brown	Sandy	6.60×10^5
PKB16	39.4988, −0.3967	2–5	8.00	Dark	Sandy	4.60×10^5
PKB17	39.4975, −0.3969	0–2	8.33	Dark	Sandy-loam	3.00×10^5
PKB18	39.4979, −0.3968	0–2	8.28	Bright	Sandy	4.90×10^5
PKB19	39.4978, −0.3961	2–5	8.09	Dark	Sandy-loam	1.20×10^6
PKB20	39.4972, −0.3974	0–2	7.89	Dull	Sandy	5.60×10^5
PKB21	39.4966, −0.3963	0–2	7.63	Light	Sandy	1.20×10^5
PKB22	39.4975, −0.3965	5–10	9.24	Dull	Sandy-loam	9.30×10^5
PKB23	39.4990, −0.3975	2–5	7.85	Dark	Sandy	4.30×10^4
PKB24	39.4980, −0.3978	0–2	7.55	Brownish dark	Sandy-loam	6.80×10^5
PKB25	39.4983, −0.3978	2–5	8.11	Brownish dark	Sandy-loam	4.10×10^6
PKB26	39.4969, −0.3959	2–5	8.06	Dull orange	Sandy	7.20×10^5
PKB27	39.4976, −0.3976	2–5	7.47	Grey	Sandy-loam	3.30×10^6
PKB28	39.4969, −0.3970	0–2	9.00	Dull orange	Sandy-loam	7.30×10^5

Microorganisms living in soil are in constant competition with other microbiota that share their niche. To evaluate the ability of soil bacteria to inhibit the growth of other bacteria, inhibition tests were performed [33]. After a 24-hour incubation period, calibrated suspensions of safe bacterial strains, including *E. coli*, *S. aureus*, *K. pneumoniae*, *Salmonella* sp., *P. fluorescens*, and *Ent. cloacae*, were spread on TSA plates using sterile swabs. The grown microorganisms from the mother grid plate were then individually replicated on these plates using sterile toothpicks to check for antibiotic effects (Table 2). A Kriging representation of these results is shown in Figure 4.

Among the 560 bacteria analyzed, a variable amount of antibiosis was detected, with *Salmonella* sp. and *K. pneumoniae* having the highest number of microorganisms with antibiotic effects. On the other hand, antibiosis against *P. fluorescens* and *Ent. cloacae* were almost anecdotal.

However, urban parks are home to a large number of microorganisms that can be sources of enzymes with industrial applications. A set of four biochemical tests was performed to detect possible enzyme producers of interest with DNase, lipase, amylase, and protease activities among all isolated bacteria (Table 3). Between 10 and 24 isolates tested positive. For the distribution of the results (Kriging representation), see Figure 5.

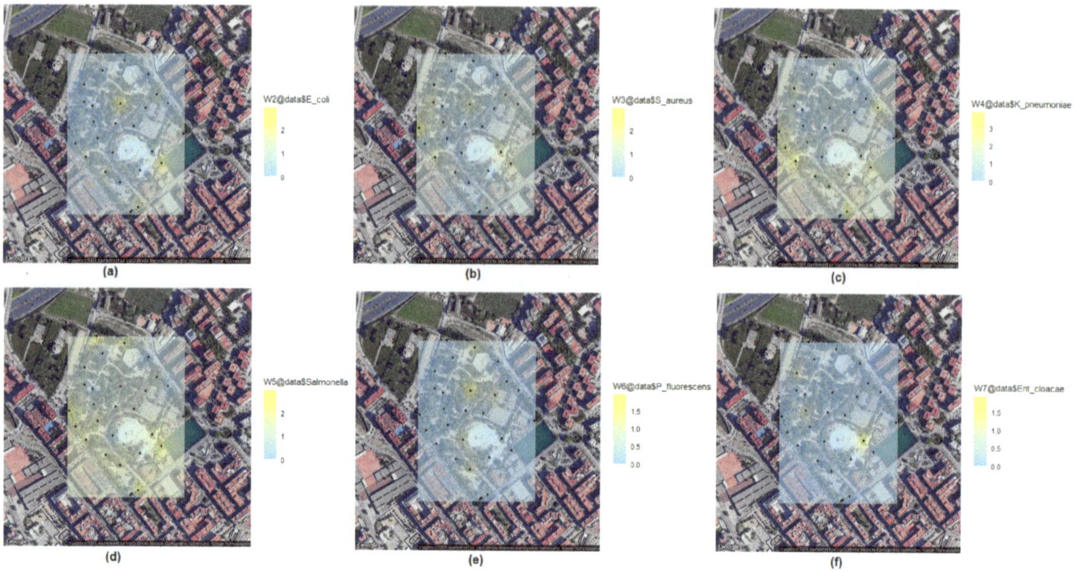

Figure 4. Kriging mapping of inhibition against different pathogens: (**a**) *E. coli*, (**b**) *S.aureus*, (**c**) *K. pneumoniae*, (**d**) *Salmonella* sp., (**e**) *P. fluorescens*, and (**f**) *Ent. cloacae*.

Table 2. Inhibition against relative safe pathogens (number of positive isolates).

Sample	*E. coli*	*S. aureus*	*K. pneumoniae*	*Salmonella* sp.	*P. fluorescens*	*Ent. cloacae*
PKB01	0	0	0	0	0	0
PKB02	0	3	3	3	0	0
PKB03	0	0	0	0	0	0
PKB04	1	1	0	1	0	0
PKB05	0	0	3	2	0	0
PKB06	1	2	4	2	0	0
PKB07	0	0	0	0	0	0
PKB08	1	1	2	2	0	0
PKB09	3	2	3	3	0	0
PKB10	0	1	3	0	0	0
PKB11	0	0	0	0	0	0
PKB12	2	1	4	3	0	0
PKB13	1	1	0	1	0	0
PKB14	1	0	1	1	0	0
PKB15	0	0	2	2	1	0
PKB16	0	0	1	2	1	0
PKB17	0	0	0	1	0	0
PKB18	3	2	1	1	2	0
PKB19	0	0	0	1	1	0
PKB20	0	0	1	2	0	0
PKB21	0	0	2	1	0	0
PKB22	0	0	1	1	0	0
PKB23	1	1	0	3	0	0
PKB24	0	2	1	1	0	0
PKB25	0	1	2	1	0	1
PKB26	0	2	1	2	0	2
PKB27	0	1	0	1	0	1
PKB28	0	0	0	0	1	0
Total	14	21	35	37	6	4

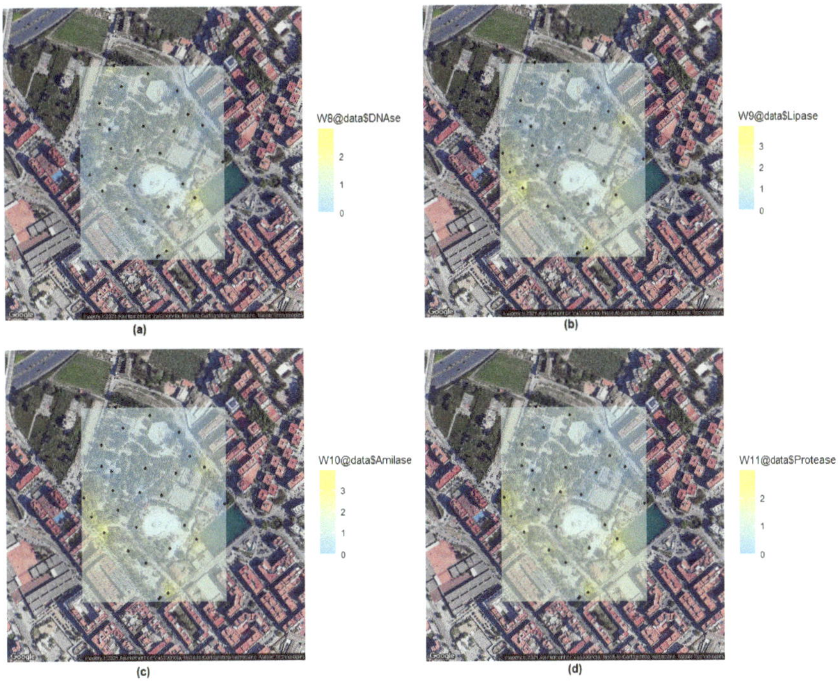

Figure 5. Kriging mapping of positive activities in different cultures: (**a**) DNase, (**b**) lipase, (**c**) amylase, and (**d**) Protease.

Table 3. Positive activities in different media cultures.

Sample	DNase	Lipase	Amylase	Protease
PKB01	0	0	0	0
PKB02	0	3	3	3
PKB03	0	0	0	0
PKB04	1	1	0	1
PKB05	0	0	3	2
PKB06	1	2	4	2
PKB07	0	0	0	0
PKB08	1	1	2	2
PKB09	3	2	3	3
PKB10	0	1	3	0
PKB11	0	0	0	0
PKB12	2	1	4	3
PKB13	1	2	0	1
PKB14	1	0	1	1
PKB15	0	1	1	1
PKB16	1	0	1	1
PKB17	1	0	0	2

Table 3. *Cont.*

Sample	DNase	Lipase	Amylase	Protease
PKB18	1	0	1	1
PKB19	1	0	1	0
PKB20	1	0	1	1
PKB21	1	0	1	1
PKB22	1	0	1	0
PKB23	2	0	1	1
PKB24	1	0	0	1
PKB25	1	2	1	1
PKB26	0	1	1	1
PKB27	1	1	1	1
PKB28	0	0	1	1
Total	10	12	24	19

Typically, a DNase reaction is an indication of pathogenicity for staphylococci and other pathogenic bacteria in clinical assays [34]. However, some soil-isolated bacteria produce transparent halos. Organisms other than staphylococci, *Serratia*, and aeromonads can produce DNase. The presence of extracellular DNA in soil or the availability of plasmid or chromosomal DNA from other soil bacteria that share the same ecological niche can be useful for specialized microbial populations with potential industrial applications, as described previously. [35].

4. Conclusions

Parks are indispensable elements of urban environments. Their soils, and the microbiota that inhabit them, have significant physical, chemical, and biological influences on their characteristics. Thus, urban parks play an important ecological role in preserving microbial diversity, giving the soil the capacity to self-purify pathogenic microorganisms, regulate the greenhouse effect, and perform other functions [36,37]. The ecological functions of the soil are directly linked to the vital activities of bacterial communities that thrive in them. For this reason, the study of these urban parks is of paramount importance.

In this study, we investigated the microbial communities existing in one of the most emblematic urban parks in the city of València. The park is a green space and serves as a recreational area, so microbial exchanges must be frequent. As it is surrounded by busy streets in a completely urban environment, the park's ecological function could be under threat due to continuous use for many generations. To this end, we determined the microbial concentration, particularly bacterial, at various points in the park by conducting a globalized study using the Kriging strategy.

Despite the soils being a mixture of various origins, the amount of bacteria detected is fairly uniform, with no levels higher than 1.20×10^7 in any of the samplings carried out. The population density allowed us to reach a number of global conclusions. On the one hand, in a significant number of sampling points, the isolation of bacteria capable of inhibiting the growth of control microorganisms has been obtained. This led us to believe that the microorganisms that inhabit the park are in intimate contact with the human population. Moreover, the isolated individuals possessed different enzymatic activities, perhaps in response to the environment, allowing them to degrade different compounds of interest.

Author Contributions: Both authors prepared the manuscript. All authors have read and agreed to the published version of the manuscript.

Funding: This research was partially funded by the Spanish Research Agency via grant no. TED2021-131040B-C33 in the program "Strategic Projects for Ecologic and Digital Transition 2021" by MCIN/AEI/10.13039/501100011033 and the European Union NextGenerationEU/PRTR.

Data Availability Statement: Not applicable.

Acknowledgments: The authors are indebted to the degree students for the sampling. The authors acknowledge A. Ruvira and D. Ruiz (Spanish Type Culture Collection, CECT, Paterna, Spain) for the MALDI-TOF MS analyses. The authors are indebted to E. Carbó (Universitat de València, València, Spain), and M.M. Jordán (Universitas Miguel Hernández, Elx, Spain) for the kind suggestions regarding the soil characterization.

Conflicts of Interest: The authors declare no conflict of interest.

References

1. Al-Delaimy, W.; Webb, M. Community Gardens as Environmental Health Interventions: Benefits Versus Potential Risks. *Curr. Environ. Health Rep.* **2017**, *4*, 252–265. [CrossRef]
2. Mafiz, A.I.; Perera, L.N.; He, Y.; Zhang, W.; Xiao, S.; Hao, W.; Sun, S.; Zhou, K.; Zhang, Y. Case study on the soil antibiotic resistome in an urban community garden. *Int. J. Antimicrob. Agents* **2018**, *52*, 241–250. [CrossRef] [PubMed]
3. Gaston, K.; Warren, P.; Thompson, K.; Smith, R. Urban domestic gardens (IV): The extent of the resource and its associated features. *Biodivers. Conserv.* **2005**, *14*, 3327–3349. [CrossRef]
4. Loram, A.; Tratalos, J.; Warren, P.; Gaston, K. Urban domestic gardens (X): The extent & structure of the resource in five major cities. *Landsc. Ecol.* **2007**, *22*, 601–615. [CrossRef]
5. Mathieu, R.; Freeman, C.; Aryal, J. Mapping private gardens in urban areas using object-oriented techniques and very high-resolution satellite imagery. *Landsc. Urban Plan.* **2007**, *81*, 179–192. [CrossRef]
6. Perera, L.N.; Mafiz, A.I.; Amarasekara, N.R.; Chang, E.; Rao, V.B.K.; Zhang, Y. Antimicrobial-resistant E. coli and Enterococcus spp. Recovered from urban community gardens. *Food Control* **2020**, *108*, 106857. [CrossRef]
7. Yuan, W.; Zhang, Y.; Riaz, L.; Yang, Q.; Du, B.; Wang, R. Multiple antibiotic resistance and DNA methylation in Enterobacteriaceae isolates from different environments. *J. Hazard. Mater.* **2021**, *402*, 123822. [CrossRef]
8. Matthiessen, L.; Bergström, R.; Dustdar, S.; Meulien, P.; Draghia-Akli, R. Increased momentum in antimicrobial resistance research. *Lancet* **2016**, *388*, 865. [CrossRef]
9. Mao, D.; Yu, S.; Rysz, M.; Luo, Y.; Yang, F.; Li, F.; Hou, J.; Mu, Q.; Alvarez, P. Prevalence and proliferation of antibiotic resistance genes in two municipal wastewater treatment plants. *Water Res.* **2015**, *85*, 458–466. [CrossRef]
10. Rodriguez-Mozaz, S.; Chamorro, S.; Marti, E.; Huerta, B.; Gros, M.; Sànchez-Melsió, A.; Borrego, C.; Barceló, D.; Balcázar, J. Occurrence of antibiotics and antibiotic resistance genes in hospital and urban wastewaters and their impact on the receiving river. *Water Res.* **2015**, *69*, 234–242. [CrossRef]
11. Yuan, W.; Tian, T.; Yang, Q.; Riaz, L. Transfer potentials of antibiotic resistance genes in Escherichia spp. strains from different sources. *Chemosphere* **2019**, *246*, 1–9. [CrossRef] [PubMed]
12. Neiderud, C.J. How urbanization affects the epidemiology of emerging infectious diseases. *Afr. J. Disabil.* **2015**, *5*. [CrossRef] [PubMed]
13. Cooley, J.D.; Wong, W.C.; Jumper, C.A.; Straus, D.C. Correlation between the prevalence of certain fungi and sick building syndrome. *Occup. Environ. Med.* **1998**, *55*, 579–584. [CrossRef] [PubMed]
14. Nicolaou, N.; Siddique, N.; Custovic, A. Allergic disease in urban and rural populations: Increasing prevalence with increasing urbanization. *Allergy Eur. J. Allergy Clin. Immunol.* **2005**, *60*, 1357–1360. [CrossRef]
15. Gilbert, J.A.; Stephens, B. Microbiology of the built environment. *Nat. Rev. Microbiol.* **2018**, *16*, 661–670. [CrossRef]
16. ECDC. ECDC Ears-net Database. Available online: https://www.ecdc.europa.eu/en/antimicrobial-resistance/surveillance-and-disease-data/data-ecdc (accessed on 24 August 2021).
17. Danko, D.; Bezdan, D.; Afshin, E.E.; Ahsanuddin, S.; Bhattacharya, C.; Butler, D.J.; Chng, K.R.; Donnellan, D.; Hecht, J.; Jackson, K.; et al. A global metagenomic map of urban microbiomes and antimicrobial resistance. *Cell* **2021**, *184*, 3376–3393.e17. [CrossRef]
18. Arnal, D.; Moya, C.; Filippelli, L.; Segura-Garcia, J.; Maicas, S. Bacteria spatial tracking in Urban Park soils with MALDI-TOF Mass Spectrometry and Specific PCR. *Biodata Min.* **2023**, *16*, 1. [CrossRef]
19. Thompson, J.; Pollio, A.; Turk, P. Comparison of munsell soil color charts and the globe soil color book. *Soil Sci. Soc. Am. J.* **2013**, *77*, 2089–2093. [CrossRef]
20. Porta-Casanellas, J. *Técnicas y Experiments en Edafología*; Colegio Oficial de Ingenieros Agrónomos de Cataluña: Barcelona, Spain, 1986.
21. Alden, A. *Sand, Silt, and Clay Soil Classification Diagram*; ThoughtCo.: New York, NY, USA, 2020.
22. Jorge, N.F.; Clark, J.; Cárdenas, M.L.; Geoghegan, H.; Shannon, V. Measuring soil colour to estimate soil organic carbon using a large-scale citizen science-based approach. *Sustainability* **2021**, *13*, 11029. [CrossRef]
23. Arahal, D.R.; Sánchez, E.; Macián, M.C.; Garay, E. Value of *recN* sequences for species identification and as a phylogenetic marker within the family "*Leuconostocaceae*". *Int. Microbiol.* **2008**, *11*, 33–39. [CrossRef]
24. Small World Initiative. Crowdsourcing Antibiotic Discovery. Available online: https://www.smallworldinitiative.org/about (accessed on 27 December 2019).
25. Altschul, S.F.; Gish, W.; Miller, W.; Myers, E.W.; Lipman, D.J. Basic local alignment search tool. *J. Mol. Biol.* **1990**, *215*, 403–410. [CrossRef] [PubMed]
26. Maier, T.; Klepel, S.; Renner, U.; Kostrzewa, M. Fast and reliable MALDI-TOF MS-based microorganism identification. *Nat. Methods* **2006**, *3*, i–ii. [CrossRef]

27. Isaaks, E.-H.; Srivastava, R.M. *An Introduction to Applied Geostatistics*; Oxford University Press: New York, NY, USA, 1989.
28. Cressie, N. *Statistics for Spatial Data*; John Wiley: New York, NY, USA, 1993.
29. Madigan, M.; Bender, D.; Buckley, D.; Sattley, W.; Stahl, D. *Brock Biology of Microorganisms*, 15th ed.; Pearson: London, UK, 2018.
30. Jordán, M.M.; Almendro-Candel, M.B.; Navarro-Pedreño, J.; Pardo, F.; García-Sánchez, E.; Bech, J. Bioavailability, mobility and leaching of phosphorus in a Mediterranean agricultural soil (ne Spain) amended with different doses of biosolids. *Environ Geochem. Health* **2022**, *44*, 7–14. [CrossRef]
31. Jordán-Vidal, M.; Montero-Martínez, M.A.; García-Sánchez, E.; Martinez-Poveda, A. Metal contamination of agricultural soils amended with biosolids (sewage sludge) at a ceramic production area in NE Spain: A 10-year resampling period. *Soil Use Manag* **2021**, *37*, 307–318.
32. Maicas, S.; Fouz, B.; Figàs-Segura, A.; Zueco, J.; Rico, H.; Navarro, A.; Carbó, E.; Segura-Garcia, J.; Biosca, E. Implementation of Antibiotic Discovery by Student Crowdsourcing in the Valèncian Community Through a Service Learning Strategy. *Front Microbiol.* **2020**, *11*, 564030. [CrossRef] [PubMed]
33. Maicas, S.; Segura-Garcia, J. Spatial distribution of antibiotic-producing bacteria in urban areas. A case study in València (Spain) *IEEE/Acm Trans. Comput. Biol. Bioinform.* **2020**. [CrossRef] [PubMed]
34. Kateete, D.; Kimani, C.; Katabazi, F.; Okeng, A.; Okee, M.; Nanteza, A.; Joloba, M.; Najjuka, F. Identification of Staphylococcus aureus: DNase and Mannitol salt agar improve the efficiency of the tube coagulase test. *Ann. Clin. Microbiol. Antimicrob.* **2010**, *9*, 1–7. [CrossRef] [PubMed]
35. Ten, L.; Liu, Q.M.; Im, W.T.; Lee, M.; Yang, D.C.; Lee, S.T. *Pedobacter ginsengisoli* sp. nov., a DNase-producing bacterium isolated from soil of a ginseng field in South Korea. *Int. J. Syst. Evol. Microbiol.* **2006**, *56*, 2565–2570. [CrossRef]
36. Nannipieri, P.; Ascher, J.; Ceccherini, M.T.; Landi, L.; Pietramellara, G.; Renella, G. Microbial diversity and soil functions. *Eur. J. Soil Sci.* **2003**, *54*, 655–670. [CrossRef]
37. Lysak, L.V.; Lapygina, E.V. The Diversity of Bacterial Communities in Urban Soils. *Eurasian Soil Sci.* **2018**, *51*, 1050–1056. [CrossRef]

land

MDPI

Article

Investigating the Spatial Heterogeneity and Influencing Factors of Urban Multi-Dimensional Network Using Multi-Source Big Data in Hangzhou Metropolitan Circle, Eastern China

Jing Zhang [1], Lei Li [1], Congmou Zhu [2,*], Qi Hao [1], Xinming Chen [3], Zhoulu Yu [4], Muye Gan [1] and Wuyan Li [5]

1 Institute of Applied Remote Sensing and Information Technology, College of Environmental and Resource Sciences, Zhejiang University, Hangzhou 310058, China; zj1016@zju.edu.cn (J.Z.); 22114167@zju.edu.cn (L.L.); 22014175@zju.edu.cn (Q.H.); ganmuye@zju.edu.cn (M.G.)
2 Collaborative Innovation Center of Statistical Data Engineering Technology & Application, Zhejiang Gongshang University, Hangzhou 310018, China
3 Territorial Consolidation Center in Zhejiang Province, Department of Natural Resources of Zhejiang Province, Hangzhou 310007, China; 0922664@zju.edu.cn
4 Zhejiang Digital Governance Space Planning and Design Co., Ltd., Hangzhou 310000, China; yuzl@zju.edu.cn
5 The Institute of Land and Urban-Rural Development, Zhejiang University of Finance and Economics, Hangzhou 310018, China; liwuyan@zufe.edu.cn
* Correspondence: congmouzhu@zjgsu.edu.cn

Abstract: Exploring the spatial heterogeneity of urban multi-dimensional networks and influencing factors are of great significance for the integrated development of metropolitan circle. This study took Hangzhou metropolitan circle as an example, using multi-source geospatial big data to obtain urban population, transportation, goods, capital, and information flow information among sub-cities. Then, spatial visualization analysis, social network analysis, and geographical detector were applied to analyze the differences in spatial structure of multiple urban networks and influencing factors in Hangzhou metropolitan circle, respectively. The results showed that (1) the network connections of population, traffic, goods, and capital flows transcended geographical proximity except that of information flow, and population and traffic flow networks were found to be more flattened in Hangzhou metropolitan circle than in other urban networks; (2) the comprehensive urban network of Hangzhou metropolitan circle was imbalanced across sub-cities, presenting hierarchical and unipolar characteristics; and (3) the influence of traffic distance on the network spatial structure of Hangzhou metropolitan was stronger than the geographical distance, and the interactions between traffic distance and socioeconomic factors would further enhance the regional differentiation of the network spatial structure. This study could provide scientific reference for constructing a coordinated and integrated development pattern in a metropolitan circle.

Keywords: urban network; spatial structure; influencing factor; multi-source big data; Hangzhou metropolitan circle

Citation: Zhang, J.; Li, L.; Zhu, C.; Hao, Q.; Chen, X.; Yu, Z.; Gan, M.; Li, W. Investigating the Spatial Heterogeneity and Influencing Factors of Urban Multi-Dimensional Network Using Multi-Source Big Data in Hangzhou Metropolitan Circle, Eastern China. *Land* **2023**, *12*, 1808. https://doi.org/10.3390/land12091808

Academic Editors: Marco Maretto and Jūratė Sužiedelytė-Visockienė

Received: 22 August 2023
Revised: 7 September 2023
Accepted: 18 September 2023
Published: 19 September 2023

1. Introduction

A metropolitan circle is a regional network with a megacity as the core where cities are spatially and functionally connected [1,2]. It has become an important spatial development strategy for competition in the global economy to reconstruct the metropolitan area across cities or provincial boundaries through the construction of intercity infrastructure and regional cooperation [3]. Its purpose is not only to expand the scale of population and urban space but also to increase the mobility of population, goods, capital, and information to strengthen intercity connections and improve resource allocation efficiency [2]. The spatial structure of a city or a region refers to the arrangement of a certain space in the urban form composed of population, materials, information, and a series of relationships generated by their interactions [4,5]. With the rapid progress of information technology and the internet, the complex interaction between real space and virtual networks has enormously reshaped the metropolitan circles' spatial structure [6,7]. Castells [8] put forward the key concept of the "flow space", indicating that regional development increasingly depends on network interactions of various element flows. China had long adopted a strategy of giving priority to the development of urban agglomerations [6], but the construction of metropolitan circles had been neglected. This phenomenon has led to a series of development constraints in metropolitan circles, such as insufficient links between cities, the weak influence of core cities, and obstacles to the implementation of urban agglomeration planning [2,9]. According to *Outline of the 14th Five-Year Plan (2021–2025) for National Economic and Social Development and Vision 2035 of the People's Republic of China* (the 14th Five-Year Plan), China was striving to build a system of metropolitan areas centered on provincial capitals or central cities in order to improve the quality of urbanization and promote regional integrated development during the 14th Five-Year Plan period. Thus, it is of great significance to decipher the heterogeneity of spatial structure and influencing factors of metropolitan circles from the perspective of various "flow space", which are critical for optimizing the spatial structure of metropolitan circle and promoting regional integration and coordinated development.

The traditional view holds that the spatial structure of a city or region reflects regional differences and similarities in socioeconomic, political, or cultural attributes [10,11]. Early studies on regional spatial structure aimed at distinguishing clustering areas with similar attributes (e.g., natural and economic status) by using factor analysis and cluster analysis [12]. It allows us to reveal regional differences and identify specific morphologies of regional spatial structure, such as single-core, sectorial, or multi-core configurations [13]. The second view concentrates more on the spatial agglomeration and dispersion of regional structure, cartography, and spatial analysis, such as spatial auto-correlation analysis, geographically weighted modeling, and hotspot detection, are usually used [14,15]. Compared with the traditional views, this viewpoint simultaneously focuses on the attributes of specific spatial units and their spatial dependencies with adjacent units, thereby identifying the core and edge structure of an area [16]. However, it tends to measures local connections, but ignores the structural configuration of the entire area, without focusing on the spatial association between the target unit and nonadjacent unit [17]. In the era of globalization and information, regional interactive relationships have transformed from focusing on neighboring space to flow space. From the perspective of "flow space", the spatial structure of a city or region is viewed as an interconnected network, and more attention is paid to the nodes, backbone, and connections of the network [7,18]. It provides a new understanding of regional spatial structure, that is, city or sub-city divisions are nodes, and various element flows are links [19,20]. Based on the network perspective, the spatial relationships between two units are determined by a network composed of multiple element flows, rather than predetermined by the similarity of attributes or proximity of geographic distances [21,22].

There are various networks in regional spatial development, including natural connections, economic connections, cultural links, and administrative links [23,24], which form a complex connected network. Recent network-based studies concentrate on the spatial structure of a single network in a city or a region, such as the population flow network [25], goods flow network [26], information flow network [27], and traffic flow network [28]. These empirical studies have explored the specific roles of particular nodes [11], the spatial arrangement of various links, and the network organization of node city on the urban agglomeration or national scale [29–31]. A series of methods including network visualization and mapping technology, gravity models, and social network analysis have been developed [22,23,32]. However, existing related studies pay less attention to the scale of the metropolitan circle, and neglect the spatial heterogeneity of urban multi-dimensional flow networks [27,31]. Regional spatial structure is determined by various flow networks [33]. By only relying on the measurement of a single flow network, we cannot accurately position the role and competitiveness of a city in a regional network. Moreover, it is not conducive for providing effective decision support for formulating spatial structure optimization and development strategies in the metropolitan circle. Many studies apply prefecture-level city network data to explore the spatial structure of a large region, such as the intercity transportation infrastructure, relational data for company headquarters and branches, intercity linkages for population, and goods mobility [34]. However, it is not adequately accurate to investigate the large-scale regional network structure using prefecture-level cities as spatial units, which may ignore the variations in the connections among sub-cities. Especially in China, the administrative scope of a city often includes several districts and counties. In the era of information, multi-source big data with high-resolution locational information are emerging, such as social media data, mobile phone signal data, and POI data [6,35,36]. These big data provide various flow network information among sub-cities that can contribute to the analysis of county-scale spatial structure.

The spatial structure of urban multi-dimensional networks in metropolitan circles is comprehensively influenced by the social economy and proximity factors [37]. In socioeconomic terms, the development level of economic, urbanization rate and population size are the core factors that affect the interactions between sub-cities [25,38]. Industrial development and financial investment affect financial and economic exchanges and cooperation between cities [39], further influencing the city's influence and service level in the surrounding areas. Meanwhile, the level of social consumption is an important intermediary and complementary factor for the interaction between cities [31]. These socioeconomic factors significantly influence the role of a city in regional network structure to a certain extent. The first law of geography states that all things are related, and things in close proximity are more closely related to each other, so that cities that are close to each other are more frequently connected [40,41]. Furthermore, the improvement in transportation has shortened the "space–time" distance between cities, and the proximity of industries also may strengthen the network links between cities, which can be considered the "industrial" distance [39]. However, to our knowledge, existing studies have usually focused on socioeconomic factors [39,42], and the influence of multi-dimensional distance on regional spatial network connections is easily ignored [43]. This is a clear and critical research gap, especially as the rapid progress of information technique and transportation in China promotes closer network connections between cities. Investigating the determining factors of the spatial network structure of various flows could provide policy supports for the integrated development of metropolitan circles.

To address these research gaps, this study attempts to consider the Hangzhou metropolitan circle, an area with early formation and rapid development, as an example, and then uses multiple internet data application platforms and web crawler technology to obtain five kinds of representative flow network information in Hangzhou metropolitan circle, including population, traffic, goods, information, and capital flows. Spatial mapping analysis, social network analysis, and geographical detector are applied to systematically explore the spatial networks of the metropolitan circle and the determining factors. The remaining part of this article is organized as follows. The second part presents the study area and data sources, the third part describes the various research methods adopted in this study, the fourth part describes the major findings, the fifth part shows the discussion and policy implications, and the sixth part concludes the paper.

2. Materials and Methods

2.1. Study Area

The Hangzhou metropolitan area is composed of four prefecture-level cities that are Hangzhou, Shaoxing, Jiaxing, and Huzhou (Figure 1), which is located in the south of the Yangtze River Delta urban agglomeration and in the northern part of Zhejiang Province. The research objective is to be attained at the county level, including four urban districts and 20 sub-cities in four prefecture-level cities (Figure 1). The district is the most important economic and political center of a prefecture-level city, so it is regarded as the central city of a prefecture-level city, and the others are regarded as sub-cities. The division of Hangzhou's central city is based on the Hangzhou 14th Five-Year Plan. Hangzhou metropolitan circle covers an area of 34,585 km^2, in the proportion of 33.21% of the total area of Zhejiang Province; it has a resident population of 23.28 million, which is 39.81% of the total population of Zhejiang Province; its GDP reached CNY 3.07 trillion in 2020 which is 47.5% of the total provincial GDP; and its per capita GDP was CNY 129,000, which was 1.2 times the provincial average. In 2007, the first joint meeting of mayors of the Hangzhou Metropolitan Economic Circle was held, marking the start of the construction of the Hangzhou Metropolitan Circle. In 2014, the National Development and Reform Commission (NDRC) approved a comprehensive reform pilot project for the economic transformation and upgradation of the Hangzhou Metropolitan Circle. Owing to long-term regional cooperation and coordinated development, the co-urbanization effect of these four cities has become increasingly significant. Hangzhou metropolitan circle has become one of the most populous and open regions in China. Spatial network elements such as population flow, goods flow, and information flow are closely related in this metropolitan circle. However, it also suffers from some development constraints, such as widening regional development gaps, weak regional division and cooperation, and a lack of regional coordination and consultation mechanisms [44]. Therefore, investigating the spatial heterogeneity and influencing factors of urban multi-dimensional networks in Hangzhou metropolitan circle is of strategic significance to promote the integration and coordinated development among sub-cities and form an orderly cooperative relationship within the region.

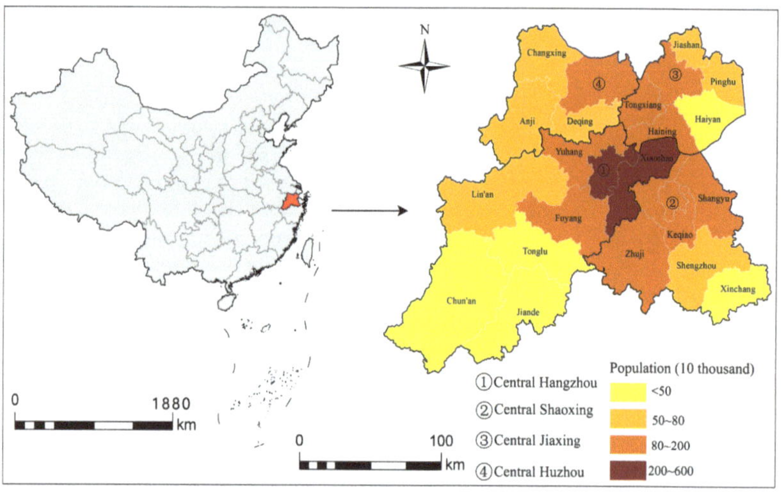

Figure 1. Location of four prefecture-level cities and 24 sub-city divisions in the Hangzhou metropolitan circle, China.

2.2. Data Sources and Preprocessing

In this study, the following five urban big data were used to identify the spatial networks of the metropolitan circle.

(1) Human flow: Population flow is the most frequent connection between sub-cities. Communication is the demand of human daily behavior, and Baidu's migration platform could map the population flow trajectory and spatial distribution through the positioning information of Internet users based on the positioning service technology. The population flow data in this platform has been widely used in China's population flow research [45]. In our study, the population flow data were collected from 1 October 2020 to 18 January 2021. Considering that the data reflects only the relative value of in-migration and out-migration, this paper used the travel index to calculate the floating population matrix among four prefecture cities in Hangzhou metropoli-

tan circle. Then, the population proportions within the 24 sub-cities were counted to construct a 24 × 24 population proportion matrix. Finally, the relative mobile population value between sub-cities was calculated to construct a 24 × 24 human flow matrix [46].

(2) Goods flow: Goods flow refers to the process of transferring goods from the supply location to the receiving location. In this study, the POI points of all logistics expressed in the region were used to construct the logistics matrix. The data were collected in 2020 from 10 major logistics companies in the study area, including ShunFeng, Zhongtong, Yunda, Jingdong, Tiantian, Shentong, Yuantong, Deppon, Baishi, and Jutu. In this study, the method proposed by Taylor [47] was used to construct a 24 × 24 goods flow matrix.

(3) Capital flow: Similar to the goods flow data, all branch institutions of five banking companies in the Hangzhou metropolitan circle, including the Industrial and Commercial Bank of China, Bank of China, Agricultural Bank, Construction Bank, and Postal Savings Bank, were selected for this study [33]. Based on the chain model in urban network analysis, bank POI points were divided into five categories and assigned different values to construct a 24 × 24 capital flow matrix [48].

(4) Information flow: With the development of internet technology, information connection has become an increasingly important method of communication between sub-cities, and sub-cities within metropolitan circles are organically connected through networks.58.com (accessed on 1 September 2023) website is the largest classified information website in China, and has the unique advantage of mutual search among sub-cities [46]. Therefore, this study used the average daily search volume of 58.com in 2020 by using the selenium library in Python to construct a 24 × 24 information flow matrix.

(5) Traffic flow: Traffic flow reflects the connection intensity of sub-cities in the transportation network. The railway is the most important and popular transportation mode for Chinese residents. In this study, the information on the number of station locations and passing trains was obtained from the China Railway Customer Service Center, which is the largest railway information publishing website in China. The distance traveled by car between sub cities was obtained from Gaode Map. These two kinds of data were jointly applied to construct a 24 × 24 traffic flow matrix [49].

2.3. Methods

2.3.1. Strength Analysis of Multiple Urban Flow Networks

Based on the above matrix information of multiple flow networks between sub-cities, the flow space network measuring method proposed by Taylor et al. (2004) was applied to measure the connection strengths of various urban flow networks [47]. Furthermore, we normalized the flow connection strength and summed these data to measure the strength of the comprehensive flow network of the metropolitan circle. The specific formula was as follows:

$$\begin{aligned} S &= R_{ij} \times R_{ji} \\ S' &= (S - S_{\min})/(S - S_{\max}) \\ F_i &= \sum_j S'_{ij} \end{aligned} \tag{1}$$

where S denotes the connection strength of the flow network between sub-city i and j, R_{ij} denotes the amount of "flow data" from sub-city i to sub-city j, and R_{ji} denotes the amount of "flow data" from sub-city j to sub-city i. S' is the normalized value of the network connection strength, S_{min} is the minimum value of the network connection strength, and S_{max} is the maximum value of the network connection strength. F_i is the connection strength of the comprehensive flow network of sub-city i, and S'_{ij} is the connection strength between sub-city i and j. Considering that various urban flow networks are equally important for the integration development of metropolitan circles, each element flow [31] is given a weight of 0.2.

2.3.2. Social Network Structure Analysis

(1) Dominant flow analysis

In this study, the dominant flow analysis method was employed to identify the general spatial structural patterns of flow networks according to the maximal flow rank relationship [50]. The higher value indicates the leading position of a sub-city in the regional flow network. This study selected the top two orders (first and second) to identify the regional main backbone network. The specific formula was as follows:

$$PC = RC_i / \sum_{n} RC_i \quad (2)$$

where PC is the relative centrality index, and RC_i is the strength of the flow connection under the i-th order.

(2) Connectivity influence degree

This method comprehensively measures the total influence of a sub-city in regional spatial structure. The higher the value is, the stronger the node effect of a sub-city, and the wider the network interaction range. This index has been widely used to reflect the influence of a sub-city in regional spatial networks [51]. The specific formula was as follows:

$$PI = \frac{\sum_{i=1}^{5} RC_i}{V_{AD} + 3V_{SD}} \quad (3)$$

where PI is the influence index of a sub-city in regional flow networks, and V_{AD} and V_{SD} are the average value and standard deviation value of RC_i, respectively, and RC_i is the flow connection strength between sub-cities under the i-th order.

(3) Dominant structural index

Based on the spatial structure algorithm of regional poly-centricity proposed by Hanssens et al. [51], this paper employed a spatial structure dominance index to reflect the network dominance level. The index is in the range of 0 to 1, where 0 suggests a distinct unipolar development trend in the regional spatial structure, and 1 suggests obvious multiple polar features in regional spatial structure. The specific formula was as follows:

$$SSI = \begin{cases} (2 - SD/SD_{rc})/2, SD \leq SD_{rc} \\ SD_{rc}/2SD, SD > SD_{rc} \end{cases} \quad (4)$$

where SSI is the regional spatial structure index, SD is the standard deviation value of the connection strength of flow in a sub-city, and SD_{rc} is the standard deviation value of the connection strength of flow in all sub-cities after ranking.

(4) Visual analysis of combined geographic–topological networks

In this study, we visualized the flow connection by using the chord diagram, the length of the arc was used to represent the size of the element attribute, and the thickness of the connections among the nodes was used to reflect the strength of the connections, which can directly reveal the topological interactions among unit nodes. Meanwhile, the distribution patterns of multiple flow strength were superimposed with geographic space, and the structure of flow connection was displayed to explore the spatial networks' mechanism of the metropolitan circle by using the ArcGIS spatial analysis platform.

2.4. Geographical Detector

(1) Model construction

In this study, we applied the geographical detector to identify the factors that influence the spatial structure of urban multi-dimensional networks in the metropolitan circle. The principle of geographic detectors is to assume that the independent variable and the dependent variable are similar in spatial distribution, and then the independent variable has a significant impact on the dependent variable [52,53]. This study used the factor detector and interactive detector modules to explore the formation mechanism of the urban multi-dimensional networks in the metropolitan circle. The detailed formula was as follows:

$$q = 1 - \frac{1}{N\sigma^2} \sum_{n=1}^{L} N_h \sigma_h^2 \tag{5}$$

where q measures the interpretative power of the independent variable, h is the stratification of the dependent or independent variable, and N_h and σ_h denote the number of sub-cities and variance in strata h, respectively. The q value is in the range of 0 to 1. A higher q value represents a stronger explanatory power of the independent variable on the dependent variable.

Furthermore, the interactive detector was used to measure the interactions between driving factors and to reveal the impact of the joint effects of explanatory variables on the explanatory power of dependent variable. The judgment types of interaction between two factors can be classified into univariate nonlinear weakening, nonlinear weakening, bivariate enhancement, and independent and nonlinear enhancement [54].

(2) Selection of driving factors

Based on the above analysis, this study selected twelve representative factors as drivers along four dimensions: population, economy, society, and distance. Population size (x1), size of working-aged population (x2), and population urbanization rate (x3) were selected to reflect the demographics of a sub city; GDP per capita (x4), gross output value of enterprise above the designated size (x5), and total retail sales of social consumer goods (x6) were selected to measure the economic development of a sub-city; number of hospital beds (x7), road density (x8), and social security expenditure and employment fiscal expenditure (x9) were selected to measure the social development of a sub-city; and geographic distance (x10), traffic distance (x11), and industrial distance (x12) were selected to measure the distances between sub-cities. Specifically, geographical distance was measured by the linear distance between the administrative centers of sub-cities; traffic distance was calculated by the shortest railway transportation time; industry distance was represented by industrial structural similarity index [39,43].

3. Results

3.1. The Strength of Multiple Urban Flow Networks in the Metropolitan Circle

Figure 2 shows the connection strengths of multiple flow networks of 24 sub-cities within four (prefecture-level) cities in Hangzhou metropolitan circle. As shown, the connection strengths of the population and goods flow networks were generally higher than the others, and the connection strength of information flow network was the lowest. In terms of the five flow networks, the spatial connections of population, traffic, goods, and capital flows showed a network radial distribution pattern with Central Hangzhou as the core, indicating that the four flow networks transcended geographical proximity. The information flow connections between sub-cities were mainly distributed within the city, indicating that the information flow did not form an open network spatial structure within Hangzhou metropolitan area. In terms of the connection strength, the top three sub-cities with the strongest connection strength of population, information, and capital flow were the Central Hangzhou, Xiaoshan, and Yuhang, indicating that these sub-cities were the main concentrated areas of population, information, and capital in Hangzhou metropolitan area. Regarding the traffic flow, the connection strength in Central Shaoxing was stronger, it was attributed to the fact that Shaoxing city is the central hub of the Hangzhou and Ningbo metropolitan areas in Eastern China, and has complete high-speed and railway transportation. In terms of the goods flow, Zhuji, one of the important production bases of small commodities in Zhejiang Province, became one of the top three sub-cities with the strongest connection strength, indicating that the rapid growth of online retail industry has greatly enhanced the position of Zhuji in the goods network.

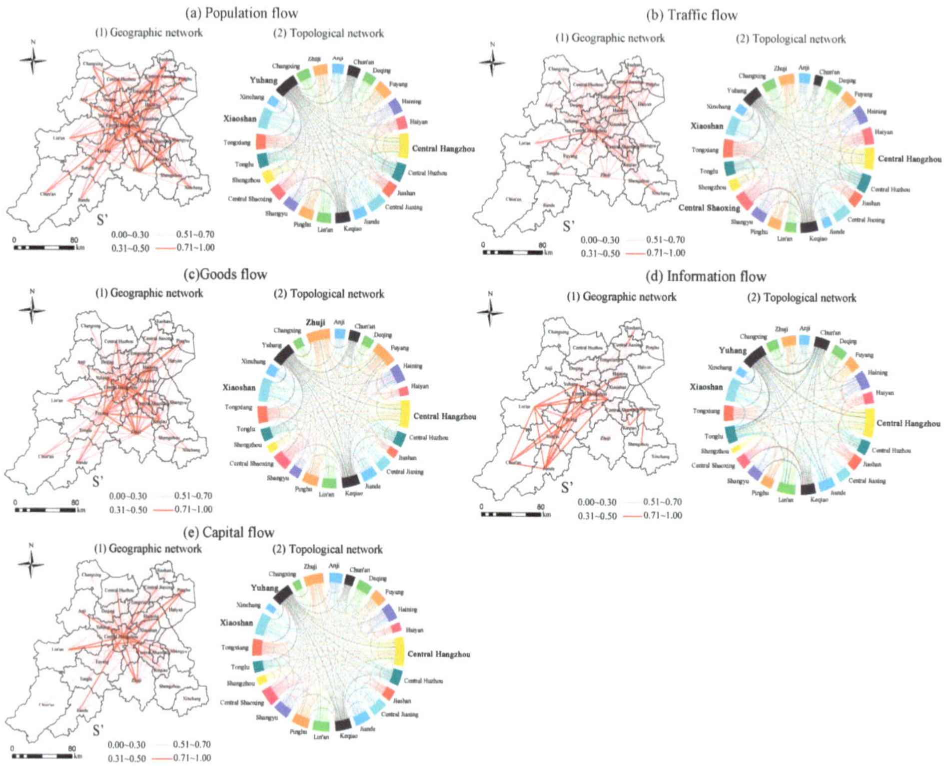

Figure 2. Strength of multiple spatial networks of Hangzhou metropolitan circle.

3.2. *The Hierarchical Structure of Multiple Urban Flow Networks in the Metropolitan Circle*

We used the connectivity influence degree to measure the connection influence of flow networks among sub-cities (Figure 3a–e). The high-value areas of connection influence of population and information flow networks were in Central Hangzhou, Xiaoshan, and Yuhang, indicating that the three sub-cities were dominant in the population and information flow networks of the Hangzhou metropolitan circle. Remote areas, such as Chun'an, Jiande, Xinchang, and Anji, showed low connectivity influence degree of regional population and information flow networks. The connection influence of traffic flow network in Central Hangzhou and Xiaoshan was at a high level, indicating that these two sub-cities took the lead in regional traffic flow network. While the low-value areas of traffic flow network were mainly distributed in the northwestern and southwestern areas. In terms of the goods network, the high-value areas of connection influence were in Central Hangzhou, Fuyang, and Zhuji, indicating that these three sub-cities had a greater influence on regional goods flow network. Regarding the capital network, Central Hangzhou, Xiaoshan, and Zhuji were the high-value areas of connection influence, indicating that the three sub-cities were in a leading position in regional capital network. While the low-value areas were located in the periphery of the metropolitan circle. Furthermore, the dominant structural index was used to identify the spatial hierarchical structure of multiple flow networks. The SSI value of the goods network was the smallest, at 0.295, followed by the capital flow network, indicating that the two flow networks presented a more pronounced unipolar development trend. The SSI value of traffic flow networks was higher than 0.50, indicating that its structure was relatively discrete.

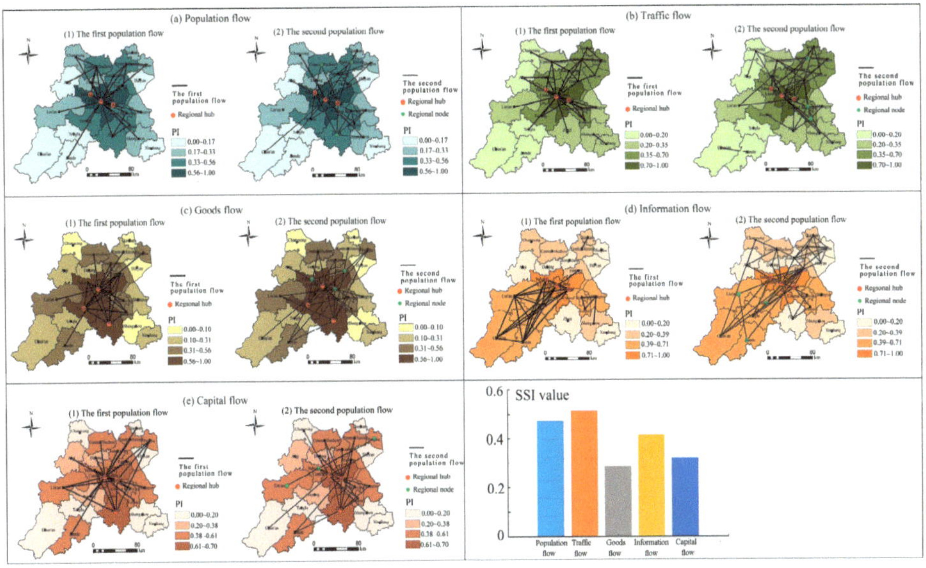

Figure 3. The connection influence and structure of multiple urban networks in Hangzhou metropolitan circle.

To further identify the structural model of multiple flow networks, the dominant flow analysis method was used to identify the first- and second-largest dominant flow networks (Figure 3). Specifically, the first-largest dominant networks of population, traffic, goods, and capital flow presented a radial network axis with Central Hangzhou, Xiaoshan, Yuhang, Zhuji, and Fuyang as the hubs. While the first-largest network of information flow was mainly concentrated in Hangzhou city, and considered Central Hangzhou, Xiaoshan, and Yuhang as the hubs, confirming that the influence of information flow has not yet overcome the restrictions of administrative boundaries. In terms of the second-largest dominant network, different element flows presented distinct patterns of spatial structure. The second-largest dominant networks of population and traffic flow were mainly distributed in the east and north of Hangzhou metropolitan circle, considering some sub-cities of Huzhou and Jiaxing as nodes. The second-largest dominant network of goods and information flow connected most sub-cities in a decentralized form, taking some sub-cities of Hangzhou as nodes. The second-largest dominant network of capital flow mainly distributed in Hangzhou and Shaoxing cities, and took Lin'an, Yuhang, and Pinghu as nodes.

3.3. Spatial Pattern of the Comprehensive Urban Flow Network in the Metropolitan Circle

The strength and pattern of the urban comprehensive flow network in Hangzhou metropolitan circle is described in Figure 4. The distribution of connectivity influence was imbalanced across sub-cities, presenting a hierarchical structure in the Hangzhou metropolitan circle. Three sub-cities, from Hangzhou, including Central Hangzhou, Yuhang, and Xiaoshan, had the highest values of connectivity influence on the comprehensive flow network, with the other sub-cities showing a decreasing trend in distance from Central Hangzhou. This indicated that the comprehensive flow network of the Hangzhou metropolitan circle presented an obvious spatial structure of core–periphery. These imbalanced distributions of comprehensive flow network were may be due to the varying locations, population, and economic sizes of sub-cities. In terms of the structural pattern of the comprehensive flow network, its first-largest dominant network considered Central

Hangzhou, Xiaoshan, and Yuhang as hubs, and was mainly distributed in Hangzhou and Shaoxing cities. This indicated that the connection between Hangzhou and Shaoxing was more close. The second-largest dominant network considered Fuyang, Keqiao, and Central Shaoxing as nodes, suggesting that the three sub-cities served as a "bridge" connecting the center and the peripheries.

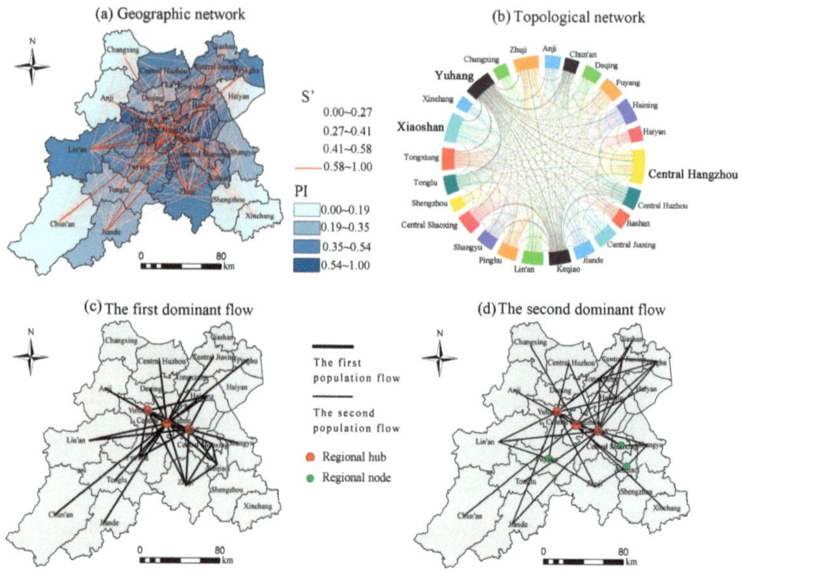

Figure 4. The strength and pattern of the comprehensive flow network in Hangzhou metropolitan circle.

3.4. Factors Influencing Comprehensive Flow Network of Metropolitan Circle

We used a geographic detector to identify the dominant factors on the distribution pattern of the comprehensive flow network in metropolitan circle. In this study, the dependent variable was the connection strength of the comprehensive flow network in Hangzhou metropolitan circle, and the independent variables included the twelve above-mentioned representative factors from four dimensions of population, economy, society, and distance. Significant influencing factors and their interpretative powers (q-values) are shown in Table 1.

Table 1. Influencing factor detection of spatial differentiation of the comprehensive spatial network.

Variables	q Statistics	Rank	Variables	q Statistics	Rank
X_1	0.8398 ***	2	X_8	0.6562 **	7
X_3	0.8353 ***	3	X_9	0.6094 **	8
X_5	0.7302 ***	6	X_{10}	0.7561 ***	5
X_6	0.8535 ***	1	X_{11}	0.8275 ***	4

Note: ***: $p < 1\%$, **: $p < 5\%$, where p represents the level of significance, $p < 5\%$ means significant difference, and $p < 1\%$ means significant difference.

From a horizontal comparison point of view, the spatial differentiation of the comprehensive flow network was influenced by a combination of socioeconomic and geographic distance. The top four factors influencing the spatial differentiation of the comprehensive flow network were, in order, the total retail sales of social consumer goods, population urbanization rate, size of working-aged population, and traffic distance, which can be regarded as the dominant factors. This indicated that regional economic development, population growth, and transportation connections were the dominant factors affecting the connection strength of the comprehensive flow network of Hangzhou metropolitan circle.

An interactive detector was used to reveal the interpretative power of the interaction between two influencing factors on the spatial heterogeneity of the comprehensive flow network. Compared with the influence of a single factor, the interpretative power of most two-factor interactions was stronger (Table 2), indicating that the two-factor interaction enhanced the spatial heterogeneity of the comprehensive flow network of the Hangzhou metropolitan circle. Specifically, the interaction type of the population size and the other significant factors was bivariate enhanced, indicating that the interaction between population increase and other socioeconomic factors enhanced the spatial differentiation of the comprehensive flow network. The interaction between population size and gross output value of enterprise above the designated size was stronger than that with other factors. In the interaction between population urbanization rate and the other factors, their interaction type was bivariate enhanced, except for the interaction between population urbanization rate and road density. The interaction between population urbanization rate and total retail sales of social consumer goods was stronger, and the interaction between population urbanization rate and road density was the weakest. The main interaction type between the gross output value of enterprise above the designated size, the total retail sales of social consumer goods, and other factors was the bivariate enhancement type, indicating that the dual effect of economic development and other socioeconomic factors promoted the spatial differentiation of the comprehensive flow network. The interaction type between gross output value of enterprise above the designated size above the designated size and road density was single-factor nonlinear attenuation with the weakest status, indicating that the interaction between them weakened the regional differentiation of the comprehensive flow network. The interaction type of road density, social security and employment fiscal expenditure, and other factors was bivariate enhanced, and the interaction between road density and social security and employment fiscal expenditure was the weakest. The interaction type between geographic distance and traffic distance was also bivariate enhancement, indicating that the interaction between them enhanced the spatial differentiation of the comprehensive flow network.

Table 2. Interactive influence results of spatial differentiation of the comprehensive spatial network.

$X_i \cap X_j$	q ($X_i \cap X_j$)	Interaction Type	$X_i \cap X_j$	q ($X_i \cap X_j$)	Interaction Type	$X_i \cap X_j$	q ($X_i \cap X_j$)	Interaction Type
$X_1 \cap X_3$	0.9125	Bivariate enhancement	$X_3 \cap X_9$	0.8801	Bivariate enhancement	$X_6 \cap X_9$	0.9013	Bivariate enhancement
$X_1 \cap X_5$	0.9250	Bivariate enhancement	$X_3 \cap X_{10}$	0.8925	Bivariate enhancement	$X_6 \cap X_{10}$	0.9048	Bivariate enhancement
$X_1 \cap X_6$	0.8621	Bivariate enhancement	$X_3 \cap X_{11}$	0.8675	Bivariate enhancement	$X_6 \cap X_{11}$	0.9301	Bivariate enhancement
$X_1 \cap X_8$	0.8603	Bivariate enhancement	$X_5 \cap X_6$	0.9626	Bivariate enhancement	$X_8 \cap X_9$	0.7605	Bivariate enhancement
$X_1 \cap X_9$	0.9018	Bivariate enhancement	$X_5 \cap X_8$	0.7129	Single-factor Nonlinear attenuation	$X_8 \cap X_{10}$	0.8742	Bivariate enhancement
$X_1 \cap X_{10}$	0.9240	Bivariate enhancement	$X_5 \cap X_9$	0.9581	Bivariate enhancement	$X_8 \cap X_{11}$	0.8320	Bivariate enhancement
$X_1 \cap X_{11}$	0.9152	Bivariate enhancement	$X_5 \cap X_{10}$	0.8202	Bivariate enhancement	$X_9 \cap X_{10}$	0.8250	Bivariate enhancement
$X_3 \cap X_5$	0.9176	Bivariate enhancement	$X_5 \cap X_{11}$	0.9266	Bivariate enhancement	$X_9 \cap X_{11}$	0.9139	Bivariate enhancement
$X_3 \cap X_6$	0.9208	Bivariate enhancement	$X_6 \cap X_8$	0.8592	Bivariate enhancement	$X_{10} \cap X_{11}$	0.8409	Bivariate enhancement
$X_3 \cap X_8$	0.8187	Single-factor nonlinear attenuation						

4. Discussion

Under the tide of globalization and information, the interactive relationships between cities have transformed from spatial proximity to a connection network dominated by time proximity [55]. Compared with the traditional statistical data-based spatial structure model, this study used the geospatial big data and social network analysis to identify the various urban flow networks of metropolitan circle, which can provide a new perspective and method to comprehensively evaluate the spatial development of the metropolitan circle in the context of single flow-led studies [7]. Furthermore, considering that the connection strength between sub-cities is significantly affected by location conditions and socioeconomic development, this study used the geographical detector to identify the determining factors of the flow networks in the metropolitan circle, which contributes to providing references for the integration development of the metropolitan circle.

The results of our study indicated that the connection strengths of the population and traffic flow networks were higher, and this finding is confirmed in other related studies of metropolitan circles and urban agglomerations [25,26]. They pointed out that population and goods flows were the most frequent in the flow networks between cities Furthermore, this study found that there were significant differences in the connection strength of different types of flow networks in Hangzhou metropolitan circle, especially the information flow, and its connection strength between sub-cities was limited by the administrative boundaries of prefecture-level cities. This phenomenon indicated that information network in Hangzhou metropolitan circle was not strong, reflecting the limited development of metropolitan circle [33]. Furthermore, the results showed that the flow connections between sub-cities from Hangzhou, Jiaxing, and Shaoxing was stronger than other sub-cities. This is because the local government proposed the development strategies of Hangzhou–Shaoxing and Hangzhou–Jiaxing integration development in 2018 and 2020, respectively. The construction of intercity transportation and public service facilities and the coordinated division of industries have strengthened the connection of multiple flows between these sub-cities [30,56].

Regarding the structural characteristics of various urban flow networks in Hangzhou metropolitan circle, the spatial urban network of the metropolitan circle presented hierarchical and unipolar characteristics. This characteristic is quite different from the dual-center or multi-center structure exhibited by urban agglomerations [39,55]. Corresponding with the Hangzhou metropolitan circle, three sub-cities from Hangzhou city, including central Hangzhou, Xiaoshan, and Yuhang, had the strongest connection influence in the flow networks, which means that various flow networks reinforce the existing urban system. This finding indicated that there is a positive correlation between flow network connections and the hierarchy of sub-cities with a metropolitan circle, where sub-cities with higher administrative levels would play stronger roles in urban network connections [57]. Additionally, this study also discovered that some sub-cities with advantageous industries played important roles as nodes in regional flow network. For example, Zhuji was regarded as the hub of goods network in Hangzhou metropolitan circle. This is due to the fact that there are many small commodity manufacturers in Zhuji, and the rapid development of the online retail industry has greatly enhanced its status in the goods network.

In the analysis of influencing factors, the dominant factors influencing the spatial differentiation of the network structure were population, social consumption, and traffic distance. This result is consistent with the findings of Zheng et al. [39]. Earlier work noted that in highly urbanized cities, social and economic activities are more prosperous, and various element flows tend to be more frequent [58]. With the continuous improvement of traffic infrastructure, the impact of traffic distance on the spatial differentiation of the regional flow network structure was found to be stronger than that of geographical distance, which is in line with Chen et al. [59]. Furthermore, the results also suggest that the interactions between socioeconomic and traffic distance have an influence on reinforcing the status of cities. This is because that in the process of regional development, the interactions between socioeconomic activities and transportation infrastructure can accelerate the flow and aggregation of various elements, promoting the status of a city in the regional network structure [39,60].

Based on the above study results, some policy implications to promote the integrated development of metropolitan areas could be suggested. First, using multi-source big data, this study found that the connection strengths of the population, goods, and traffic flow networks between sub-cities were strong, while the connections of the information and capital flow networks were relatively weak. Therefore, the government ought to increase investment in information network infrastructure and apply information technologies such as artificial intelligence, cloud computing, and big data to accelerate the development of digital financial networks [49]. Second, according to the results of this study, the spatial structure of urban network in Hangzhou metropolitan circle was characterized by a core–periphery structure. Numerous population, capital, technology, and other resources flow into Central Hangzhou, resulting in limited development of marginal areas. Therefore, in order to transform the regional spatial structure into a balanced "network–node" pattern, the government should support the free-flow enhancement of the radiation and driving role of hubs and core nodes, and focus on developing the linkage ability of peripheral sub-cities to connect with extra-regional networks [33,51]. Finally, the results showed that traffic distance and geographical distance were found to have significant impacts on the spatial network structure of metropolitan circles, and the interactions of traffic distance with population size and economic development would reinforce their influence. Thus, local governments could promote the development of transportation infrastructure represented by high-speed railways, expressways, and intercity intelligent transportation, which can greatly shorten the space-time distance between sub-cities. Additionally, provincial government could innovatively develop more cross-regional and closely related group areas to rationally allocate population and industries, information infrastructure and public service facilities can be improved to accelerate the flow and connection of regional resources. These series of cross-regional construction measures contribute to promoting the polycentric development of regional spatial structure.

The contributions of this study to the analysis of regional flow networks include two aspects. First, this study employed multi-source big data to systematically investigate the connections and structures of various flow networks between sub-cities in the metropolitan circle, which is conducive to analyze the spatial structure of metropolitan circles and its differences from the perspective of different element flow. Most studies identify regional spatial structure by using data from a single flow category [22,55], and fail to reveal the multi-dimensional spatial network structure. The second contribution is that this study quantitatively explored the distance and socioeconomic factors that influence the spatial differentiation of the connection strength of comprehensive flow network, which could provide policy inspiration for promoting the integrated development of metropolitan circles. It is worth pointing out that as far as we know, there are few studies on the analysis of the determinants of regional flow network connections combining geographic, demographic, social, and economic dimensions as a result of the limitation of big data and methods. This study is an attempt to fulfil these research gaps in this regard.

Overall, this study also faces two limitations. First, although the big data used in this study are for the sake of data availability and convenience, the accuracy of the results is still slightly insufficient. For example, the population flow ignores the travel data of private cars, and the goods flow does not consider large-scale material transportation. Thus, future research needs to refine the comprehensiveness and accuracy of the data. Second, this study does not consider the temporal evolution of the spatial network structure of the metropolitan circle, and it is essential to reveal the internal mechanism of regional spatial structure from a time-across perspective [61]. Future research needs to investigate the spatiotemporal characteristics of the spatial network structure of metropolitan circles based on spatiotemporal panel data.

5. Conclusions

The connection strengths and spatial structure of multiple flow networks between sub-cities in metropolitan circles is a critical basis to evaluate the integrated development of metropolitan circles. Considering the Hangzhou metropolitan circle as a case study area, this study examined the spatial network structure of the metropolitan circle based on multiple big data, and determining factors influencing the spatial differentiation of the comprehensive flow network between sub-cities in metropolitan circle were identified using a geographical detector.

The following are the main findings of this study. The connection strengths and spatial distribution vary in different types of elements in the metropolitan circle. Central Hangzhou, Yuhang, and Xiaoshan from Hangzhou city as well as some sub-cities with characteristic industrial advantages were identified as regional nodes in different flow networks. The distribution of the connectivity influence on comprehensive flow network presented hierarchical and unipolar characteristics in the metropolitan circle, which is quite different from the multi-center structure of urban agglomerations. The spatial differentiation of the connection strength of comprehensive flow network was strongly affected by distance and socioeconomic factors in Hangzhou metropolitan circle, especially the traffic distance, and its interactions with socioeconomic factors would strongly enhance the spatial differentiation of flow network.

Author Contributions: Conceptualization, J.Z. and C.Z.; methodology, L.L. and Q.H.; software, Z.Y.; formal analysis, W.L.; investigation, X.C.; writing—original draft preparation, M.G. and C.Z. All authors have read and agreed to the published version of the manuscript.

Funding: This research was funded by the Basic scientific research business cost project of Zhejiang Gongshang University, grant number XT202209, the National Natural Science Foundation of China, grant number 41701171, and the Open Fund of Key Laboratory of Urban Land Resources Monitoring and Simulation, Ministry of Natural Resources, grant number KF-2020-05-073.

Data Availability Statement: Not applicable.

Conflicts of Interest: The authors declare no conflict of interest.

References

1. Hall, P.G.; Pain, K. *The Polycentric Metropolis: Learning from Mega-City Regions in Europe*; Routledge: London, UK, 2006.
2. Fang, C. China's urban agglomeration and metropolitan area construction under the new development pattern. *Econ. Geogr.* **2021**, *41*, 1–7. (In Chinese)
3. Li, X.; Zhou, J.; Huang, Y. Understanding the Guangdong-Hong Kong-Macao Greater Bay Area from the Perspective of Mega-City Region. *Prog. Geogr.* **2018**, *37*, 1609–1622. (In Chinese)
4. Rodrigue, J.-P.; Comtois, C.; Slack, B. *The Geography of Transport Systems*; Routledge: London, UK, 2009.
5. Wang, Y. Analysis on the Evolution of Spatial Relationship Between Population and Economy in the Beijing-Tianjin-Hebei and Shandong Region of China. *Sustain. City Soc.* **2022**, *83*, 103948.
6. Tu, W.; Zhu, T.; Xia, J.; Zhou, Y.; Lai, Y.; Jiang, J. Portraying the Spatial Dynamics of Urban Vibrancy Using Multisource Urban Big Data. *Comput. Environ. Urban Syst.* **2020**, *80*, 101428. [CrossRef]
7. Takeuchi, H.; Kozu, H. Spatial structure of railroad station areas in metropolitan areas:applying a gravity model and rank-size model. *Stud. Reg. Sci.* **2015**, *45*, 73–86. [CrossRef]
8. Castells, M. Space for Flows, Space of Places: Materials for a Theory of Urbanism in the Information Age. In *The City Reader*; Routledge: London, UK, 2015; pp. 263–274.
9. Niu, X.; Wang, Y.; Liu, J.; Feng, Y. Special Structure of Shanghai Conurbation Area from Perspective of Inter City Functional Links. *Urban Plan. Forum* **2018**, *5*, 80–87.
10. Fischer, M.M. Regional Taxonomy: A Comparison of Some Hierarchic and Nonhierarchic Strategies. *Reg. Sci. Urban Econ.* **1980**, *10*, 503–537. [CrossRef]
11. Feng, J.; Zhong, Y. Restructuring of Social Space in Beijing from 2000 to 2010. *Acta Geogr. Sin.* **2018**, *73*, 711–737. (In Chinese)
12. Yang, J.; Song, G.; Lin, J. Measuring Spatial Structure of China's Megaregions. *J. Urban Plann. Dev.* **2014**, *141*, 04014021. [CrossRef]
13. Harris, C.D.; Ullman, E.L. The Nature of Cities. *Ann. AAPSS* **1945**, *242*, 7–17. [CrossRef]
14. Anselin, L. *Spatial Econometrics: Methods and Models*; Springer: Dordrecht, The Netherlands, 1988; Volume 4.
15. Wang, L.; Zhao, P. From Dispersed to Clustered: New Trend of Spatial Restructuring in China's Metropolitan Region of Yangtze River Delta. *Habitat Int.* **2018**, *80*, 70–80. [CrossRef]
16. Derudder, B.; Taylor, P.J.; Hoyler, M.; Ni, P.; Liu, X.; Zhao, M.; Witlox, F. Measurement and Interpretation of Connectivity of Chinese Cities in World City Network, 2010. *Chin. Geogr. Sci.* **2013**, *23*, 261–273. [CrossRef]
17. Zhang, P.; Zhao, Y.; Zhu, X.; Cai, Z.; Xu, J.; Shi, S. Spatial Structure of Urban Agglomeration Under the Impact of High-Speed Railway Construction: Based on the Social Network Analysis. *Sustain. Cities Soc.* **2020**, *62*, 102404. [CrossRef]
18. Taylor, P.J.; Derudder, B. *World City Network: A Global Urban Analysis*; Routledge: Abingdon, UK, 2015.
19. Meijers, E. From Central Place to Network Model: Theory and Evidence of a Paradigm Change. *Tijdschr. Voor Econ. En Soc. Geogr.* **2007**, *98*, 245–259. [CrossRef]
20. Du, W.; Zhang, Q.; Chen, Y.; Ye, Z. An Urban Short-Term Traffic Flow Prediction Model Based on Wavelet Neural Network with Improved Whale Optimization Algorithm. *Sustain. City Soc.* **2021**, *69*, 102858. [CrossRef]

21. Abbasi, A.; Altmann, J.; Hossain, L. Identifying the Effects of Co-authorship Networks on the Performance of Scholars: A Correlation and Regression Analysis of Performance Measures and Social Network Analysis Measures. *J. Informetrics* **2011**, *5*, 594–607. [CrossRef]
22. Hui, E.; Li, X.; Chen, T.; Lang, W. Deciphering the Spatial Structure of China's Megacity Region: A New Bay Area—The Guangdong-Hong Kong-Macao Greater Bay Area in the Making. *Cities* **2020**, *105*, 102168. [CrossRef]
23. Chao, H.; Ma, X.; Li, G. Characteristics and Changes of the Polycentric Spatial Structure in the Pearl River Delta Region. *Areal Res. Dev.* **2014**, *33*, 12–18. (In Chinese)
24. Guo, D.; Jin, H.; Gao, P.; Zhu, X. Detecting Spatial Community Structure in Movements. *Int. J. Geogr. Inf. Sci.* **2018**, *32*, 1326–1347. [CrossRef]
25. Zhang, W.; Chong, Z.; Li, X.; Nie, G. Spatial Patterns and Determinant Factors of Population Flow Networks in China: Analysis on Tencent Location Big Data. *Cities* **2020**, *99*, 102640. [CrossRef]
26. Zhao, P.; Liu, X.; Shi, W.; Jia, T.; Li, W.; Chen, M. An Empirical Study on the Intra-Urban Goods Movement Patterns Using Logistics Big Data. *Int. J. Geogr. Inf. Sci.* **2018**, 1520236. [CrossRef]
27. Zhang, W.; Fang, C.; Zhou, L.; Zhu, J. Measuring Megaregional Structure in the Pearl River Delta by Mobile Phone Signaling Data: A Complex Network Approach. *Cities* **2020**, *104*, 102809. [CrossRef]
28. Demšar, U.; Reades, J.; Manley, E.; Batty, M. Revisiting the Past: Replicating Fifty-Year-Old Flow Analysis Using Contemporary Taxi Flow Data. *Ann. Am. Assoc. Geogr.* **2018**, *108*, 811–828. [CrossRef]
29. Alessandra, F.; Vicente, R. Migration Flows and Quality of Life in a Metropolitan Area: The Case of Barcelona, Spain. *Appl. Res. Qual. Life* **2010**, *5*, 241–259.
30. Zhang, W.; Thill, J. Mesoscale Structures in World City Networks. *Ann. Am. Assoc. Geogr.* **2019**, *109*, 887–908. [CrossRef]
31. Liu, D.; Shi, Y. The Influence Mechanism of Urban Spatial Structure on Urban Vitality Based on Geographic Big Data: A Case Study in Downtown Shanghai. *Buildings* **2022**, *12*, 569. [CrossRef]
32. De Montis, A.; Barthélemy, M.; Chessa, A.; Vespignani, A. The Structure of Interurban Traffic: A Weighted Network Analysis. *Environ. Plan. B Plan. Des.* **2007**, *34*, 905–924. [CrossRef]
33. Wang, S.; Gao, S.; Wang, Y. Spatial Structure of the Urban Agglomeration Based on Space of Flows: The Study of the Pearl River Delta. *Geogr. Res.* **2019**, *38*, 1849–1861. (In Chinese)
34. Cao, X.; Ouyang, S.; Yang, W.; Luo, Y.; Liu, D. Transport Accessibility and Spatial Connections of Cities in the Guangdong-Hong Kong-Macao Greater Bay Area. *Chin. Geogr. Sci.* **2019**, *29*, 820–833. [CrossRef]
35. Jendryke, M.; Balz, T.; McClure, S.C.; Liao, M. Putting People in the Picture: Combining Big Location-Based Social Media Data and Remote Sensing Imagery for Enhanced Contextual Urban Information in Shanghai. *Comput. Environ. Urban Syst.* **2017**, *62*, 99–112. [CrossRef]
36. Jiang, J.; Li, Q.; Tu, W.; Shaw, S.-L.; Yue, Y. A Simple and Direct Method to Analyze the Influences of Sampling Fractions on Modeling Intra-City Human Mobility. *Int. J. Geogr. Inf. Sci.* **2019**, *33*, 618–644. [CrossRef]
37. Liu, C.; Guan, M.; Duan, D. Spatial Pattern and Influential Mechanism of Interurban Technology Transfer Network in China. *Acta Geogr. Sin.* **2018**, *73*, 1462–1477. (In Chinese)
38. Gong, P.; Yang, J. Spatial Structure of Urban Agglomerations in China Based on Microblog Data. *Urban Dev. Res.* **2020**, *27*, 1 8, 61, 181.
39. Zheng, L.; Gu, W.; Long, F.; Zhang, S. The Urban Network Structure and Influential Mechanism of Guizhou Province from the Perspective of Various Flows. *Sci. Geogr. Sin.* **2020**, *40*, 939–947. (In Chinese)
40. Cordon, P.; Richardson, H.; Wong, H. The Distribution of Population and Employment in a Polycentric City: The Case of Los Angeles. *Environ. Plan. A Econ. Space* **1986**, *18*, 161–173. [CrossRef]
41. Jin, B.; Yan, L.; Sha, J.; Yao, Z. The Spatial Difference of the Intensity of Information and Capital Flow in Chinese Provincial Region: A Case Study of Mobile Phone Transactions in Jingdong Mall in 2015. *Sci. Geogr. Sin.* **2018**, *38*, 223–232. (In Chinese)
42. Li, Y.; Wu, Q.; Zhang, Y.; Zhang, H.; Fang, S.; Jin, S. Spatial Structure and Formation Mechanism of Expressway Traffic Flow Network Based on Space of Flows: A Case Study of Guangdong Province. *Geogr. Res.* **2021**, *40*, 2204–2219. (In Chinese)
43. He, C.; Jin, L.; Liu, Y. How Does Multi-Proximity Affect the Evolution of Export Product Space in China. *Geogr. Res.* **2017**, *36*, 1613–1626.
44. Lian, J.; Zeng, G.; Teng, T. Research on the Difference of Function Diffusion of the Central City Based on the Economic Connection Intensity: A Case of the Central City of Hangzhou Metropolitan Area and Its Compact Cities. *Areal Res. Dev.* **2017**, *36*, 54–58. (In Chinese)
45. Ye, Q.; Zhang, L.; Peng, P.; Huang, J. The Network Characteristics of Urban Agglomerations in the Middle Reaches of the Yangtze River Based on Baidu Migration Data. *Econ. Geogr.* **2017**, *37*, 53–59.
46. Zhang, J.; Hao, Q.; Chen, X.; Zhu, C.; Zhang, L.; Hong, M.; Wu, J.; Gan, M. Exploring Spatial Network Structure of the Metropolitan Circle Based on Multi-Source Big Data: A Case Study of Hangzhou Metropolitan Circle. *Remote Sens.* **2022**, *14*, 5266. [CrossRef]
47. Taylor, P.; Evans, D.; Pain, K. Application of the Interlocking Network Model to Mega-City-Regions: Measuring Polycentricity within and Beyond City-Regions. *Reg. Stud.* **2008**, *42*, 1079–1093. [CrossRef]
48. Xing, L.; Du, S.; Sun, G.; Chen, Q. Analysis on Network Structure Characteristics and Its Influencing Factors in Hubei Province Based on the Perspective of Multi-Dimensional Feature Flow. *Resour. Environ. Yangtze Basin* **2022**, 1–13. Available online: http://kns.cnki.net/kcms/detail/42.1320.X.20220822.1731.004.html (accessed on 23 August 2022).

49. Qiu, J.; Liu, Y.; Chen, H.; Gao, F. Urban Network Structure of Guangdong-Hong Kong-Macao Greater Bay Area with the View of Space of Flows: A Comparison Between Information Flow and Transportation Flow. *Econ. Geogr.* **2019**, *39*, 7–15. (In Chinese)
50. Nystuen, J.D.; Dacey, M.F. A Graph Theory Interpretation of Nodal Regions. *Papers Reg. Sci. Assoc.* **1961**, *7*, 29–42. [CrossRef]
51. Hanssens, H.; Derudder, B.; Witlox, S. Assessing the Functional Polycentricity of the Mega-City-Region of Central Belgium Based on Advanced Producer Service Transaction Links. *Reg. Stud.* **2013**, *48*, 1939–1953. [CrossRef]
52. Wang, J.; Xu, C. Geodetector: Principle and Prospective. *Acta Geograph. Sin.* **2017**, *72*, 116–134.
53. Wang, J.; Zhang, T.; Fu, B. A Measure of Spatial Stratified Heterogeneity. *Ecol. Indic.* **2016**, *67*, 250–256. [CrossRef]
54. Hou, M.; Deng, Y.; Yao, S. Coordinated Relationship Between Urbanization and Grain Production in China: Degree Measurement, Spatial Differentiation, and Its Factors Detection. *J. Clean. Prod.* **2022**, *331*, 129957. [CrossRef]
55. Fang, C.; Yu, X.; Zhang, X.; Fang, J.; Liu, H. Big Data Analysis on the Spatial Networks of Urban Agglomeration. *Cities* **2020**, *102*, 102735. [CrossRef]
56. Su, S.; Zhao, C.; Zhou, H.; Li, B.; Kang, M. Unraveling the Relative Contribution of TOD Structural Factors to Metro Ridership: A Novel Localized Modeling Approach with Implications on Spatial Planning. *J. Transport Geogr.* **2022**, *100*, 103308. [CrossRef]
57. Liu, Y.; Zhang, X.; Pan, X.; Ma, X.; Tang, M. The Spatial Integration and Coordinated Industrial Development of Urban Agglomerations in the Yangtze River Economic Belt, China. *Cities* **2020**, *104*, 102801. [CrossRef]
58. Wang, Y.; Niu, X.; Song, X. Research Progress of Regional Spatial Structure under the Perspective of Space of Flow. *Urban Plan. Int.* **2017**, *32*, 27–33. [CrossRef]
59. Chen, W.; Liu, W.; Ke, W.; Wang, N. The Spatial Structures and Organization Patterns of China's City Networks Based on the Highway Passenger Flows. *J. Geogr. Sci.* **2018**, *28*, 477–494. [CrossRef]
60. Zhou, J.; Bi, X.; Zou, W. Driving Mechanism of Urban-Rural Integration in Huaihai Economic Zone: Based on the Space of Flow. *J. Nat. Resour.* **2020**, *35*, 1881–1896. (In Chinese)
61. Zhou, M.; Yue, Y.; Li, Q.; Wang, D. Portraying Temporal Dynamics of Urban Spatial Divisions with Mobile Phone Positioning Data: A Complex Network Approach. *ISPRS Int. J. Geo Inf.* **2016**, *5*, 240. [CrossRef]

land

MDPI

Article

Earth Observation Data and Geospatial Deep Learning AI to Assign Contributions to European Municipalities Sen4MUN: An Empirical Application in Aosta Valley (NW Italy)

Tommaso Orusa [1,2,*], Annalisa Viani [3] and Enrico Borgogno-Mondino [1]

1 Department of Agricultural, Forest and Food Sciences (DISAFA), GEO4Agri DISAFA Lab, Università degli Studi di Torino, Largo Paolo Braccini 2, 10095 Grugliasco, Italy
2 IN.VA spa and Earth Observation Valle d'Aosta—eoVdA, Località L'Île-Blonde 5, 11020 Brissogne, Italy
3 Institute for Piedmont, Liguria, Aosta Valley (IZS PLV) S.C Valle d'Aosta—CeRMAS (National Reference Center for Wildlife Diseases), Località Amerique 7/G, 11020 Quart, Italy
* Correspondence: tommaso.orusa@unito.it

Abstract: Nowadays, European program Copernicus' Sentinel missions have allowed the development of several application services. In this regard, to strengthen the use of free satellite data in ordinary administrative workflows, this work aims to evaluate the feasibility and prototypal development of a possible service called Sen4MUN for the distribution of contributions yearly allocated to local municipalities and scalable to all European regions. The analysis was focused on the Aosta Valley region, North West Italy. A comparison between the Ordinary Workflow (OW) and the suggested Sen4MUN approach was performed. OW is based on statistical survey and municipality declaration, while Sen4MUN is based on geospatial deep learning techniques on aerial imagery (to extract roads and buildings to get real estate units) and yearly Land Cover map components according to European EAGLE guidelines. Both methods are based on land cover components which represent the input on which the financial coefficients for assigning contributions are applied. In both approaches, buffers are applied onto urban class (LC_b). This buffer was performed according to the EEA-ISPRA soil consumption guidelines to avoid underestimating some areas that are difficult to map. In the case of Sen4MUN, this is applied to overcome Sentinel sensor limits and spectral mixing issues, while in the case of OW, this is due to limits in the survey method itself. Finally, a validation was performed assuming as truth the approach defined by law as the standard, i.e., OW, although it has limitations. MAEs involving LC_b, road lengths and real estate units demonstrate the effectiveness of Sen4MUN. The developed approach suggests a contribution system based on Geomatics and Remote sensing to the public administration.

Keywords: Sen4MUN; geomatics for public administration; Sentinel-1 & Sentinel-2; AGEA orthophoto; ArcGIS Pro; land cover; money assignment to local entities; Europe; Italy; Alpine region

Citation: Orusa, T.; Viani, A.; Borgogno-Mondino, E. Earth Observation Data and Geospatial Deep Learning AI to Assign Contributions to European Municipalities Sen4MUN: An Empirical Application in Aosta Valley (NW Italy). *Land* **2024**, *13*, 80. https://doi.org/10.3390/land13010080

Academic Editor: Jūratė Sužiedelytė-Visockienė

Received: 4 December 2023
Revised: 8 January 2024
Accepted: 9 January 2024
Published: 10 January 2024

1. Introduction

The European Space program Copernicus, with its Sentinels missions, has allowed the creation of many research projects and prototypes aiming at developing several services, many of them already available based on earth observation data [1–6]. Unfortunately, there are still few available or prototype services based on an applied use of geomatics and remote sensing, despite an expansion of Space Economy. The distribution systems of regional contributions to municipalities are overwhelmingly based on territorial components. Therefore, the geospatial component and the technological transfer offered by the growing development of geomatics and remote sensing in this sector would be enormous, allowing for important impacts on administrative processes currently carried out with systems that are often onerous, as well as poorly efficient and inaccurate. To achieve real digitalization in public administration is necessary to develop services capable of responding to ordinary questions in a modern way, guaranteeing scalability and a certain degree of standardization. In fact, nowadays there is a growing need of standardization and validation procedures of geographical and earth observation data/product/services with special concerns about the expected roles they can have within monitoring/control actions from institutional subjects (e.g., CAP controls. natural hazards, etc.).

One of these, still under development, and one of the few that aims to use remote sensing in a highly economic key is the Sen4CAP system for the payment of contribution and control in agriculture, according to the Common Agricultural Policy (CAP) in light of EU regulation (N. 746/2018) [7–9]. The EU's Common Agricultural Policy (CAP) seeks to raise agricultural productivity in Europe in a sustainable manner while maintaining a respectable level of living for EU farmers. The CAP, which has an annual budget of about EUR 59 billion, uses a variety of strategies such as direct payments, market reforms, and rural development, to improve the sustainability and competitiveness of agriculture in Europe [10]. The Integrated Administration and Control System (IACS) oversees and manages the majority of the CAP budget with the goal of protecting the program's finances and assisting farmers in submitting their declarations. [11]. Earth observation (EO) satellites are regarded to play a bigger part in the CAP reform satellite's enhancement and cost-effectiveness of the IACS. It is important to note that the CAP reform has established the role of EO data as being of utmost importance. Their adoption is expected to be required in 2024–2025, and may also extend to future planned missions like IRIDE, which the Italian Space Agency (ASI) and the European Space Agency (ESA) are working on together. Sentinel missions—Sentinel-1 and Sentinel-2 (hereinafter called Sentinels) in particular—play a major role. In the years after 2020, the initiative will give special consideration to demonstrating how data obtained from Sentinels may help modernize and streamline the CAP. Sen4CAP was established by the ESA in response to requests from European payment agencies like other EO services [12–14].

Also, other space agencies worldwide like NASA (US), JAXA (Japan), CNES (France), DLR (Germany), and more recently the ASI (Italy) with the ambitious program IRIDE, have developed or have planned to strengthen their services addressed to the private and public sector, not only for research purposes, but to promote technological transfer in many sectors [15], but at the present time fewer are related to contribution systems in different sectors which play a huge role. The space race has opened up new frontiers of investment for individuals and enterprises at different levels, not only the large economic giants of the ICT and geospatial services such as Planet, SatVu, Albedo, Maxar, e-Geos, etc. Despite several services being available from forest management monitoring to precision agriculture, to the management of migratory flows passing through the management of urban areas and planning of smart cities and much more [16], almost none are related to contributions system. The reason is related to the fact that, on one hand, many services and products offered are often too general and not specific, or worse, without solid scientific validation. On the other hand, public administration is too slow to embrace the new opportunities offered by applied sciences. It is often incapable of understanding them because they are not transferred in the correct way or suitable for use as a support and then a replacement for old methods without creating big changes (especially when it comes to money). Added to all this is bureaucracy and the need to legislate on new approaches so that they become operational and can complement and replace conventional ones. It is therefore not surprising that more bureaucratic countries are often slower to assimilate the news. Added to this is the need for foresight on the part of the political sector and of regional managers and technicians capable of understanding and getting involved, abandoning safe but often obsolete paths to support new ones that require hard work, experimentation, and continuous understanding [17].

Therefore, this work aims to create a tentative service based on European EO data and geospatially based processing about the distribution of contributions intended for municipalities called Sen4MUN, where Sen means Sentinels for municipalities. It is worth noting that in Italy, as in other EU countries, the municipalities receive income from different types according to different criteria [18]. Current incomes of taxes, contributions, and equalization nature are made up of four income items: taxes, duties, and similar income; tax sharing; equalization funds from central administrations; equalization funds from the region or autonomous province. Generally, equalization funds from central administrations and regions or autonomous provinces are the most important and the main core of the approach suggested [19,20]. Concerning contribution delivered by regions to municipalities, they are assigned following two possible approaches; the first one is based on the estimated expenditure, taking into account the previous year (the more you spend the more you receive money for local development within a budget defined at the regional and governmental level); the second is a little more rigorous, based on indicators of territorial development (in which, at least in Italy, the more the urbanized area develops, roads, housing units, factories and more in general areas for urban use, the higher the revenues) as a function of the resident population without going into the criticisms of the present approaches, which certainly should be reviewed from the point of view of environmental sustainability dealing with SDGs goals [21]. Since the second approach permits an exploitation of the possibility offered by EO data and GIS updated data; a system similar to the idea at the base of Sen4CAP has been suggested in this work. Urbanized components, like other territorial bio-physical surfaces, can potentially be mapped to the temporal resolution of Sentinels by translating the information produced by a medium-high resolution land cover into a datum that can be used by administrations at an economic level [22]. To date, the monitoring of land cover is carried out only from an environmental point of view (to map the territory or quantify land consumption or for other research purposes), but without, up to now, deriving any form of economic quantification. To try to bridge this gap, the Sen4MUN has been designed.

Sen4MUN aims to create a single and standardized approach for the item concerning government revenue or, as in the case of this study, regional revenues for local authorities considering also the environmental issues. Sen4MUN aims to replace the approach currently based on the estimation of territorial indicators obtained from statistical analyzes or very rough estimates used up to now by suggesting a more rigorous, efficient, and objective approach, based on the technology transfer offered by EO data trying to suggest new approaches by space economy and the actual and expected roles of geomatics within the next generation EU framework from science to public services.

In this regard, therefore, to strengthen the use of free European satellite data in Ordinary administrative workflows such as Sen4CAP, this work has been focused to assess the feasibility and prototypal development of a possible service called Sen4MUN for the distribution of contributions yearly allocated to local municipalities and scalable to all European regions.

The analysis was focused on Valle d'Aosta Region, North West Italy, considered complex in land cover mapping because of its geo-morphology and Sentinels geometrical resolution limits in alpine areas linked to SAR distortions [23] and spectral mixing, as well as phenology detection in pixels located in high slope degree in multispectral remote sensed data [24].

A comparison between the ordinary methodology based on the estimation of territorial indicators (OW), with the buffer zone retrieved with statistical surveys and the Sen4MUN approach, which is based on yearly land cover classification according to EAGLE guidelines, has been performed. Finally, due to the fact that some roads cannot be correctly mapped by Sentinels due to a GSD limiting factor, updated GIS geodatabases were included in the Sen4MUN prototypal approach.

2. Materials and Methods

2.1. Study Area

Sen4MUN was realized and tested within the Aosta Valley Region in the North West of Italy. Aosta Valley is a region in northwestern Italy which hosts the highest peaks in the whole alpine chain and in general in Europe. The region borders France in the Western part and Switzerland in the Northern. Figure 1 has reported Aosta Valley's location in respect to Italy. Though it is the smallest in Italy, its geomorphology makes it one of the most complex. It is situated in the Western Alps [25].

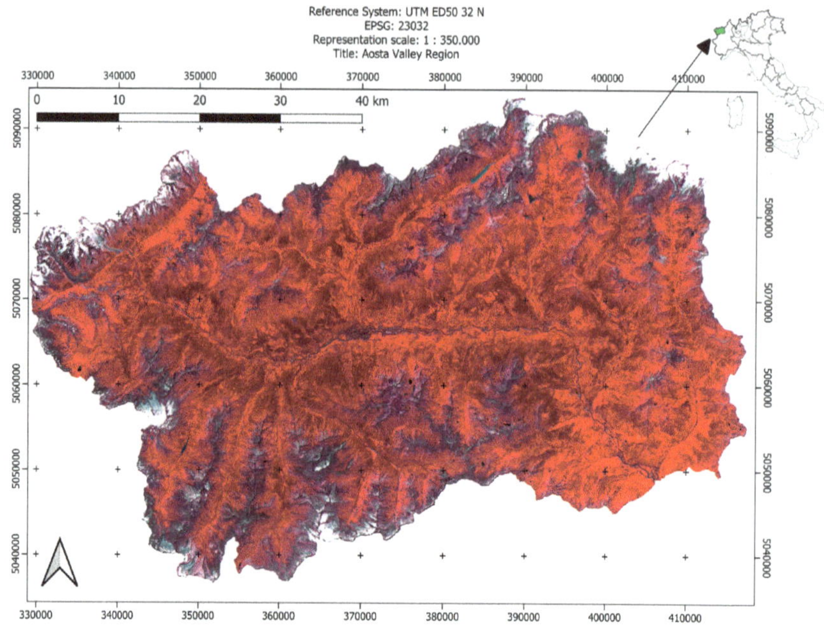

Figure 1. Area of study, involving the Region Valle d'Aosta, NW Italy. False color Sentinel-2 imagery 2022 meteorological summer composite (NIR, Red, Green).

2.2. Ordinary Workflow

This approach adopts municipality declaration and statistical data obtained from regional or national offices that starts from building practices or territorial surveys performed within a time range (for example every 5 years [26]. In the most likely cases, data comes from GIS databases (like the cadastre of buildings real estate units and municipalities streets lengths), and the land cover areas divided into three marco-classes derived from municipality declarations. These three land cover components are: urban areas which are the most important, semi-anthropic areas, and sterile areas, buffered 10–20 or 30 m (over-estimated) due to inhomogeneity data collection through the different times [22]. Nowadays, Ordinary Workflow (hereinafter called OW) does not use deep learning processing to update the GIS geodatabase and earth observation data. In fact, almost the entire OW data collection is based on municipality declaration and statistical surveys at different level with yearly temporal gaps. This is the case of the Aosta Valley region that follows the Cerutti's approach since 1979, which is entirely based on statistical surveys and declarations [27–29]. This approach considered the following patterns: the municipal area, the urban and anthropic areas, the semi-anthropic areas (that includes all vegetated areas cultivated and not cultivated by humans), the sterile areas (that means all land unproductive surfaces, like water bodies and courses, snow and ice, rocks), roads lengths, and finally real estate units. All these components are properly weighted and normalized according to financial criteria defined at regional, national, and European levels. The financial criteria are based on the resources made available by the annual regional budget on the basis of a legal resolution, the expenses made in the current year, the actual resident population and the municipal area. These indicators, based on state and regional tax formulas, which vary over time because they are the result of political economy choices, define the coefficients to be applied to the territorial development components covered by this study and which characterize the contributions paid. In fact, the financial criteria are defined and regulated by national law and regional regulations, leaving greater freedom in their calculation to regions with special statutes such as the Valle d'Aosta Autonomous Region.

Therefore, OW is based on data collection campaigns for statistical purposes and declarations from municipalities. The municipal offices are required to communicate certain parameters such as the municipal area, resident population, roads belonging to the municipality and real estate units present in the municipality. The poor communication between offices at various levels often makes these data not always representative and homogeneous in qualitative terms for all municipalities.

It is worth noting that OW is based on a simple weighted sum of factors that sum to total land cover as reported in financial land components workflow in Figure 1 which is used also in Sen4MUN using a different collection data method. In this regard, the role of remote sensing and GIS is to properly map these areas instead of using periodic surveys as it happens nowadays and in the issue previously discussed. Equations (12) and (13) are the core of this approach, and each variable cab be obtained weighing each component (representing the input variable) properly mapped. Table 1 shows how each variable was obtained for the only purpose of complete clarity, although the procedure is as described previously, a simple weighing of weighed remote sensed variables.

Table 1. Ordinary Workflow equations adopted to retrieve territorial inputs for the financial contribution counting. It is worth to note that these equations are adopted also in Sen4MUN approach.

Description	Algorithm
Land Cover Areas (LCA)	$d = e + g + i$ (1) where, e = urban and anthropic area g = semi-anthropic areas i = sterile areas d = municipality's administrative boundaries
LCA Weights	$f = e \times \alpha$ (2) where, e = urban and anthropic area α = weight (in this case = 3) f = urban weighted area $h = g \times \beta$ (3) where, g = semi-anthropic area β = weight (in this case = 1.5) h = semi-anthropic weighted area $l = i \times \gamma$ (4) where, i = sterile area γ = weight (in this case = 0.5) l = sterile weighted area
Weighted areas	$m = f + h + l$ (5) where, f = urban weighted area h = semi-anthropic weighted area l = sterile weighted area $n = \frac{m}{\sum_{i=1}^{n} m} \times 100$ (6) where, m = conventional municipality area $\sum_{i=1}^{n} m$ = sum of all the municipalities in the regional areas n = conventional weighted municipality area

Table 1. *Cont.*

Description	Algorithm
Weighted areas	$o = n \times \delta$ (7) where, n = conventional weighted municipality area δ = weight (in this case = 50%) o = sub-conventional weighted municipality area
Roads length	$q = \frac{p}{\sum_{i=1}^{n} p} \times 100$ (8) where, p = roads length $\sum_{i=1}^{n} p$ = sum of all the municipalities in the regional areas q = roads weighted length $r = q \times \varepsilon$ (9) where, q = roads weighted length ε = weight (in this case = 30%) r = sub-conventional weighted roads length
Real estate units	$t = \frac{s}{\sum_{i=1}^{n} s} \times 100$ (10) where, s = real estate units $\sum_{i=1}^{n} s$ = sum of all the municipalities real estate units t = real estate weighted units $u = t \times \zeta$ (11) where, t = real estate weighted units ζ = weight (in this case = 20%) u = sub-conventional weighted real estate units

The sub-conventional parameters, respectively obtained by Equations (7), (9) and (11) are used to obtain algorithms reported in Equations (12) and (13), respectively. These are adopted by the regional Aosta Valley Autonomous offices to assess the municipality contributions according to the financial budget foreseen annually.

$$v = o + r + u \tag{12}$$

where,

v = sum of all sub-conventional weighted parameters,

$$z = v \times \eta \tag{13}$$

where,

η = a financial weight in this case 11.50%.

2.3. *Sen4MUN*

The Sen4MUN approach is based on remote sensing and, in particular, yearly land cover, GIS-updated geodatabases, and deep learning to assess the parameters necessary as input in the OW.

2.3.1. Earth Observation Data and Processing

In this regard, urban and anthropic areas, semi-anthropic areas, and sterile areas are computed from Aosta Valley yearly land cover classes aggregation available at SCT regional geoportal and eoVdA webpages. Road lengths are obtained by the GIS viability monthly update geodatabase, as well as real estate units also including deep learning (this part will be further discussed). The reference year 2020 was considered. Land cover (LC) complies with EAGLE guidelines and is realized according to the following methods [30] and it is based on Sentinels missions (S1–S2) [31–34] adopting Aosta Valley land cover [25,30]. It is worth to note that, to assist the semantic and technological foundation of a European harmonized information management capability for land monitoring, the EAGLE Group has been working on a solution and proof of concept since 2008. Mostly, but not only, in their capacities as Eionet members, land monitoring specialists from many European Environment Agency (EEA) member nations have formed the self-initiated and public EAGLE Group. As a result, the EAGLE Group uses a bottom-up methodology to compile information and experiences from current land cover (LC) and land use (LU) categorization methodologies and projects. The Copernicus Land Monitoring Service recognizes EAGLE as a key and necessary element to facilitate a broad change in emphasis from categorization to characterization. This has resulted in the enforcement of EAGLE compliance in new CLMS products. The EAGLE guidelines envisage dividing the land cover components according to a pyramidal hierarchical approach, starting from macro classes to reach detailed components.

Since LC classes are more detailed than those reported in the Cerutti's approach (which are the same at EAGLE at macro-level), they were aggregated according to the Cerutti's description to integrate the Sen4MUN approach into the OW as follows in Table 2.

Table 2. Comparison between Land Cover EAGLE and Cerutti's classes.

Land Cover EAGLE Class	Cerutti Class
Urban and anthropic areas	urban and anthropic area
Shrubland and transitional woods	semi-anthropic areas
Woody crops	semi-anthropic areas
Water surfaces	sterile areas
Water courses	sterile areas
Needle-leaved forests	semi-anthropic areas
Broad-leaved forests	semi-anthropic areas
Mixed forests and moors	semi-anthropic areas
Permanent snow and ice	sterile areas
Natural grasslands and alpine pastures	semi-anthropic areas
Lawn pastures	semi-anthropic areas
Bare rocks	sterile areas
Discontinuous herbaceous vegetation of medium-low altitude	semi-anthropic areas
Sparse herbaceous vegetation at high altitudes	semi-anthropic areas
Alpine wetlands	sterile areas

2.3.2. Geospatial Deep Learning Data and Processing

Roads and real estate units were extracted both from cadastral maps and deep learning adopting open-source libraries and Python scripts integrated with ESRI ArcGIS Pro v.2.9 for object detection and classification [35–38]. Roads and building footprints [39–43] were extracted using Convolutional Neural Network (CNN) techniques [44–47] onto the AGEA (Agency for Disbursements in Agriculture) 2020 ortho-rectified imagery, yearly available at the national level. In particular, ArcGIS pretrained deep learning models were adopted to extract roads and real estate units (that has been assigned from cadastre and municipal declaration of habitability to building footprint extracted using these models). Concerning road extraction, the deep learning model to extract roads from high resolution satellite imagery named: Road Extraction—Global in ESRI ArcGIS Pro v.2.9 was adopted. The implementation is based on the Sat2Graph model by [48]. Sat2Graph relies on a novel encoding scheme. It is capable of creating a three-dimensional tensor from the road network graph. This capability is named as Graph Tensor Encoding (GTE). By combining the benefits of segmentation-based and graph-based techniques, this graph-tensor encoding scheme enables the training of a basic, non-recurrent neural network model to directly translate the input satellite/aerial images into the road network graph (i.e., edges and vertices). In the case of a road network graph G = {V, E} covering a region measuring W meters by

H meters, GTE employs a $W^{-\lambda} \times H^{-\lambda} \times (1 + 3 \cdot D_{max})$ 3D-tensor (represented as T) to hold the graph's encoding. In this case, D_{max} is the maximum number of edges that can be encoded at each $\lambda \times \lambda$ grid, and λ is the spatial resolution, which limits the encoded graph so that no two vertices can be co-located within a $\lambda \times \lambda$ grid. The two spatial axes in the two-dimensional plane are represented by the first two dimensions of T. To encode the graph information, we utilize the vector at each spatial point, $u_{x,y} = [T_{x,y,1}, T_{x,y,2},\ldots T_{x,y}, (1 + 3\cdot D_{max})]^T$. Its first element (vertexness), $p_v \in [0, 1]$, specifies the likelihood that a vertex would exist at point (x, y). D_{max} 3-element groups, which encode the data of a possible outgoing edge from location (x, y), come after the initial element. The first element $p_{ei} \in [0, 1]$ (edgeness) of the *i*-th 3-element group encodes the likelihood of having an outgoing edge toward (dx_i, dy_i), that is, an edge pointing from (x, y) to (x + dx_i, y + dy_i). Since vertices with degrees greater than six are extremely uncommon in road network graphs, we have set D_{max} to six in this instance. GTE only use the *i*-th 3-element group to encode edges pointing toward a 360 D_{max}-degree sector from $(i-1) \cdot 360$ D_{max} degrees to $i \cdot 360$ D_{max} degrees in order to minimize the number of possible distinct isomorphic encodings of the same input graph. It is simple to encode a road network graph into GTE The encoding algorithm first interpolates the straight road segment in the road network graph for the purpose of road network extraction. To keep the distance between consecutive points under d meters, it chooses the bare minimum of equally spaced intermediate spots By controlling the edge vector's length in GTE, this interpolation technique stabilizes the training process.

Interpolation for stacked roads may result in vertices from two overlapped road segments at the same location. When this occurs, the pre-trained model in ArcGIS permits to move the endpoint vectors of the two edges using an iterative conflict-resolution process. The objective is to ensure that the separation between any two vertices (derived from the two edges that overlap). The GTE decoding technique returns a graph's anticipated GTE, which is frequently noisy, to the standard graph format (G = {V,E}). There are two steps in the decoding algorithm: (1) vertex extraction and (2) edge connection. It has been considered vertices and edges with a probability larger than a threshold (referred to as p_{thr}), since both the edgeness and verticeness predictions are real numbers between 0 and 1.

During the vertex extraction phase, the decoding algorithm locates the local maxima of the vertexness map in order to extract possible vertices. The algorithm takes into consideration just those local maxima whose vertexness exceeds p_{thr}. The decoding algorithm joins the outgoing edges of each candidate vertex $v \in V$ to other vertices in the edge connection stage. The following distance function (reported in Equation (14)) is used by the algorithm to calculate the distance of the *i*-th edge of vertex $v \in V$ to all other neighboring vertices u.

$$\begin{aligned} d(v,i,u) = {} & ||(v_x + dx_i, v_x + dy_i) - (u_x, u_y)||+ \\ & +w\cdot \cos_{dist}((dx_i, dy_i), (u_x - v_x, v_y - v_y)) \end{aligned} \tag{14}$$

where w is the weight of the cosine distance in the distance function and $\cos_{dist}$(v1, v2) is the cosine distance of the two vectors. In this case, we set w to a high value, like 100, to prevent erroneous connections. Once this distance has been calculated, the decoding algorithm inserts an edge between v and u 0 and selects a vertex, u 0, that minimizes the distance function d(v, i, u). In order to prevent false edges from being added to the graph when there are no suitable candidate vertices nearby, we specified a maximum distance criterion of 15 m. Finally, Sat2Graph uses cross-entropy loss (denoted as L_{CE}) and L2-loss. The vertexness channel (p_v) and edgeness channels are subjected to the cross-entropy loss.

p_{ei} $i \in \{1, 2,\ldots, D_{max}\}$), and the edge vector channels ((dx_i, dy_i) $i \in \{1, 2,\ldots, D_{max}\}$) get the L2-loss. GTE varies across extended road sections.

In this instance, distinct ground truth labels in the GTE format can be mapped to the same road structure. We only calculate the losses for edgeness and edge vectors at position (x, y) when there is a vertex at that location in the ground truth due to this discrepancy. The overall loss function is displayed below in Equation (15) (ground truth is represented by $\hat{T}$ $\hat{p}_v$, $\hat{p}_{ei}$, $\hat{dx}_i$, and $\hat{dy}_i$).

$$L(T,\hat{T}) = \sum_{(x,y)\in[1..W]\times[1..H]} (L_{CE}(p_v, \hat{p}_v) + \\ +\hat{T}_{x,y,1} \times \left(\sum_{i=1}^{D_{max}} (L_{CE}(p_{ei}, \hat{p}_{ei}) + L_2\left((dx_i, dy_i), \left(\hat{dx}_i, \hat{dy}_i\right)\right)\right)) \tag{15}$$

Concerning building footprints, these were obtained using the deep learning model to extract building footprints from high-resolution aerial and satellite imagery named: Building Footprint Extraction—New Zealand. In this last case, real estate extraction is based on the Mask R-CNN model architecture implemented using ArcGIS API for Python [49].

Developed on top of Faster R-CNN, Mask R-CNN is a state-of-the-art model for instance segmentation. A region-based convolutional neural network called Faster R-CNN [50] provides bounding boxes together with a confidence score for each object's class identification. Mask R-CNN works on two-stage mainly based on Faster R-CNN architecture: (a) (phase 1): there are two networks in the first stage; a region proposal network and a backbone network (ResNet, VGG, Inception, etc.) to provide a collection of region suggestions, these networks execute once for each image. The feature map's regions that contain the object are called region proposals; (b) (phase 2): The network predicts object classes and bounding boxes for every suggested region that was acquired in phase one. While fully linked layers in the networks always need a constant size vector to produce predictions, each proposed region might have a variable size. The RoIAlign technique or RoI pool, which is quite similar to MaxPooling, are used to fix the size of these proposed regions. When segmenting scenes or objects with irregular borders, segmentation models may produce boundaries that are too smooth and may not be accurate. A point-based rendering neural network module named PointRend has been included as an upgrade into the pre-trained model in order to obtain a clear segmentation border. This module provides the segmentation problem with a rendering perspective, utilizing techniques from classical computer graphics. Labels are frequently predicted by image segmentation models on a regular grid with low resolution, such as 1/8th of the input. To upscale the forecasts to the original resolution in these models, interpolation is used. PointRend, on the other hand, upscales the predictions using an iterative subdivision technique by having a trained tiny neural network predict the labels of points at certain places. This technique efficiently produces output with great resolution [51].

In this work, in order to improve the quality of the results obtained and, in particular, the extraction of a building footprint to get real estate units, PointRend has been adopted into the pre-trained model. To enable PointRend within Mask R-CNN in the ESRI ArcGIS front-end or scripting console, the following parameters has been set as reported in Equation (16):

$$\text{model} = \text{MaskRCNN}(\text{data} = \text{data}, \text{pointrend} = \text{True}) \tag{16}$$

An 8-bit, 4-band high resolution (0.20 m GSD) aerial AGEA imagery of 2020 of the whole Aosta Valley was adopted. Due to the high computation necessary to extract the features in the whole region, the processing was conducted in a workstation with the following main characteristics: 64 GB CPU memory; 2 TB SSD storage and with a graphic card Nvidia RTX A4000 16 GB GDDR6.

Then, from the roads extracted, they were checked with the viability geodatabase, and only municipality roads and areas were considered (as specifically reported in the regional laws) for the road length computation. Then, the land cover areas and roads were buffered 20 m for the following reasons: some sparse urban areas are difficult to map due to the GSD limits of Sentinels [52–59], then this threshold is adopted both in OW procedure and ISPRA Land Units in the case of soil consumption estimation [29]. Finally, the areas obtained were used as inputs in the OW equations reported in Table 1.

2.4. Validation

The validation of the Sen4MUN approach and in particular of the surface estimated adopting GIS [60–64] and earth observation data to retrieve the parameters adopted in the OW was realized by computing the Mean Absolute Error (MAE) according to Equation (17) here reported:

$$MAE = \frac{\sum_{i=1}^{n} |p_i - o_i|}{n} \quad (17)$$

where p_i is the prediction (Sen4MUN component area), o_i is the OW component areas estimated without remote sensing methods, and n is the number of municipalities in the Aosta Valley autonomous area equal to 74).

It is worth noting that the EAGLE classification starts from the same macro-classes identified by Cerutti's. The unique difference in the two approaches is represented by the names adopted (despite the description is perfectly the same). Therefore, the EAGLE approach starts from macro to arrive to detail with sub-classes while Cerutti's stops to macro-level.

This is a key point because it permits comparability between the two methods even if different technical approaches are followed.

In order to sum up the suggested newer approach, a workflow of the Sen4MUN approach is provided in Figure 2.

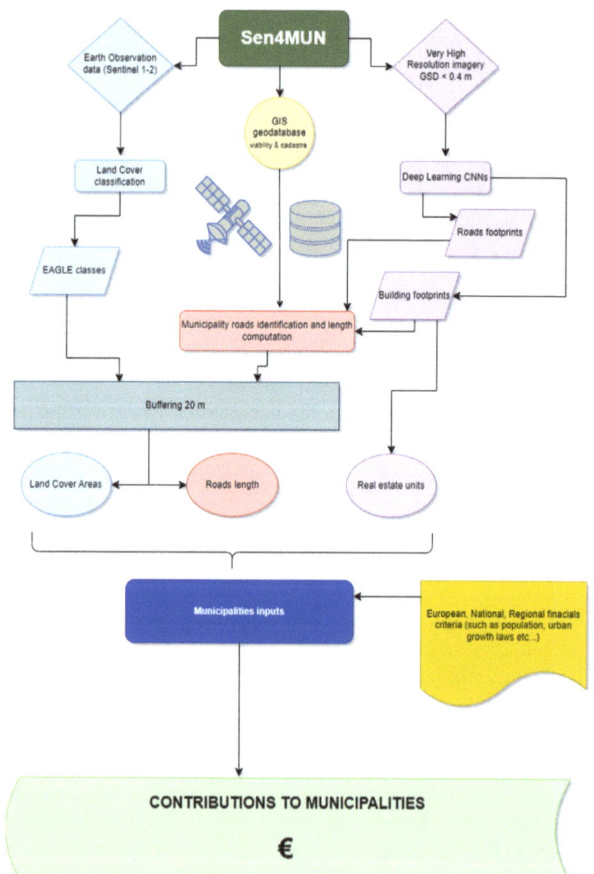

Figure 2. Sen4MUN workflow.

3. Results

The Sen4MUN approach was validated by computing the MAE compared to the surfaces calculated with the traditional method from statistical surveys. It is worth noting that, for the reference year 2020, the main surfaces and inputs were computed both with OW and Sen4MUN for each municipality within the Aosta Valley. Furthermore, also deep learning pre-trained models concerning on roads and real estate units have been evaluated. The pre-trained ESRI ArcGIS model Building Footprint Extraction—New Zealand has produced an F1 score of 86.20 all over the region, with an overall precision of 0.885 and a recall of 0.861. Despite many building footprints being available in ArcGIS, this is one that works better in geomorphological complex areas, while the pre-trained ESRI ArcGIS model Road Extraction—Global showed a precision score of 0.804 in the area of study. However, including results without PointRend as reported in Equation (16), it had a score of 0.873. Excluding false positives due to rocks artifacts and after a manual refining due to photointerpretation, the final layer had an overall accuracy of 0.961. It is worth noting that a 4-bands aerial imagery was adopted to perform the extraction. In particular, to refine the accuracy of the models in ArcGIS Pro in order to reach a threshold score of 0.80, hyperparameter optimization has been performed, fine-tuning the hyperparameters of the model, such as learning rate, batch size, and optimizer settings, to improve the training process and enhance the model's performance. Moreover, a repeat cycle approach has been followed by iteration through the training, evaluation, and refinement steps multiple times until the model's performance reaches an acceptable level. Finally, MAEs were computed for land cover components, road length and real estate units.

Since urban areas play a major role in the present regional regulation in terms of township incomes, MAEs involving each municipality were computed also considering their road length and real estate units. The results obtained are reported in Table 3 and a general Table A1 in Appendix A and in the graphs concerning on the MAEs reported in Figure 3.

Table 3. Overall MAEs and areas computed with both OW and Sen4MUN.

Urban & Anthropic Areas (km^2)			Road Length (km)			Real Estate Units		
Sen4MUN	OW	MAE	Sen4MUN	OW	MAE	Sen4MUN	OW	MAE
103.9	92.1	0.16	1686.7	1626.4	0.81	303,049	293,214	133

Table 3 reports the overall MAE computed considering the same component retrieved with OW and Sen4MUN.

Figure 3 shows the MAEs for the three territorial components used for the calculation of municipal contributions by the Valle d'Aosta Autonomous Region. In particular, the following are reported in the section: (A) the MAEs for urban and anthropic areas expressed in square km; (B) the MAEs for the length of municipal roads expressed in km; (C) the MAEs for real estate units expressed in units. In all three sections of Figure 3, it is interesting to note that the largest MAEs generally occur in municipalities with fragmented urban areas with scattered villages and houses and roads in steep areas due to the limitations of the sensor and processing techniques or errors inherent in the traditional method, such as to cause the imbalances as illustrated below.

As reported in Table 3, an overall MAE of 0.16 km^2 was obtained involving urban and anthropic areas, while 0.81 km was obtained for road length and 11 units in the case of real estate units. An overall MAE of 0.82 km^2 was computed involving all EAGLE Land cover classes, respecting those computed with Cerutti's approach. All the values obtained are significant and seem to suggest a validity of Sen4MUN. The values obtained in Table 3 for the two approaches are not very distant as indicated by the MAEs. It should be highlighted that the standard for assigning contributions is currently OW because it is regulated at a legislative level. It is worth noting that the errors came from Cerutti's approach which overestimate some areas due to limits in this approach itself. Furthermore, it is interesting to note that in all the components, for example urbanized areas, road lengths and real estate units, MAEs are associated with municipalities that underwent more changes in at least one of the components in the reference year and with sparse villages. Unlike what can be derived from the MAEs here computed, the Sen4MUN system is more effective in monitoring territorial changes because it is based on satellite and GIS data updated at a high temporal frequency than the ordinary system. The MAEs obtained allow the transferability of the new approach and possible replacement of the ordinary one even in an Alpine reality such as the one investigated. In fact, MAEs are generally higher in mountainous areas than those obtainable in lowland areas due to the limiting factors affecting some remote sensing applications (land cover mapping in particular).

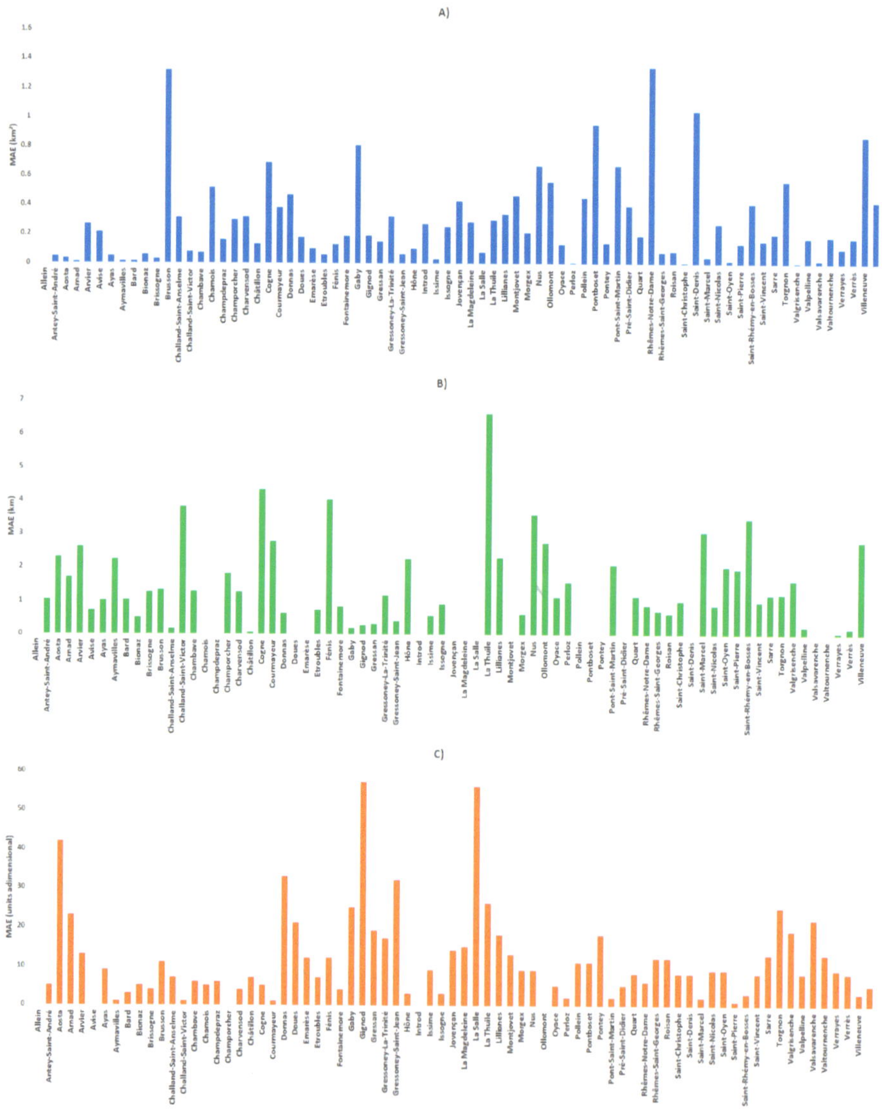

Figure 3. MAEs computed per each municipality in Aosta Valley Autonomous Region comparing Sen4MUN with OW considering (**A**) Urban and anthropic areas; (**B**) Road length; (**C**) Real estate units.

4. Discussions

The results show the consistency of the methods. Furthermore, the procedure adopts free worldwide coverage data. The creation of land covers on an annual, biennial, or monthly basis or even on a time scale of Sentinels (potentially every 5 days) allows the continuous monitoring of territorial dynamics. Moreover, this activity is part of the actions that each European country carries out for the monitoring of soil consumption and that the regions or provinces carry out to offer detailed products according to the guidelines of the European Environmental Agency. Furthermore, the availability for the administration of a regional air flight with very high resolution imagery (like an AGEA flight), albeit for agricultural purposes, allows this type of data to be integrated into other service chains. Such as in the case of Sen4MUN, to extract the footprint of buildings and roads thus makes public spending more efficient through the expansion of derivable products and services. Certainly, it is complex to create universally valid models, and locally suitable models are always preferable. But it must also be said that to create standards such as the Sen4CAP adopted by the European Union, compromises must be reached. Sen4MUN was specifically developed in a complex area such as to make it scalable in a much simpler way to plain and hilly areas which are the majority at a European level. The fact that the approach returns interesting results in an alpine and geomorphologically complex context is certainly encouraged from a scalability perspective. It should be remembered that the largest MAEs occur in municipalities that have more scattered settlements with isolated houses mixed with woods. Areas by which their nature are complex to map at Sentinel scale are not so numerous. Finally, it should also be remembered that many of the differences are linked not to a lack of accuracy and precision of Sen4MUN, which from a purely photo-interpretative analysis proved to be more performing than the traditional OW approach, but precisely as a result of errors in the dataset from statistical investigations of OW. This dataset, despite the errors it suffers from, was taken as a reference base as current regulations require this to be the data collection tool for the purposes of assigning contributions. During the analyses in some municipalities with high MAEs, it was found that Sen4MUN, on the contrary, mapped components that had escaped OW.

Naturally, the critical points in the Sen4MUN approach result in a correct mapping and knowledge of geomatics and remote sensing and their limits, such as the most suitable approach (hierarchical, single-directed, data fusion, etc.), the classification algorithms, as well as the input data that best respond to the components to be mapped, etc. Furthermore, the definition of the optimal number of training areas validation sets according to the area to be mapped is crucial [65–76] in obtaining high accuracies and minimization of errors [77–87]. Despite this work, a ready land cover has been adopted with an overall accuracy upper than 0.94, the whole procedure considering all these issues has been tackled in previous scientific literature involving the Aosta Valley territory [88]. In fact, in the case of a stand-alone and not prototype of this service at an application level not only on a regional or national scale, but on a European scale given the characteristics of the continent, the optimal solution would be to implement a regulated procedure, but which gives a certain degree of freedom in the algorithms and training sets (providing only thresholds) so that chains capable of responding best according to the area to be mapped are developed. Sen4CAP has been excessively standardized also in the algorithms making it underperform in the mountain areas [89]. The need to standardize as much as possible a procedure that moves economic contributions is crucial for large-scale regulation [47], but this is performed with the support of researchers, technicians, and academic experts and after several experiments in UE, to better define the operational leans and freedoms useful for developing consistent products and services for each European reality without creating disparities and differences. As far as the authors are aware, there are no applications in scientific literature for the monetarization of land cover and usage or AI geospatial deep learning at a national, regional, or local contribution scale for services provided by the public administration, therefore a comparison of the results obtained to date is extremely difficult. At the same time, fewer are scientific works on monetary attribution to the components of the coverage for the estimation of ecosystem services and nature based solutions [90,91], but with the intent of estimates of mere research and not of actual transfer of technological application to the public sector. Furthermore, the works themselves are on a large and non-municipal scale and involve geographical areas different from those under study. Therefore, we hope that this work will stimulate the Italian, European and international scientific and technical community to deepen and explore the application potential of geomatics in the operational workflows of public administration by offering new services and comparing the results and approaches obtained for the purpose of continuous improvement.

The use of the prototypal Sen4MUN Aosta Valley has made it possible to make the procedure for assigning contributions to municipalities more objective through the use of Sentinel data and updated GIS geodatabases, favoring technology transfer and the implementation of a possible new service within the Copernicus program and other future programs, such as an IRIDE capable of offering increasingly high spatial and temporal resolution data [37] useful for direct applications to the public sector [92–96].

5. Conclusions

Sen4MUN can be used as a standard procedure for assigning contributions to municipalities. The procedure is consistent and in line with the ordinary territorial workflow based on statistical surveys without the use of earth observation data, deep learning, and geodatabase. Although the system has been tested and used on a prototype level in the Aosta Valley, it can be scaled up to other Italian and European regions. The hope is that this service will become operational and spread to all member countries of the European Union as well as other national realities. Sen4MUN could join other services; coming from various Earth observation programs such as Copernicus, favoring an ever more massive technology transfer to the public sector by rationalizing activities and processes with plural activities with a view to the ever-increasing importance of geomatics in management flows and implementation planners of local, national and European policies. Finally, Sen4MUN seems to be capable of providing useful data for contribution purposes to public administration.

Author Contributions: Conceptualization, T.O.; methodology, T.O. and A.V.; software, T.O.; validation, T.O.; formal analysis, T.O.; investigation, T.O.; resources, T.O.; data curation, T.O. and A.V.; writing—original draft preparation, T.O.; writing—review and editing, T.O. and A.V.; visualization, T.O.; supervision, T.O. and E.B.-M.; project administration, T.O. and E.B.-M.; funding acquisition, T.O. All authors have read and agreed to the published version of the manuscript.

Funding: This research received no external funding.

Data Availability Statement: Data can be accessed through request at the Enti locali della Regione Autonoma Valle d'Aosta offices.

Acknowledgments: Thanks to the Regione Autonoma Valle d'Aosta and INVA spa and eoVdA for having made this work possible, and, in particular, the Regional Cartographic Office in particular Chantal Trèves, and to Enti Locali e Ufficio finanza e contabilità degli Enti locali della Regione Autonoma Valle d'Aosta and in particular, Tiziana Vallet, and also Stefania Fanizzi, Nicoletta Berno, Emanuela Oro, Alessandra Sibona and more in general, everyone working in the Aosta Valley offices. Finally, the Geo4Agri DISAFA Lab colleagues for the great support.

Conflicts of Interest: The authors declare no conflicts of interest.

Appendix A

The data used to realize the histogram available in Figure 3 are reported here below in Table A1.

Table A1. MAEs and areas computed with both OW and Sen4MUN (S4M).

Aosta Valley Municipalities	Urban Areas (km²)			Roads Length (km)			Real Estate Units		
	S4M	OW	MAE	S4M	OW	MAE	S4M	OW	MAE
Allein	0.28	0.33	0.05	16.37	15.32	1.05	643	638	5
Antey-Saint-André	1.04	1.00	0.04	17.98	15.67	2.31	3250	3208	42
Aosta	9.58	9.57	0.01	106.26	104.55	1.71	65,073	65,050	23
Arnad	1.68	1.41	0.27	29.34	26.73	2.61	2614	2601	13
Arvier	0.87	0.66	0.21	22.73	22.02	0.71	1970	1971	1
Avise	0.45	0.50	0.05	12.59	11.58	1.00	899	890	9
Ayas	2.84	2.85	0.01	21.28	19.04	2.24	10,022	10,021	1
Aymavilles	1.37	1.39	0.01	31.05	30.03	1.02	3349	3346	3
Bard	0.26	0.20	0.06	3.04	2.56	0.49	320	315	5
Bionaz	0.55	0.58	0.03	15.22	13.95	1.26	626	622	4
Brissogne	2.11	0.79	1.32	19.78	18.46	1.32	1510	1499	11
Brusson	1.55	1.86	0.31	28.91	28.75	0.16	5229	5222	7
Challand-Saint-Anselme	0.85	0.93	0.08	19.30	15.49	3.81	3231	3230	1
Challand-Saint-Victor	0.63	0.70	0.07	16.93	15.66	1.27	1776	1770	6
Chambave	1.39	0.88	0.52	24.36	25.13	0.77	1804	1799	5
Chamois	0.17	0.33	0.16	3.15	3.15	0.00	568	562	6
Champdepraz	0.94	0.65	0.29	17.60	15.78	1.81	1411	1412	1
Champorcher	0.69	1.00	0.31	22.97	21.72	1.25	2355	2351	4
Charvensod	1.55	1.42	0.13	9.95	9.90	0.05	3897	3890	7
Châtillon	2.97	2.28	0.69	47.63	43.32	4.31	7515	7510	5
Cogne	2.10	1.72	0.38	25.21	22.44	2.78	5528	5527	1
Courmayeur	3.37	2.90	0.47	53.52	52.90	0.62	15,334	15,301	33
Donnas	1.90	1.72	0.18	26.89	28.21	1.32	4069	4048	21
Doues	0.60	0.70	0.10	24.41	27.03	2.62	1329	1317	12
Emarèse	0.29	0.34	0.06	13.57	12.85	0.72	984	977	7
Etroubles	0.70	0.57	0.13	18.90	14.89	4.01	1314	1302	12
Fénis	1.47	1.28	0.18	27.13	26.30	0.83	3409	3405	4
Fontainemore	0.66	1.47	0.81	26.91	26.74	0.16	1560	1535	25
Gaby	0.61	0.80	0.19	9.51	9.25	0.26	1394	1337	57
Gignod	1.37	1.23	0.15	30.44	30.14	0.30	2692	2673	19
Gressan	2.84	2.52	0.32	27.43	26.26	1.17	8330	8313	17
Gressoney-La-Trinité	0.82	0.88	0.06	4.18	3.80	0.38	1944	1912	32
Gressoney-Saint-Jean	1.86	1.97	0.10	18.56	16.32	2.25	5350	5347	3
Hône	1.16	0.89	0.27	13.00	14.51	1.52	2226	2226	0
Introd	0.56	0.53	0.03	13.87	13.33	0.54	1444	1435	9
Issime	0.77	1.02	0.25	10.47	9.56	0.91	1263	1260	3
Issogne	1.50	1.08	0.42	26.35	30.83	4.48	2305	2291	14
Jovençan	0.72	0.43	0.28	10.45	13.32	2.87	974	959	15
La Magdeleine	0.28	0.36	0.08	5.46	5.62	0.16	1157	1101	56
La Salle	2.25	1.95	0.29	42.81	36.21	6.59	7445	7419	26
La Thuile	1.77	1.44	0.33	27.23	24.94	2.28	7223	7205	18
Lillianes	0.52	0.98	0.46	19.48	22.28	2.80	1160	1147	13
Montjovet	1.62	1.42	0.21	39.41	38.81	0.60	3271	3262	9
Morgex	2.33	1.66	0.67	22.97	19.41	3.57	6982	6973	9
Nus	2.44	1.88	0.56	54.40	51.67	2.73	5125	5125	0
Ollomont	0.42	0.55	0.13	8.80	7.68	1.12	965	960	5
Oyace	0.27	0.26	0.01	3.50	1.93	1.57	475	473	2
Perloz	0.46	0.91	0.45	19.45	21.95	2.51	1248	1237	11
Pollein	2.11	1.16	0.95	10.27	14.13	3.86	2217	2206	11
Pontboset	0.33	0.47	0.14	10.31	11.32	1.01	809	791	18
Pontey	1.17	0.50	0.67	6.80	4.72	2.07	1296	1294	2
Pont-Saint-Martin	2.00	1.61	0.39	21.39	21.49	0.10	5165	5160	5
Pré-Saint-Didier	1.10	0.91	0.19	20.58	19.46	1.13	5925	5917	8
Quart	3.90	2.56	1.34	59.69	58.83	0.86	6495	6489	6
Rhêmes-Notre-Dame	0.31	0.39	0.08	6.28	5.58	0.70	719	707	12
Rhêmes-Saint-Georges	0.31	0.39	0.08	6.19	5.57	0.62	820	808	12

Table A1. *Cont.*

Aosta Valley Municipalities	Urban Areas (km^2)			Roads Length (km)			Real Estate Units		
	S4M	OW	MAE	S4M	OW	MAE	S4M	OW	MAE
Roisan	0.66	0.66	0.01	13.44	12.45	0.99	1455	1447	8
Saint-Christophe	2.98	1.93	1.04	44.18	44.32	0.15	5519	5511	8
Saint-Denis	0.49	0.53	0.04	9.98	6.92	3.06	1163	1161	2
Saint-Marcel	1.46	1.19	0.27	34.99	34.13	0.86	2308	2299	9
Saint-Nicolas	0.46	0.48	0.02	19.19	17.17	2.02	1184	1175	9
Saint-Oyen	0.35	0.22	0.14	7.68	5.72	1.96	556	555	1
Saint-Pierre	2.22	1.82	0.41	40.67	37.22	3.45	5225	5222	3
Saint-Rhémy-en-Bosses	0.88	0.73	0.15	19.48	18.51	0.98	1274	1266	8
Saint-Vincent	2.32	2.13	0.20	40.48	39.29	1.19	8956	8943	13
Sarre	2.74	2.18	0.56	41.53	40.33	1.21	6616	6591	25
Torgnon	1.10	1.10	0.00	21.17	19.56	1.61	4332	4313	19
Valgrisenche	0.36	0.54	0.17	14.00	13.78	0.22	859	851	8
Valpelline	0.60	0.62	0.02	13.58	14.25	0.67	1322	1300	22
Valsavarenche	0.48	0.66	0.18	8.46	9.29	0.83	1036	1023	13
Valtournenche	2.76	2.66	0.10	37.56	37.51	0.05	13,870	13,861	9
Verrayes	1.76	1.59	0.17	26.29	26.12	0.17	3218	3210	8
Verrès	2.14	1.27	0.87	13.50	10.74	2.77	3956	3953	3
Villeneuve	1.43	1.01	0.42	28.31	30.04	1.73	2192	2187	5
TOTAL	103.9	92.1	0.16	1686.7	1626.4	0.81	303,049	293,214	11

References

1. Matevosyan, H.; Lluch, I.; Poghosyan, A.; Golkar, A. A Value-Chain Analysis for the Copernicus Earth Observation Infrastructure Evolution: A Knowledgebase of Users, Needs, Services, and Products. *IEEE Geosci. Remote Sens. Mag.* **2017**, *5*, 19–35. [CrossRef]
2. Žlebir, S. Copernicus Earth Observation Programme. In Proceedings of the 40th COSPAR Scientific Assembly, Moscow, Russia, 2–10 August 2014.
3. Schroedter-Homscheidt, M.; Arola, A.; Killius, N.; Lefèvre, M.; Saboret, L.; Wandji, W.; Wald, L.; Wey, E. The Copernicus Atmosphere Monitoring Service (CAMS) Radiation Service in a Nutshell. In Proceedings of the 22nd SolarPACES Conference, Abu Dhabi, United Arab Emirates, 11–14 October 2016.
4. Thépaut, J.-N.; Dee, D.; Engelen, R.; Pinty, B. The Copernicus Programme and Its Climate Change Service. In Proceedings of the IGARSS 2018—2018 IEEE International Geoscience and Remote Sensing Symposium, Valencia, Spain, 22–27 July 2018; pp. 1591–1593.
5. Szantoi, Z.; Strobl, P. *Copernicus Sentinel-2 Calibration and Validation*; Taylor & Francis: Abingdon, UK, 2019; Volume 52.
6. Peuch, V.-H.; Engelen, R.; Rixen, M.; Dee, D.; Flemming, J.; Suttie, M.; Ades, M.; Agustí-Panareda, A.; Ananasso, C.; Andersson, E.; et al. The Copernicus Atmosphere Monitoring Service: From Research to Operations. *Bull. Am. Meteorol. Soc.* **2022**, *103*, E2650–E2668. [CrossRef]
7. Koetz, B.; Defourny, P.; Bontemps, S.; Bajec, K.; Cara, C.; de Vendictis, L.; Kucera, L.; Malcorps, P.; Milcinski, G.; Nicola, L.; et al. SEN4CAP Sentinels for CAP Monitoring Approach. In Proceedings of the 2019 JRC IACS Workshop, Valladolid, Spain, 10–11 April 2019.
8. Sarvia, F.; Petris, S.D.; Orusa, T.; Borgogno-Mondino, E. MAIA S2 Versus Sentinel 2: Spectral Issues and Their Effects in the Precision Farming Context. In *International Conference on Computational Science and Its Applications*; Springer: Berlin/Heidelberg, Germany, 2021; pp. 63–77.
9. Koontz, T.M. Money Talks? But to Whom? Financial versus Nonmonetary Motivations in Land Use Decisions. *Soc. Nat. Resour.* **2001**, *14*, 51–65.
10. Sishodia, R.P.; Ray, R.L.; Singh, S.K. Applications of Remote Sensing in Precision Agriculture: A Review. *Remote Sens.* **2020**, *12*, 3136. [CrossRef]
11. Lupia, F.; Antoniou, V. Copernicus Sentinels Missions and Crowdsourcing as Game Changers for Geospatial Information in Agriculture. *GEOmedia* **2018**, *22*, 32–35.
12. Carella, E.; Orusa, T.; Viani, A.; Meloni, D.; Borgogno-Mondino, E.; Orusa, R. An Integrated, Tentative Remote-Sensing Approach Based on NDVI Entropy to Model Canine Distemper Virus in Wildlife and to Prompt Science-Based Management Policies. *Animals* **2022**, *12*, 1049. [CrossRef] [PubMed]
13. Viani, A.; Orusa, T.; Borgogno-Mondino, E.; Orusa, R. Snow Metrics as Proxy to Assess Sarcoptic Mange in Wild Boar: Preliminary Results in Aosta Valley (Italy). *Life* **2023**, *13*, 987. [CrossRef]
14. Ippoliti, C.; Candeloro, L.; Gilbert, M.; Goffredo, M.; Mancini, G.; Curci, G.; Falasca, S.; Tora, S.; Di Lorenzo, A.; Quaglia, M.; et al. Defining Ecological Regions in Italy Based on a Multivariate Clustering Approach: A First Step towards a Targeted Vector Borne Disease Surveillance. *PLoS ONE* **2019**, *14*, e0219072. [CrossRef]

15. Amani, M.; Ghorbanian, A.; Ahmadi, S.A.; Kakooei, M.; Moghimi, A.; Mirmazloumi, S.M.; Moghaddam, S.H.A.; Mahdavi, S.; Ghahremanloo, M.; Parsian, S.; et al. Google Earth Engine Cloud Computing Platform for Remote Sensing Big Data Applications: A Comprehensive Review. *IEEE J. Sel. Top. Appl. Earth Obs. Remote Sens.* **2020**, *13*, 5326–5350. [CrossRef]
16. Orusa, T.; Mondino, E.B. Landsat 8 Thermal Data to Support Urban Management and Planning in the Climate Change Era: A Case Study in Torino Area, NW Italy. In *Remote Sensing Technologies and Applications in Urban Environments IV*; International Society for Optics and Photonics: Strasbourg, France, 2019; Volume 11157, p. 111570O.
17. Gascon, F.; Cadau, E.; Colin, O.; Hoersch, B.; Isola, C.; Fernández, B.L.; Martimort, P. *Copernicus Sentinel-2 Mission: Products, Algorithms and Cal/Val. In Earth Observing Systems XIX*; International Society for Optics and Photonics: Bellingham, WA, USA, 2014; Volume 9218, p. 92181E.
18. Louvin, R. Flessibilità Fiscale e Zone Franche. Profili Giuridici e Finanziari. In *IUS Publicum Europaeum*; Nuova Serie: Aosta, Italy, 2022; Volume 9, pp. 1–240.
19. Louvin, R. Il Comitato Europeo Delle Regioni: Bilancio e Rilancio. In *Il Comitato delle Regioni, Regioni e Regioni Alpine: Riflessioni ed Esperienze sul futuro dell'Unione Europea*; Università degli Studi di Trento, Facoltà di Giurisprudenza: Trento, Italy, 2022; pp. 61–68.
20. Louvin, R. L'évolution Des Compétences Communales En Italie: Analogies et Discordances Par Rapport Au Cadre Français. In *Quelle (s) Commune (s) pour le XXIe Siècle? Approche de Droit Comparé*; L'Harmattan: Paris, France, 2018; pp. 193–216.
21. Wunder, S.; Kaphengst, T.; Frelih-Larsen, A. Implementing Land Degradation Neutrality (SDG 15.3) at National Level: General Approach, Indicator Selection and Experiences from Germany. In *International Yearbook of Soil Law and Policy 2017*; Springer: Berlin/Heidelberg, Germany, 2018; pp. 191–219.
22. Congedo, L.; Sallustio, L.; Munafò, M.; Ottaviano, M.; Tonti, D.; Marchetti, M. Copernicus High-Resolution Layers for Land Cover Classification in Italy. *J. Maps* **2016**, *12*, 1195–1205. [CrossRef]
23. Samuele, D.P.; Filippo, S.; Orusa, T.; Enrico, B.-M. Mapping SAR Geometric Distortions and Their Stability along Time: A New Tool in Google Earth Engine Based on Sentinel-1 Image Time Series. *Int. J. Remote Sens.* **2021**, *42*, 9135–9154. [CrossRef]
24. Orusa, T.; Viani, A.; Cammareri, D.; Borgogno Mondino, E. A Google Earth Engine Algorithm to Map Phenological Metrics in Mountain Areas Worldwide with Landsat Collection and Sentinel-2. *Geomatics* **2023**, *3*, 221–238. [CrossRef]
25. Orusa, T.; Cammareri, D.; Borgogno Mondino, E. A Possible Land Cover EAGLE Approach to Overcome Remote Sensing Limitations in the Alps Based on Sentinel-1 and Sentinel-2: The Case of Aosta Valley (NW Italy). *Remote Sens.* **2022**, *15*, 178. [CrossRef]
26. Ferrario, C.; Ferri, V. I Comuni Italiani e l'autonomia Finanziaria: Una Scomoda Opportunità? *Sci. Reg.* **2023**, *23*, 105–138.
27. Cerutti, P. *Uso Del Territorio e Forme Contributive*, 1st ed.; UTET: Torino, Italy, 1979; Volume 1.
28. Rosanò, A. La Riforma Della Legge Della Regione Autonoma Valle d'Aosta in Materia Di Attività Condotte Nell'ambito Delle Politiche Promosse Dall'Unione Europea. In *Quaderni AISDUE 1/2023*; Editoriale Scientifica: Naples, Italy, 2023; pp. 359–372.
29. Strollo, A.; Smiraglia, D.; Bruno, R.; Assennato, F.; Congedo, L.; De Fioravante, P.; Giuliani, C.; Marinosci, I.; Riitano, N.; Munafò, M. Land Consumption in Italy. *J. Maps* **2020**, *16*, 113–123. [CrossRef]
30. Orusa, T.; Cammareri, D.; Borgogno Mondino, E. A Scalable Earth Observation Service to Map Land Cover in Geomorphological Complex Areas beyond the Dynamic World: An Application in Aosta Valley (NW Italy). *Appl. Sci.* **2022**, *13*, 390. [CrossRef]
31. Berger, M.; Aschbacher, J. Preface: The Sentinel Missions—New Opportunities for Science. *Remote Sens. Environ.* **2012**, *120*, 1–2. [CrossRef]
32. Malenovskỳ, Z.; Rott, H.; Cihlar, J.; Schaepman, M.E.; García-Santos, G.; Fernandes, R.; Berger, M. Sentinels for Science: Potential of Sentinel 1, 2, and 3 Missions for Scientific Observations of Ocean, Cryosphere, and Land. *Remote Sens. Environ.* **2012**, *120*, 91–101. [CrossRef]
33. Bereta, K.; Caumont, H.; Daniels, U.; Goor, E.; Koubarakis, M.; Pantazi, D.-A.; Stamoulis, G.; Ubels, S.; Venus, V.; Wahyudi, F. The Copernicus App Lab Project: Easy Access to Copernicus Data. In Proceedings of the 22nd International Conference on Extending Database Technology (EDBT), Lisbon, Portugal, 26–29 March 2019; pp. 501–511.
34. Colson, D.; Petropoulos, G.P.; Ferentinos, K.P. Exploring the Potential of Sentinels-1 & 2 of the Copernicus Mission in Support of Rapid and Cost-Effective Wildfire Assessment. *Int. J. Appl. Earth Obs. Geoinf.* **2018**, *73*, 262–276.
35. Zhang, X.; Li, X.; An, J.; Gao, L.; Hou, B.; Li, C. Natural Language Description of Remote Sensing Images Based on Deep Learning. In Proceedings of the 2017 IEEE International Geoscience and Remote Sensing Symposium (IGARSS), Fort Worth, TX, USA, 23–28 July 2017; pp. 4798–4801.
36. Shakya, A.; Biswas, M.; Pal, M. Parametric Study of Convolutional Neural Network Based Remote Sensing Image Classification. *Int. J. Remote Sens.* **2021**, *42*, 2663–2685. [CrossRef]
37. Bullock, E.L.; Healey, S.P.; Yang, Z.; Houborg, R.; Gorelick, N.; Tang, X.; Andrianirina, C. Timeliness in Forest Change Monitoring: A New Assessment Framework Demonstrated Using Sentinel-1 and a Continuous Change Detection Algorithm. *Remote Sens. Environ.* **2022**, *276*, 113043. [CrossRef]
38. Cao, Y.; Niu, X.; Dou, Y. Region-Based Convolutional Neural Networks for Object Detection in Very High Resolution Remote Sensing Images. In Proceedings of the 2016 12th International Conference on Natural Computation, Fuzzy Systems and Knowledge Discovery (ICNC-FSKD), Changsha, China, 13–15 August 2016; pp. 548–554.
39. Ji, S.; Zhang, C.; Xu, A.; Shi, Y.; Duan, Y. 3D Convolutional Neural Networks for Crop Classification with Multi-Temporal Remote Sensing Images. *Remote Sens.* **2018**, *10*, 75. [CrossRef]

40. Baroud, S.; Chokri, S.; Belhaous, S.; Mestari, M. A Brief Review of Graph Convolutional Neural Network Based Learning for Classifying Remote Sensing Images. *Procedia Comput. Sci.* **2021**, *191*, 349–354. [CrossRef]
41. Pan, X.; Zhao, J. A Central-Point-Enhanced Convolutional Neural Network for High-Resolution Remote-Sensing Image Classification. *Int. J. Remote Sens.* **2017**, *38*, 6554–6581. [CrossRef]
42. Shirmard, H.; Farahbakhsh, E.; Heidari, E.; Beiranvand Pour, A.; Pradhan, B.; Müller, D.; Chandra, R. A Comparative Study of Convolutional Neural Networks and Conventional Machine Learning Models for Lithological Mapping Using Remote Sensing Data. *Remote Sens.* **2022**, *14*, 819. [CrossRef]
43. Cresson, R. A Framework for Remote Sensing Images Processing Using Deep Learning Techniques. *IEEE Geosci. Remote Sens. Lett.* **2018**, *16*, 25–29. [CrossRef]
44. Albawi, S.; Mohammed, T.A.; Al-Zawi, S. Understanding of a Convolutional Neural Network. In Proceedings of the 2017 International Conference on Engineering and Technology (ICET), Antalya, Turkey, 21–23 August 2017; pp. 1–6.
45. Zhong, J.; Zhong, S.; Zhang, Q.; Maia, N.; Shen, Y.; Liu, S.; Yu, Y.; Peng, Z. Vision-Based System for Simultaneous Monitoring of Shaft Rotational Speed and Axial Vibration Using Non-Projection Composite Fringe Pattern. *Mech. Syst. Signal Process.* **2019**, *120*, 765–776. [CrossRef]
46. Pickering, S.; Tanaka, S.; Yamada, K. The Impact of Municipal Mergers on Local Public Spending: Evidence from Remote-Sensing Data. *J. East Asian Stud.* **2020**, *20*, 243–266. [CrossRef]
47. Basten, C.; von Ehrlich, M.; Lassmann, A. Income Taxes, Sorting and the Costs of Housing: Evidence from Municipal Boundaries in Switzerland. *Econ. J.* **2017**, *127*, 653–687. [CrossRef]
48. He, S.; Bastani, F.; Jagwani, S.; Alizadeh, M.; Balakrishnan, H.; Chawla, S.; Elshrif, M.M.; Madden, S.; Sadeghi, M.A. Sat2graph: Road Graph Extraction through Graph-Tensor Encoding. In Proceedings of the Computer Vision–ECCV 2020: 16th European Conference, Glasgow, UK, 23–28 August 2020; Proceedings, Part XXIV 16. pp. 51–67.
49. He, K.; Gkioxari, G.; Dollár, P.; Girshick, R. Mask R-Cnn. In Proceedings of the IEEE International Conference on Computer Vision, Venice, Italy, 22–29 October 2017; pp. 2961–2969.
50. Ren, S.; He, K.; Girshick, R.; Sun, J. Faster R-Cnn: Towards Real-Time Object Detection with Region Proposal Networks. *Adv. Neural Inf. Process. Syst.* **2015**, *28*, 161–243. [CrossRef]
51. Kirillov, A.; Wu, Y.; He, K.; Girshick, R. PointRend: Image Segmentation as Rendering. *arXiv* **2020**, arXiv:1912.08193.
52. Henriksson, G.; Weitoft, G.R.; Allebeck, P. Associations between Income Inequality at Municipality Level and Health Depend on Context–A Multilevel Analysis on Myocardial Infarction in Sweden. *Soc. Sci. Med.* **2010**, *71*, 1141–1149. [CrossRef]
53. Esmailpour, N.; Dehghan Dehkordi, E.; Abdali, S.; Ilderemi, H. An Overview on Income Experiences and Procedures of Municipalities with an Emphasis on Their Sustainability. *Eur. Online J. Nat. Soc. Sci. Proc.* **2015**, *4*, 227.
54. Viani, A.; Orusa, T.; Mandola, M.L.; Robetto, S.; Belvedere, M.; Renna, G.; Scala, S.; Borgogno-Mondino, E.; Orusa, R. Tick's Suitability Habitat Maps and Tick-Host Relationships in Wildlife. A One Health Approach Based on Multitemporal Remote Sensed Data, Entropy and Meta®Population Dataset in Aosta Valley, NW Italy. In Proceedings of the GeoVet 2023 International Conference, Teramo, Italy, 19–21 September 2023.
55. Büttner, G. CORINE Land Cover and Land Cover Change Products. In *Land Use and Land Cover Mapping in Europe*; Springer: Berlin/Heidelberg, Germany, 2014; pp. 55–74.
56. Feranec, J.; Soukup, T.; Hazeu, G.; Jaffrain, G. *European Landscape Dynamics: CORINE Land Cover Data*; CRC Press: Boca Raton, FL, USA, 2016.
57. ESCAP. *Producing Land Cover Change Maps and Statistics: Step by Step Guide on the Use of QGIS and RStudio*; 2020 ESA; ESCAP: Frascati, Italy, 2020.
58. Comber, A.J.; Wadsworth, R.; Fisher, P. Using Semantics to Clarify the Conceptual Confusion between Land Cover and Land Use: The Example of 'Forest'. *J. Land Use Sci.* **2008**, *3*, 185–198. [CrossRef]
59. Orusa, T.; Orusa, R.; Viani, A.; Carella, E.; Borgogno Mondino, E. Geomatics and EO Data to Support Wildlife Diseases Assessment at Landscape Level: A Pilot Experience to Map Infectious Keratoconjunctivitis in Chamois and Phenological Trends in Aosta Valley (NW Italy). *Remote Sens.* **2020**, *12*, 3542. [CrossRef]
60. Conrad, O.; Bechtel, B.; Bock, M.; Dietrich, H.; Fischer, E.; Gerlitz, L.; Wehberg, J.; Wichmann, V.; Böhner, J. System for Automated Geoscientific Analyses (SAGA) v. 2.1.4. *Geosci. Model Dev.* **2015**, *8*, 1991–2007. [CrossRef]
61. QGIS Development Team. QGIS Geographic Information System. Open Source Geospatial Foundation Project. 2018. Available online: https://qgis.org/en/site/ (accessed on 7 January 2024).
62. Grizonnet, M.; Michel, J.; Poughon, V.; Inglada, J.; Savinaud, M.; Cresson, R. Orfeo ToolBox: Open Source Processing of Remote Sensing Images. *Open Geospat. Data Softw. Stand.* **2017**, *2*, 15. [CrossRef]
63. Inglada, J.; Christophe, E. The Orfeo Toolbox Remote Sensing Image Processing Software. In Proceedings of the 2009 IEEE International Geoscience and Remote Sensing Symposium, Cape Town, South Africa, 12–17 July 2009; Volume 4, p. IV-733.
64. Racine, J.S. *RStudio: A Platform-Independent IDE for R and Sweave*; JSTOR: New York, NY, USA, 2012.
65. Homewood, K.; Lambin, E.F.; Coast, E.; Kariuki, A.; Kikula, I.; Kivelia, J.; Said, M.; Serneels, S.; Thompson, M. Long-Term Changes in Serengeti-Mara Wildebeest and Land Cover: Pastoralism, Population, or Policies? *Proc. Natl. Acad. Sci. USA* **2001**, *98*, 12544–12549. [CrossRef] [PubMed]
66. Seto, K.C.; Kaufmann, R.K. Modeling the Drivers of Urban Land Use Change in the Pearl River Delta, China: Integrating Remote Sensing with Socioeconomic Data. *Land Econ.* **2003**, *79*, 106–121. [CrossRef]

67. Olioso, A.; Soria, G.; Sobrino, J.; Duchemin, B. Evidence of Low Land Surface Thermal Infrared Emissivity in the Presence of Dry Vegetation. *IEEE Geosci. Remote Sens. Lett.* **2007**, *4*, 112–116. [CrossRef]
68. Parker, D.C.; Manson, S.M.; Janssen, M.A.; Hoffmann, M.J.; Deadman, P. Multi-Agent Systems for the Simulation of Land-Use and Land-Cover Change: A Review. *Ann. Assoc. Am. Geogr.* **2003**, *93*, 314–337. [CrossRef]
69. Kennedy, R.E.; Townsend, P.A.; Gross, J.E.; Cohen, W.B.; Bolstad, P.; Wang, Y.; Adams, P. Remote Sensing Change Detection Tools for Natural Resource Managers: Understanding Concepts and Tradeoffs in the Design of Landscape Monitoring Projects. *Remote Sens. Environ.* **2009**, *113*, 1382–1396. [CrossRef]
70. Kennedy, R.E.; Yang, Z.; Gorelick, N.; Braaten, J.; Cavalcante, L.; Cohen, W.B.; Healey, S. Implementation of the LandTrendr Algorithm on Google Earth Engine. *Remote Sens.* **2018**, *10*, 691. [CrossRef]
71. Phiri, D.; Simwanda, M.; Salekin, S.; Nyirenda, V.R.; Murayama, Y.; Ranagalage, M. Sentinel-2 Data for Land Cover/Use Mapping: A Review. *Remote Sens.* **2020**, *12*, 2291. [CrossRef]
72. Rose, R.A.; Byler, D.; Eastman, J.R.; Fleishman, E.; Geller, G.; Goetz, S.; Guild, L.; Hamilton, H.; Hansen, M.; Headley, R.; et al. Ten Ways Remote Sensing Can Contribute to Conservation. *Conserv. Biol.* **2015**, *29*, 350–359. [CrossRef]
73. Tamiminia, H.; Salehi, B.; Mahdianpari, M.; Quackenbush, L.; Adeli, S.; Brisco, B. Google Earth Engine for Geo-Big Data Applications: A Meta-Analysis and Systematic Review. *ISPRS J. Photogramm. Remote Sens.* **2020**, *164*, 152–170. [CrossRef]
74. Brisco, B.; Brown, R.; Hirose, T.; McNairn, H.; Staenz, K. Precision Agriculture and the Role of Remote Sensing: A Review. *Can. J. Remote Sens.* **1998**, *24*, 315–327. [CrossRef]
75. Green, D.G.; Leishman, T. Computing and Complexity—Networks, Nature and Virtual Worlds. In *Philosophy of Complex Systems*; Elsevier: Amsterdam, The Netherlands, 2011; pp. 137–161.
76. Karra, K.; Kontgis, C.; Statman-Weil, Z.; Mazzariello, J.C.; Mathis, M.; Brumby, S.P. Global Land Use/Land Cover with Sentinel 2 and Deep Learning. In Proceedings of the 2021 IEEE International Geoscience and Remote Sensing Symposium IGARSS, Brussels, Belgium, 11–16 July 2021; pp. 4704–4707.
77. Agarwal, C. A Review and Assessment of Land-Use Change Models: Dynamics of Space, Time, and Human Choice. *Hum. Ecol. Risk Assess. Int. J.* **2002**, *346*, 102–118.
78. Long, H.; Li, Y.; Liu, Y.; Woods, M.; Zou, J. Accelerated Restructuring in Rural China Fueled by 'Increasing vs. Decreasing Balance' Land-Use Policy for Dealing with Hollowed Villages. *Land Use Policy* **2012**, *29*, 11–22. [CrossRef]
79. Atzberger, C. Advances in Remote Sensing of Agriculture: Context Description, Existing Operational Monitoring Systems and Major Information Needs. *Remote Sens.* **2013**, *5*, 949–981. [CrossRef]
80. Yeh, A.G.; Li, X. An Integrated Remote Sensing and GIS Approach in the Monitoring and Evaluation of Rapid Urban Growth for Sustainable Development in the Pearl River Delta, China. *Int. Plan. Stud.* **1997**, *2*, 193–210. [CrossRef]
81. Li, X.; Yeh, A.G.-O. Analyzing Spatial Restructuring of Land Use Patterns in a Fast Growing Region Using Remote Sensing and GIS. *Landsc. Urban Plan.* **2004**, *69*, 335–354. [CrossRef]
82. Tarasenko, M.V. Current Status of the Russian Space Programme. *Space Policy* **1996**, *12*, 19–28. [CrossRef]
83. Xiao, J.; Shen, Y.; Ge, J.; Tateishi, R.; Tang, C.; Liang, Y.; Huang, Z. Evaluating Urban Expansion and Land Use Change in Shijiazhuang, China, by Using GIS and Remote Sensing. *Landsc. Urban Plan.* **2006**, *75*, 69–80. [CrossRef]
84. Berger, M.M. Happy Birthday, Constitution: The Supreme Court Establishes New Ground Rules for Land-Use Planning. *Urban Lawyer* **1988**, *20*, 735–800.
85. Otukei, J.R.; Blaschke, T. Land Cover Change Assessment Using Decision Trees, Support Vector Machines and Maximum Likelihood Classification Algorithms. *Int. J. Appl. Earth Obs. Geoinf.* **2010**, *12*, S27–S31. [CrossRef]
86. Heistermann, M.; Müller, C.; Ronneberger, K. Land in Sight?: Achievements, Deficits and Potentials of Continental to Global Scale Land-Use Modeling. *Agric. Ecosyst. Environ.* **2006**, *114*, 141–158. [CrossRef]
87. Rindfuss, R.R.; Stern, P.C. Linking Remote Sensing and Social Science: The Need and the Challenges. In *People and Pixels: Linking Remote Sensing and Social Science*; National Academy of Science: Washington, DC, USA, 1998; pp. 1–27.
88. Orusa, T.; Viani, A.; Moyo, B.; Cammareri, D.; Borgogno-Mondino, E. Risk Assessment of Rising Temperatures Using Landsat 4–9 LST Time Series and Meta®Population Dataset: An Application in Aosta Valley, NW Italy. *Remote Sens.* **2023**, *15*, 2348. [CrossRef]
89. Kohler, T.; Giger, M.; Hurni, H.; Ott, C.; Wiesmann, U.; von Dach, S.W.; Maselli, D. Mountains and Climate Change: A Global Concern. *Mt. Res. Dev.* **2010**, *30*, 53–55. [CrossRef]
90. Crossman, N.D.; Bryan, B.A.; de Groot, R.S.; Lin, Y.-P.; Minang, P.A. Land Science Contributions to Ecosystem Services. *Curr. Opin. Environ. Sustain.* **2013**, *5*, 509–514. [CrossRef]
91. Hagedoorn, L.C.; Koetse, M.J.; van Beukering, P.J. Estimating Benefits of Nature-Based Solutions: Diverging Values from Choice Experiments with Time or Money Payments. *Front. Environ. Sci.* **2021**, *9*, 686077. [CrossRef]
92. Napp, S.; Petrić, D.; Busquets, N. West Nile Virus and Other Mosquito-Borne Viruses Present in Eastern Europe. *Pathog. Glob. Health* **2018**, *112*, 233–248. [CrossRef] [PubMed]
93. Caminade, C.; Morse, A.P.; Fevre, E.M.; Mor, S.; Baylis, M.; Kelly-Hope, L. Forecasting the Risk of Vector-Borne Diseases at Different Time Scales: An Overview of the CLIMate SEnsitive DISease (CLIMSEDIS) Forecasting Tool Project for the Horn of Africa. In Proceedings of the EGU23, 25th EGU General Assembly, Vienna, Austria, 23–28 April 2023.
94. Viani, A.; Orusa, T.; Divari, S.; Lovisolo, S.; Zanet, S.; Borgogno-Mondino, E.; Orusa, R.; Bollo, E. *Bartonella* spp. Distribution Assessment in Red Foxes (Vulpes Vulpes) Coupling Geospatially-Based Techniques. 2023, 76° Convegno SISVET, Bari. Available online: https://hdl.handle.net/2318/1938950 (accessed on 7 January 2024).

95. Bolton, D.K.; Gray, J.M.; Melaas, E.K.; Moon, M.; Eklundh, L.; Friedl, M.A. Continental-Scale Land Surface Phenology from Harmonized Landsat 8 and Sentinel-2 Imagery. *Remote Sens. Environ.* **2020**, *240*, 111685. [CrossRef]
96. Bhaduri, B.; Bright, E.; Coleman, P.; Dobson, J. LandScan. *Geoinformatics* **2002**, *5*, 34–37.

land

Article

Spatial and Temporal Variation Characteristics of Ecological Environment Quality in China from 2002 to 2019 and Influencing Factors

Junjie Li [1], Xiangbin Peng [1], Ruomei Tang [1], Jing Geng [2,*], Zipeng Zhang [3], Dong Xu [4] and Tingting Bai [5]

1 College of Art and Design, Nanjing Forestry University, Nanjing 210037, China; lijunjie@njfu.edu.cn (J.L.); xiangbin@njfu.edu.cn (X.P.); zousujuan@abc.edu.cn (R.T.)
2 Academician Workstation of Zhai Mingguo, University of Sanya, Sanya 572022, China
3 College of Geographical and Remote Sciences, Xinjiang University, Urumqi 830017, China; zp_zhang@xju.edu.cn
4 State Key Laboratory of Remote Sensing Science, Beijing Normal University, Beijing 100091, China; xd@mail.bnu.edu.cn
5 School of Business Administration, Northeastern University, Shenyang 110189, China; 2110438@stu.neu.edu.cn
* Correspondence: jinggeng@sanyau.edu.cn

Abstract: Since the beginning of the new century, there has been a notable enhancement in China's ecological environment quality (EEQ), a development occurring in tandem with climate change and the extensive ecological restoration projects (ERPs) undertaken in the country. However, comprehensive insights into the spatial and temporal characteristics of China's EEQ, and its responses to both climate change and human activities over the past two decades, have remained largely elusive. In this study, we harnessed a combination of multi-source remote-sensing data and reanalysis data. We employed Theil–Sen median trend analysis, multivariate regression residual analysis, and the Hurst index to examine the impacts and changing patterns of climatic factors and human activities on China's EEQ during the past two decades. Furthermore, we endeavored to forecast the future trajectory of EEQ. Our findings underscore a significant improvement in EEQ across most regions of China between 2002 and 2019, with the most pronounced enhancements observed in the Loess Plateau, Northeast China, and South China. This transformation can be attributed to the combined influence of climatic factors and human activities, which jointly accounted for alterations in EEQ across 78.25% of China's geographical expanse. Human activities (HA) contributed 3.93% to these changes, while climatic factors (CC) contributed 17.79%. Additionally, our projections indicate that EEQ is poised to continue improving in 56.70% of China's territory in the foreseeable future. However, the Loess Plateau, Tarim Basin, and Inner Mongolia Plateau are anticipated to experience a declining trend. Consequently, within the context of global climate change, the judicious management of human activities emerges as a critical imperative for maintaining EEQ in China. This study, bridging existing gaps in the literature, furnishes a scientific foundation for comprehending the evolving dynamics of EEQ in China and informs the optimization of management policies in this domain.

Keywords: EEQ; Chinese High-Resolution Ecological Quality Dataset (CHEQ); climate change; ERPs; human activities

Citation: Li, J.; Peng, X.; Tang, R.; Geng, J.; Zhang, Z.; Xu, D.; Bai, T. Spatial and Temporal Variation Characteristics of Ecological Environment Quality in China from 2002 to 2019 and Influencing Factors. *Land* **2024**, *13*, 110. https://doi.org/10.3390/land13010110

Academic Editor: Jūratė Sužiedelytė-Visockienė

Received: 18 November 2023
Revised: 8 January 2024
Accepted: 9 January 2024
Published: 19 January 2024

1. Introduction

The ecological environment serves as the fundamental underpinning of human survival and progress. Since the advent of the new century, China's trajectory towards modernization has continued its forward momentum, accompanied by rapid socio-economic growth and urbanization. These developments have exerted substantial pressure on the ecological environment [1]. To address this challenge, in November 1998, the State Council introduced the China Ecological Environment Construction Plan (CECP). This visionary plan outlines a comprehensive strategy spanning 50 years, aimed at instigating nationwide ERPs. The overarching goal is to enhance China's ecological environment and facilitate the sustainable development of both societal and natural systems [2]. A recent study has underscored the remarkable outcome of these long-term ERPs, revealing that China has contributed to 25% of global greening efforts [3]. Nevertheless, despite this achievement, a comprehensive knowledge of the ecological and environmental benefits accrued from ERPs over the past two decades remains elusive. Henceforth, it becomes imperative to illuminate the spatial and temporal evolution of ecological environment quality (EEQ) in China during the preceding 18 years. Additionally, it is essential to investigate the respective contributions of human activities and climate change to variations in EEQ. Such insights hold profound practical significance as they inform the development and implementation of future ERPs, bolster ecological conservation initiatives, and advance the realization of sustainable development goals within China.

ERPs are designed to rehabilitate and enhance ecosystems that have suffered damage through the application of artificial interventions and ecological technologies [4]. This endeavor serves several vital objectives, including the promotion of biodiversity, bolstering ecosystem stability, fortifying resilience against natural disasters and anthropogenic disruptions, and upholding ecosystem security [5]. Since 1998, China has been steadfast in its pursuit of significant advancements in land greening through large-scale initiatives. Notably, the nation has embarked on the construction of nine major ERPs. The comprehensive enhancement of afforestation in terms of both quantity and quality has been a concurrent focus. As of 2023, China has successfully accomplished afforestation on a staggering 6800×10^4 ha of land. This concerted investment in ecological restoration has yielded evident environmental benefits [6]. Noteworthy examples include the transformative effects of initiatives like converting farmland back into forests, restoring pastures to grasslands, safeguarding natural forests, and implementing soil and water conservation measures. These efforts have demonstrably improved land quality, elevated vegetation coverage, enhanced water retention, and mitigated the incidence of climatic and geological catastrophes, such as sandstorms and mudslides. Furthermore, ERPs have also conferred a diverse array of ecological functions and services. These include serving as carbon sinks, regulating climate, and managing resources. These multifaceted contributions promote the sustainable development of the ecological environment [7]. According to statistics from the State Forestry Administration, China's forest coverage increased from 16.55% in 2000 to 23.04% in 2020, and grassland vegetation coverage expanded from 44% in 2000 to 56.1% in 2020. The culmination of these extensive ecological restoration initiatives in China has played an instrumental role in elevating the quality of the country's ecological environment and has made substantial contributions to ecosystem protection [6].

In the realm of EEQ assessment, China's Ministry of Environment introduced the Ecological Environment Status Index (EI) in 2006. This comprehensive index comprises various components, including water quality, soil quality, forest coverage, and pollution levels. Nonetheless, the EI encounters challenges associated with data acquisition complexities and a laborious calculation process [8]. The rapid advancement of remote-sensing technology has significantly streamlined the monitoring and investigation of EEQ on a large scale [9,10]. In 2013, Xu et al. proposed the EEQ evaluation index, known as the Remote Sensing Ecological Index (RSEI) [11]. This index offers the distinct advantages of objectivity, simplicity, and accessibility, rendering it a widely adopted tool in EEQ assessment and research. Although some scholars have used the RSEI model to assess the EEQ in China, most of the previous studies have focused on the analysis of specific regions and the assessment of space only for specific situations, and the applicability of the model in China remains to be studied [12]. And the RSEI index lacks integration with the national EI index for accuracy verification. Moreover, it exhibits limitations, such as the incomplete coverage of evaluation criteria and the omission of dominant ecosystem service functions in specific regions when applied to diverse geographic areas [12,13]. Consequently, there is an urgent need to devise a method characterized by temporal and spatial universality to quantitatively assess EEQ across various Chinese regions.

To address these challenges, this study utilizes a combination of multi-source remote-sensing datasets and reanalysis datasets. Considering the complex climatic and geographic environment of China, we added numerous environmental variables to obtain EEQ data based on the original RESI model construction, and constructed the Chinese High-Resolution Ecological Quality Dataset (CHEQ). These data sources serve as the foundation for an in-depth assessment of the spatial and temporal characteristics of EEQ and the associated influencing mechanisms in China during the period spanning 2002 to 2019, employing residual analysis as the primary analytical tool. The specific objectives of our research encompass three key areas: (1) to generate a comprehensive, high-precision, long time-series dataset documenting EEQ in China, spanning the past two decades, and to elucidate the spatial and temporal patterns characterizing China's EEQ during this period; (2) to investigate the respective contributions of human activities and climate change to fluctuations in China's EEQ over the past 18 years; and (3) to forecast the anticipated trends and spatial distribution patterns of China's EEQ in the future.

This study introduces two principal innovations. First, it presents a high-precision, gridded ecological quality assessment model for China, verified using data from over 2000 ground stations. This model has led to the creation of China's inaugural high-resolution, high-precision long-term ecological quality dataset, offering invaluable scientific data support for scholarly research. Second, our study is the first to investigate the spatial and temporal variations in China's ecological quality and their underlying drivers, employing traditional methods as a foundation while utilizing the high-precision dataset.

2. Data

Data Sources

The data used in this study include Chinese administrative division data; MOD13A2 Normalized Vegetation Index data (NDVI); MOD17A3 Net Primary Productivity (NPP) of vegetation; data on climatic elements: precipitation (PRE), actual evapotranspiration (AET), potential evapotranspiration (PET), solar radiation (SRAD), atmospheric pressure (VAP), saturated water vapor pressure difference (VPD), wind speed (VS), surface runoff (RO), drought index (DI), soil moisture (SOIL), drought index (PDSI), water deficit (DEF), surface temperature (TEMP), maximum temperature (TMMN), and minimum temperature (TMMX) [14–16]. These climate data come from the Terraclimate dataset. A detailed description of the data used in this study is shown in Table 1.

Table 1. Detailed description of data.

Data Name	Time Period	Date Type	Spatial Resolution	Time Scale	Data Sources
Ecological Restoration Project Boundary Data	/	shp	/	/	NTPDC [a]
EI	2018	/	/	Annual	MEE [ab]
MOD13A2	2002–2019	HDF	1000 m	Annual	NASA [c]
MOD17A3	2002–2019	HDF	1000 m	Annual	NASA [c]
MOD09A1	2002–2019	HDF	1000 m	Annual	NASA [c]
MCD12Q1	2002–2019	HDF	1000 m	Annual	NASA [c]
TerraClimate	2002–2019	Necdef	4600 m	Monthly	GEE [d]

Note: [a.] National Tibet Data Center (https://data.tpdc.ac.cn/home (accessed on 2 May 2023)). [b.] MEE: Ministry of Ecology and Environment of the People's Republic of China (https://www.mee.gov.cn/ (accessed on 2 May 2023)). [c.] NASA: National Aeronautics and Space Administration (https://www.nasa.gov/ (accessed on 2 May 2023)). [d.] GEE: Google Earth Engine: (https://earthengine.google.com/ (accessed on 2 May 2023)).

In this study, the relevant data from 2002–2019 were selected and processed as follows: data cropping, scale conversion, standardization of indicators, synthesis of annual data by monthly mean temperature, and setting all data to the same spatial resolution and coordinate system.

3. Methods

Figure 1 shows the workflow of this study, including the distribution map of EEQ in China from 2002–2019, and the contribution analysis of EEQ (natural factors and human activities). The specific steps are as follows: (1) Obtain basic data through multi-source remote-sensing technology, extract, crop, synthesize, resample, standardize, and perform principal component analysis (PCA) and other preprocessing to obtain the specific situation of China's EEQ from 2002 to 2019, and then obtain the observed overall trend of China's EEQ; (2) Based on the climate dataset and observed EEQ values, perform multiple linear regression and residual trend analysis to obtain residual and simulated EEQ values; (3) Obtain China's residual and simulated EEQ trend changes through Theil–Sen median; (4) Based on regression models and climate data calculations, the contribution degree of climatic factors (CC) is obtained, and the difference between the observed EEQ and the predicted EEQ is used to represent the contribution degree of human activities (HA).

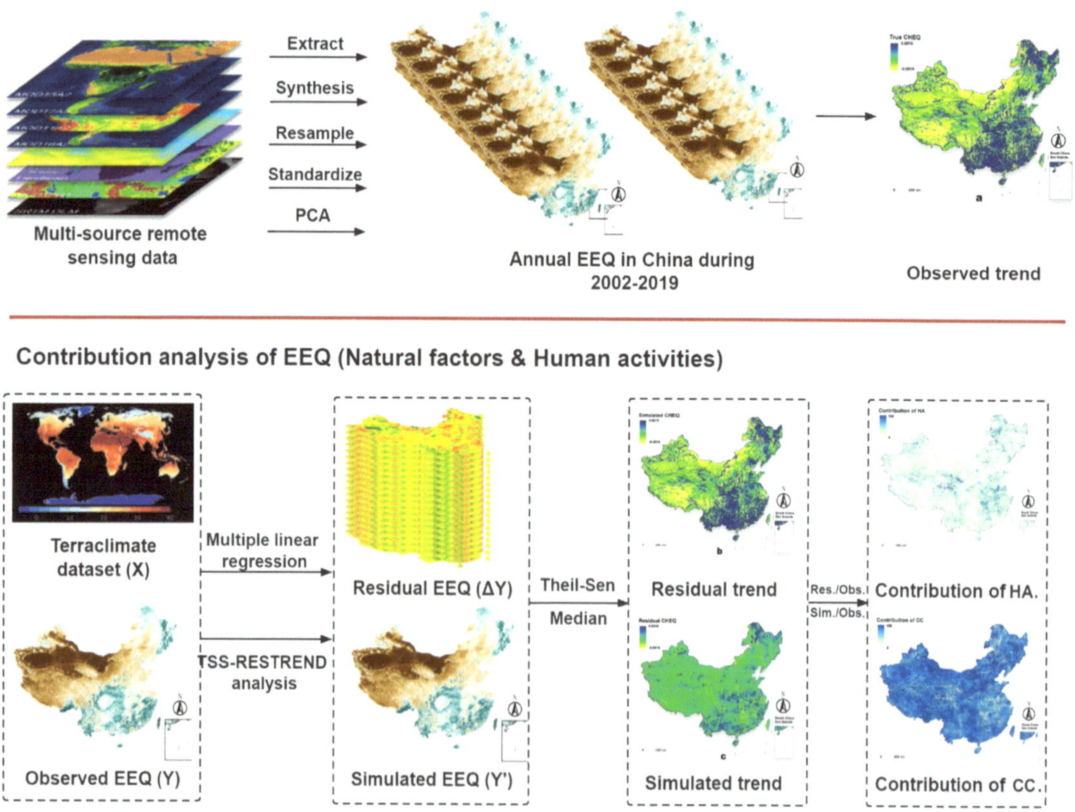

Figure 1. Research method flowchart. The methodological process consists of two parts, i.e., EEQ mapping, as well as EEQ driver analysis.

3.1. Estimation of EEQ

The building of an ecological civilization is inextricably linked to the optimization of habitat quality and the maintenance of biodiversity. Habitat quality indicates the ability of an ecosystem to provide sustained survival for species or groups of species.

In this study, we developed a new *EEQ* evaluation model based on the RSEI model proposed by Xu et al. [12] and generated high-resolution *EEQ* data in China. And we assessed the applicability of RSEI and CHEQ. Compared to the RSEI index, we additionally introduced the AI index in the calculation of CHEQ, which is derived from the "Technical Criterion for Ecosystem Status Evaluation".

$$EEQ = \frac{PC1 - PC1_{\text{min}}}{PC1 - PC1_{\text{max}}} \tag{1}$$

$$PC1 = PCA(NDVI, NDBSI, LST, WET, AI) \tag{2}$$

This study utilized PCA. PCA is a commonly used data analysis method: PCA can be used to extract the main feature components of the data and is often used for dimensionality reduction of high-dimensional data. This is a commonly used information system data-processing method that can extract the most representative principal components from multiple variables, thereby simplifying the data analysis process and improving the interpretation ability of the data. Based on the results of PCA, the variation patterns and potential factors of the data can be explained for data interpretation and application. where PC1 is the first principal component, $PC1_{\text{min}}$ is the minimum value of PC1, $PC1_{\text{max}}$ is the maximum value of PC1, NDVI is the Normalized Vegetation Index data, NDBSI is the Normalized Difference Built-Up Index, LST is the land surface temperature, WET is the humidity, and AI is the abundance index. NDVI and LST are MOD13A2 and MOD11A1 products, respectively, and both NDBSI and WET are calculated based on the MOD09A1. For the formulae of NDBSI and WET, please refer to https://www.indexdatabase.de/ (accessed on 2 May 2023). For the calculation process of AI index, please refer to https://www.mee.gov.cn/ (accessed on 2 May 2023).

3.2. Theil–Sen Median Trend Analysis

In this investigation, we primarily employed the Theil–Sen median (TSM) trend analysis method [12], a robust non-parametric statistical approach extensively applied for trend assessment. Particularly well-suited for determining the median trend within datasets, TSM minimizes the impact of outliers while simultaneously offering high computational efficiency. It finds frequent application in trend analyses concerning lengthy time-series data [17–19]. The outcomes derived through the TSM technique contribute significantly to delineating the temporal trajectory of EEQ within the time series.

$$\text{Slope} = \text{Median}\left(\frac{CHEQ_j - CHEQ_i}{j - i}\right), \ 2002 \le i < j \le 2019 \tag{3}$$

where slope, in this context, signifies the directional course of EEQ across time. Here, $CHEQ_j$ and $CHEQ_i$ denote the CHEQ values for years j and i, respectively. When the slope value is less than 0, it signifies a diminishing trend in EEQ over the designated time span. This conveys a state of decline or deterioration in EEQ during this interval. Conversely, a positive slope value greater than 0 indicates an upswing in ecosystem quality throughout the specified period. This suggests an enhancement or elevation in ecosystem quality over the same interval.

3.3. Mann–Kendall Model

The Mann–Kendall test, a rank-based nonparametric examination, is adept at scrutinizing both linear and nonlinear trends within datasets [20,21]. In the context of this study, the Mann–Kendall test served as a tool to ascertain the significance of trends in CHEQ. The resultant statistical values, denoted as S and Z_{Slope}, were calculated utilizing the following formulae:

$$S = \sum_{i=1}^{n-1} \sum_{j=i+1}^{n} sgn(CHEQ_j - CHEQ_i) \tag{4}$$

$$sgn(X_j - X_i) \begin{cases} 1, & CHEQ_j - CHEQ_i > 0 \\ 0, & CHEQ_j - CHEQ_i = 0 \\ -1, & CHEQ_j - CHEQ_i < 0 \end{cases} \tag{5}$$

$$Var(S) = \frac{n(n-1)(2n+5)}{18} \tag{6}$$

$$Z_{Slope} = \begin{cases} \frac{S-1}{\sqrt{Var(S)}}, & S > 0 \\ 0, & S = 0 \\ \frac{S+1}{\sqrt{Var(S)}}, & S < 0 \end{cases} \tag{7}$$

where $CHEQ_i$ signifies the CHEQ values corresponding to years j and i within the time series. The symbol "n" designates the length of this series, which, in this case, is 19 years. The test statistic Z_{Slope} is constrained within the range of $(-\infty, +\infty)$. At a specified significance level α, the alterations within the time series attain significance if $|Z_{Slope}|$ surpasses $Z_{1-\alpha/2}$. For this study, a significance level of $\alpha = 0.05$ is selected. Consequently, it is determined that, if Z_{Slope} exceeds 1.96 ($Z_{1-0.05/2}$), namely, $Z_{Slope} > 1.96$, a discernible trend in China's CHEQ within the 2002–2019 timeframe can be confidently deemed significant at the 0.05 confidence level.

3.4. Multiple Regression Residual Analysis

The central objective of this study is to holistically evaluate the collective ramifications of diverse climate shifts on ecosystems, while concurrently dissecting the proportional influence attributed to human interventions. Prior investigations have predominantly centered on modeling the ecological repercussions of climate change employing precipitation and temperature data [22–24]. Nonetheless, the real-world impact on ecosystems stems from an amalgamation of factors, encompassing saturated water vapor pressure, soil moisture, actual evapotranspiration, and other climatic elements. To address this complexity, the present study adopts a multiple regression residuals approach for a comprehensive quantitative examination of the mutual contributions posed by climatic factors and human activities towards shifts in ecosystem quality. The ensuing sequence delineates the specific procedural framework:

1. A multivariate linear regression model was constructed to establish the interrelation between climate factor indicators and CHEQ, utilizing CHEQ as the dependent variable. This model encompassed diverse factors—TEMP, PRE, AET, PET, DEF, RO, PDSI, VAP, VPD, SRAD, VS, and SOIL—as independent variables. Through this approach, the regression link between climate factor indicators and CHEQ was established.
2. By integrating the regression model coefficients with climate data, the anticipated CHEQ value (denoted as $CHEQ_{CC}$: CHEQ influenced solely by climate change) was computed.
3. To distinctly delineate the CHEQ impact attributable solely to human activities, the disparity between the observed CHEQ value and the projected CHEQ value was computed as the CHEQ residual (designated as $CHEQ_{HA}$).

The corresponding formula is provided below:

$$CHEQ_{CC} = \sum_{i=1}^{n} a_i x_i + b \tag{8}$$

$$CHEQ_{HA} = CHEQ_{obs} - CHEQ_{CC} \tag{9}$$

where $CHEQ_{CC}$ denotes the forecasted value of CHEQ as determined by the regression model, where x_i represents the climate factor, and a_i and b stand as parameters within the regression model equation. The disparities between the observed value $CHEQ_{obs}$ and the projected value $CHEQ_{CC}$ are referred to as $CHEQ_{HA}$ residuals.

3.5. Analysis of the Determination of the Drivers of CHEQ Changes

This study endeavors to discern the primary influencers steering alterations in China's CHEQ. This endeavor encompasses computing the contributions attributed to climatic factors and human activities regarding CHEQ adjustments, as detailed in Table 2 [25]. The precise calculation procedure is elucidated within the confines of Table 2, with the respective terminologies illustrated as follows: "CC" signifies climate factors, "HA" signifies human activities, "Slope($CHEQ_{obs}$)" represents the observed CHEQ trend, "Slope($CHEQ_{CC}$)" signifies the trend of CHEQ predictions (reflecting changes under the sway of climate factors), and "Slope($CHEQ_{HA}$)" signifies the trend of CHEQ residuals (pertaining to changes influenced by human activities) [26,27].

Table 2. Criteria for determining the drivers of CHEO change and calculation of contribution rate.

Trend (CHEQobs)	Driving Factors	Classification Criteria for Driving Factors		Contribution Rate of Driving Factors/%	
		Trend (CHEQCC)	Trend (CHEQHA)	Climatic Factors	Human Activities
Greater than 0	CC&HA	>0	>0	$\frac{Slope(CHEQ_{CC})}{Slope(CHEQ_{obs})}$	$\frac{Slope(CHEQ_{HA})}{Slope(CHEQ_{obs})}$
	CC	>0	<0	100	0
	HA	<0	>0	0	100
Less than 0	CC&HA	>0	<0	$\frac{Slope(CHEQ_{CC})}{Slope(CHEQ_{obs})}$	$\frac{Slope(CHEQ_{HA})}{Slope(CHEQ_{obs})}$
	CC	>0	>0	100	0
	HA	<0	<0	0	100

3.6. Hurst Index

The anticipation of China's future CHEQ evolution holds paramount importance for guiding forthcoming ERPs [28]. The Hurst index, widely employed across disciplines such as climatology and ecology, serves as a tool to assess the sustained patterns within extensive time-series data [29,30]. The advantages of the Hurst index lie in its self-similarity and long-term dependence attributes exhibited within measurement index time series. In this study, we adopt the R/S analysis of the Hurst index to evaluate the persistency of China's CHEQ data across an extensive time span, subsequently employing this information to prognosticate future CHEQ shifts based on the trends observed between 2002 and 2019.

The Hurst index (H) manifests in three primary forms: (1) $0.5 < H < 1$, indicating a continuous sequence within the time series. A proximity to 1 signifies enhanced continuity, implying that future changes align with past trends; (2) $H = 0.5$, denoting a stochastic sequence in CHEQ's time series, thereby suggesting a lack of long-term correlation; and (3) $0 < H < 0.5$, reflecting inverse persistence within the time series. This inversely persistent state implies a future trend contrary to past patterns. Greater proximity to 0 amplifies the strength of this inverse persistence.

4. Results and Analysis

4.1. Accuracy Verification of EEQ

In this investigation, we employ the 2018 county-level eco-index data, as provided by the Ministry of Ecology and Environment of China, to assess the validity and precision of the grid-scale CHEQ data generated within the confines of this research. As illustrated in Figure 2, a comparative analysis of accuracy between the CHEQ and RSEI models is conducted across six distinct regions in China. Upon careful examination of the figure, it becomes evident that the CHEQ model exhibits a notably superior degree of conformity when contrasted with the conventional RSEI model across various regions. It is worth noting, however, that the CHEQ model does exhibit a higher root mean square error (RMSE) compared to RSEI, with the exception of the eastern region, as well as the central and southern regions. In summary, the CEHQ model, as introduced in this study, demonstrates an enhanced level of generalizability when compared to existing approaches.

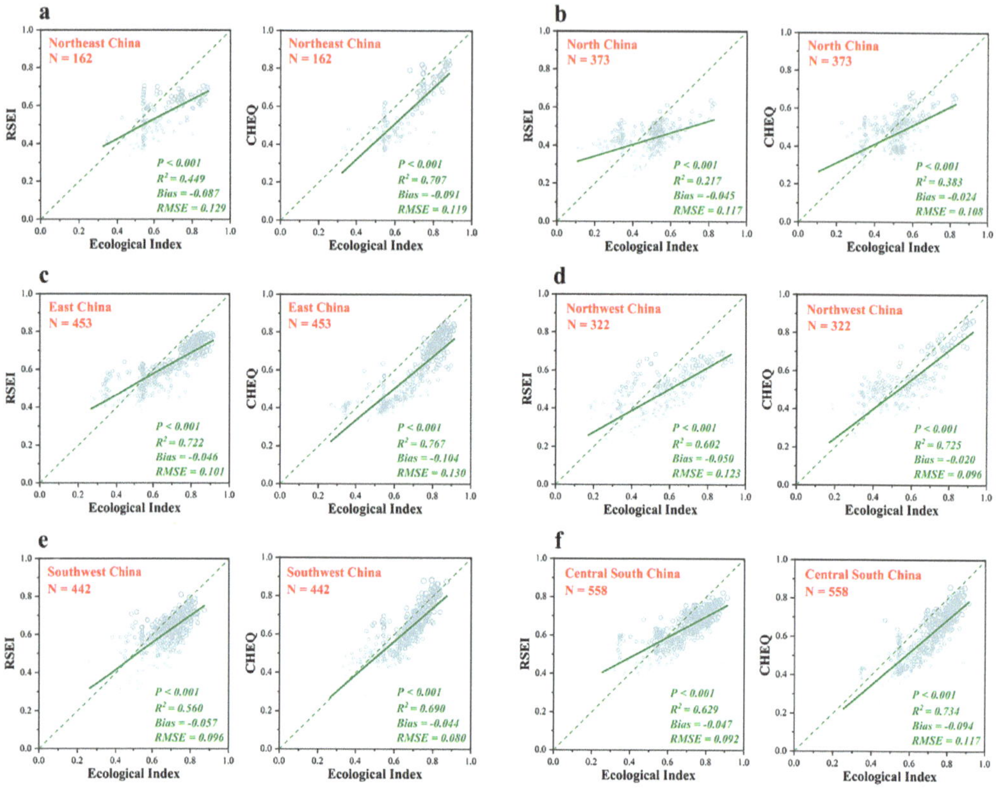

Figure 2. Comparison of the accuracy of CHEQ and RSEI models for six regions in China. China partition data from the Resources, Environment, Science and Data Centre. (**a**): Northeast China; (**b**): North China; (**c**): East China; (**d**): Northwest China; (**e**): Southwest China; (**f**): Central South China. (https://www.resdc.cn/Default.aspx (accessed on 2 May 2023)).

Furthermore, we utilized the MCD12Q1 land use data to contrast the longitudinal variations of CHEQ and RSEI across distinct land use and vegetation types. As illustrated in Figure 3, notable disparities emerge in the mean values of RSEI and CHEQ across various land use categories. Specifically, RSEI exhibits higher mean values in bare soil and built-up areas, while CHEQ displays a more consistent and gradual decline in built-up areas. Conversely, in natural areas encompassing both vegetation and water bodies, CHEQ demonstrates higher mean values compared to RSEI. These observations collectively suggest that CHEQ excels in characterizing EEQ across diverse land cover types. Of paramount significance, it is noteworthy that the standard deviation of CHEQ consistently registers lower values than that of RSEI across all land classes. This disparity signifies that CHEQ maintains superior stability and continuity when assessed on a time-series scale. In summation, CHEQ emerges as the more apt choice for applications in large-scale ecological and environmental studies in China when compared to RSEI.

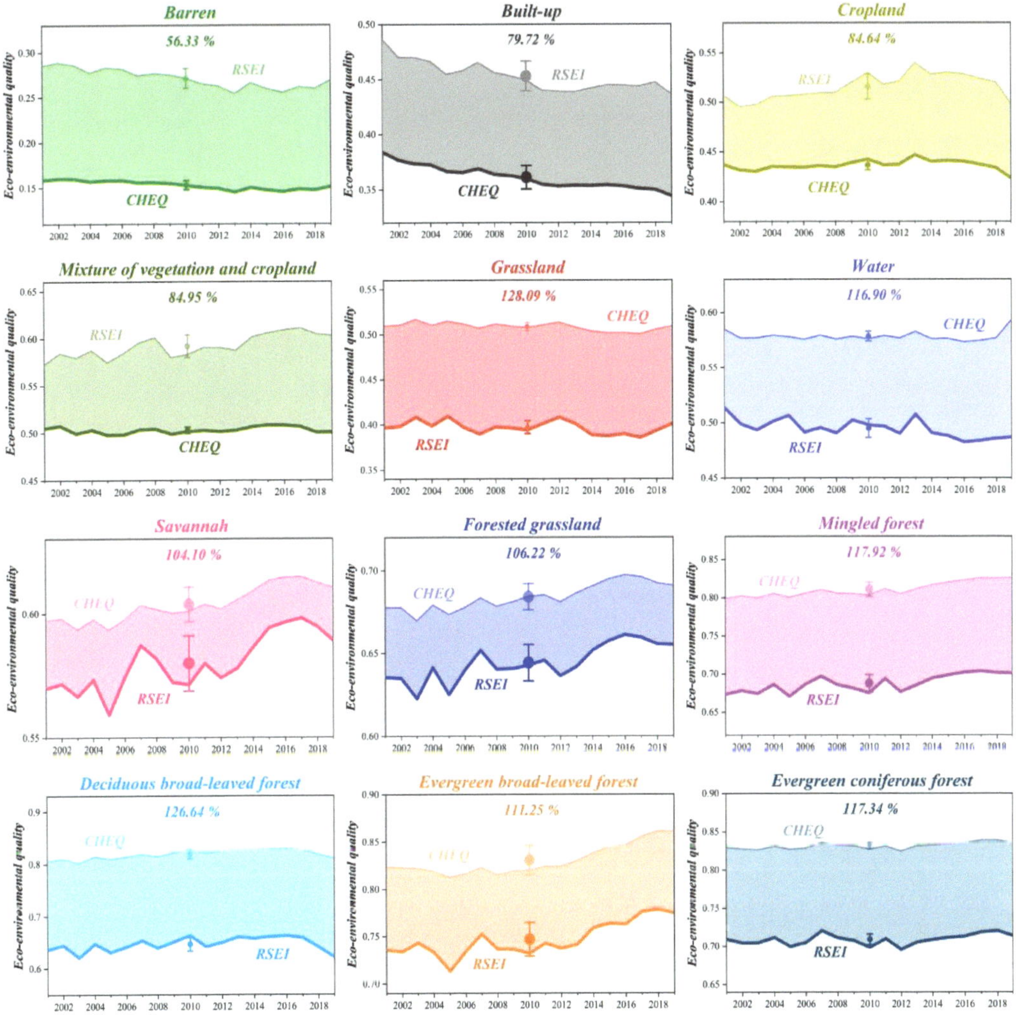

Figure 3. Characteristics of long time-series variation of CHEQ and RSEI under different land use types and vegetation types. The box plot represents the mean and standard deviation of CHEQ and RSEI over multiple years.

4.2. Characteristics of Spatial and Temporal Changes in the EEQ

In general, the trend in EEQ change across China during the period from 2002 to 2019 remains relatively stable, as depicted in Figure 4. To further elucidate the distribution of these EEQ trends during the same timeframe, Figure 5 offers a comprehensive breakdown. Figure 6 shows the trend of EEQ distribution in China from 2002–2019. Figure 5a illustrates the spatial distribution of observed EEQ trends, Figure 5b portrays the spatial distribution of EEQ trends as derived from climate indicator simulations, and Figure 5c delineates the trends in residual EEQ, representing the impact of human activities exclusively.

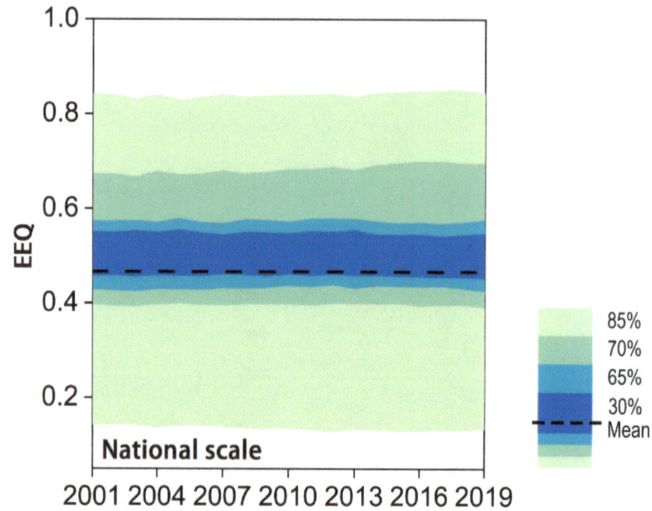

Figure 4. Temporal trends in EEQ in China during 2001–2019.

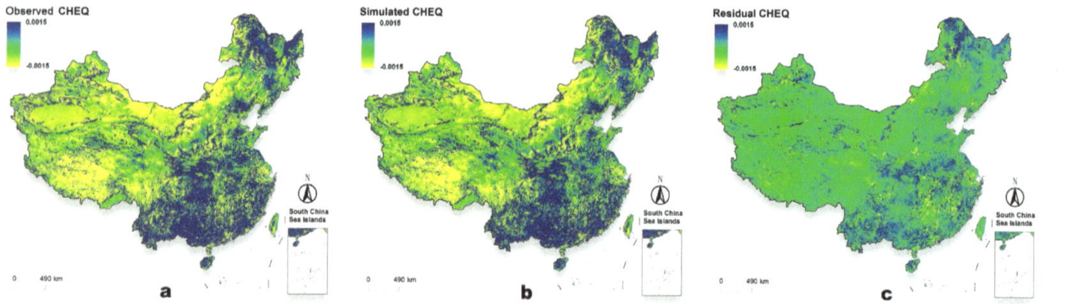

Figure 5. Trend distribution of EEQ in China from 2002 to 2019. (**a**) Spatial change trend map of original CHEQ. (**b**) Map of spatial trends in CHEQ under the influence of climate change. (**c**) Map of spatial trends in CHEQ under the influence of human activities.

As illustrated in Figure 5a, a pronounced spatial heterogeneity characterizes the trends in EEQ change across China during the period spanning from 2002 to 2019. Broadly, the majority of regions within China have exhibited improvements in their EEQ over the past 18 years, a trend in line with the findings of Liao [31]. Noteworthy upward EEQ trends are discernible in Northeast China, the northern portions of North China, Central China, the central regions of East China, South China, as well as rapid growth trends in the Loess Plateau, Sichuan, Qinghai, Gansu, and the middle and lower reaches of the Yangtze River. These patterns are closely linked to a series of ERPs and environmental protection strategies carried out in China in recent years [9]. The Loess Plateau stands as a prominent ecological restoration area in China, grappling with severe soil erosion and degradation resulting from decades of excessive cultivation and unsustainable human activities [32]. To ameliorate the local ecological environment, China initiated extensive ERPs, such as converting farmland back into forests and grasslands, and implementing water conservation measures aimed at mitigating soil erosion through vegetation restoration and land preservation. These efforts have notably enhanced the quality of the local ecological environment [33]. In another example, the construction of eco-friendly infrastructures, including wetland restoration, the rehabilitation of natural habitats, and riverbank stabilization, has been implemented in the middle and lower reaches of the Yangtze River in the past decade. These measures have contributed to improved local water quality and ecosystems, providing the necessary conditions for enhancing local EEQ [34]. Meanwhile, over the last 18 years, China has witnessed a significant decline in EEQ in certain regions, including the Yangtze River Delta, Hunan, Jiangxi, the Junggar Basin in Xinjiang, the eastern and western parts of Inner Mongolia, and parts of southern Tibet. The acceleration of industrialization in China, accompanied by expanding urban areas, has resulted in substantial emissions of industrial and urban pollutants, exacerbating ecological and environmental challenges [35]. Furthermore, the Yangtze River Delta, as the central hub of China's economic development, has adversely affected local vegetation through the extensive use of impermeable surface materials during urbanization and construction processes [36]. Conversely, the central and western regions of China, situated inland and distant from the ocean, grapple with low precipitation levels and densely populated localized areas [37]. Consequently, issues such as water scarcity and pollution are prevalent, contributing to localized ecosystem imbalances.

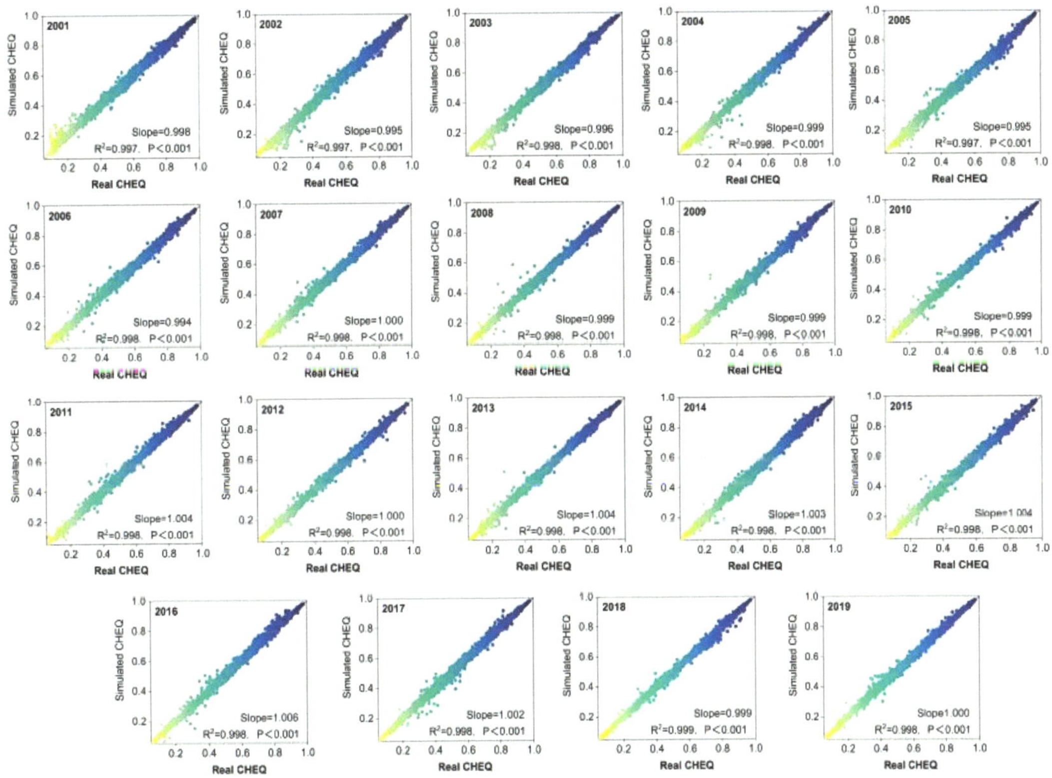

Figure 6. Accuracy assessment of simulated CHEQ during 2002–2019. The color mapping in the figure maps values from 0–1.

The spatial distribution of the simulated EEQ trend is illustrated in Figure 5b. Remarkably, the spatial distribution pattern of both observed and simulated EEQ values in China from 2002 to 2019 closely aligns. This alignment implies that climate change and natural factors have predominantly governed the enhancement of EEQ in China over the past 18 years [13]. Conversely, EEQ changes attributable to human activities exhibit a markedly distinct distribution pattern during the period from 2002 to 2019, as depicted in Figure 5c. The regions demonstrating increased EEQ are primarily situated in the Loess Plateau, Northeast China, Central China, and South China. Northeast China has implemented an array of ERPs, including the establishment of the Three-North Protective Forest System, the

Natural Forest Resource Protection Project, and initiatives focused on wetland preservation and restoration. Concurrently, they have undertaken an extensive policy of converting farmland into forests and grasslands. These efforts have revitalized vegetation cover, improved local soil quality, effectively mitigated land degradation, and, thereby, elevated local EEQ [22]. In the central and southern regions, collaborative inter-regional efforts have been initiated. This includes the implementation of projects such as the Protection Forest System in the Yangtze River Basin (Phase II and Phase III), the Protection Forest System in the Pearl River Basin, and the Comprehensive Management of Karst Rocky Desertification in Southwest China. These initiatives have effectively safeguarded local flora and natural ecological environments, reducing the impacts of industrial and agricultural discharges on local EEQ [7]. Moreover, the central region, in tandem with economic development, has embarked on soil and water conservation and river management projects to mitigate the adverse effects of land degradation and water scarcity on EEQ [38].

4.3. Contribution of Human Activities and Natural Factors

To assess the scientific robustness and applicability of the residual trend analysis model developed in this investigation, we conducted a validation exercise to gauge the accuracy of the simulated EEQ values. As evident from Figure 6, both R^2 and the slope closely approach the value of 1, indicative of the model's effectiveness in capturing the influence of climatic elements on EEQ [26]. Consequently, the TSS-RESTREND model employed in this research is substantiated, scientifically sound, and universally applicable. This model serves as a valuable tool for scrutinizing the driving forces behind the temporal variations in EEQ across China.

The contributions of climatic factors and human activities to the changes in EEQ in China between 2002 and 2019 are depicted in Figure 7. In this context, "CC" signifies the influence of climatic factors alone, "HA" represents the influence of human activities alone, and "HA&CC" denotes the combined influence of climatic factors and human activities.

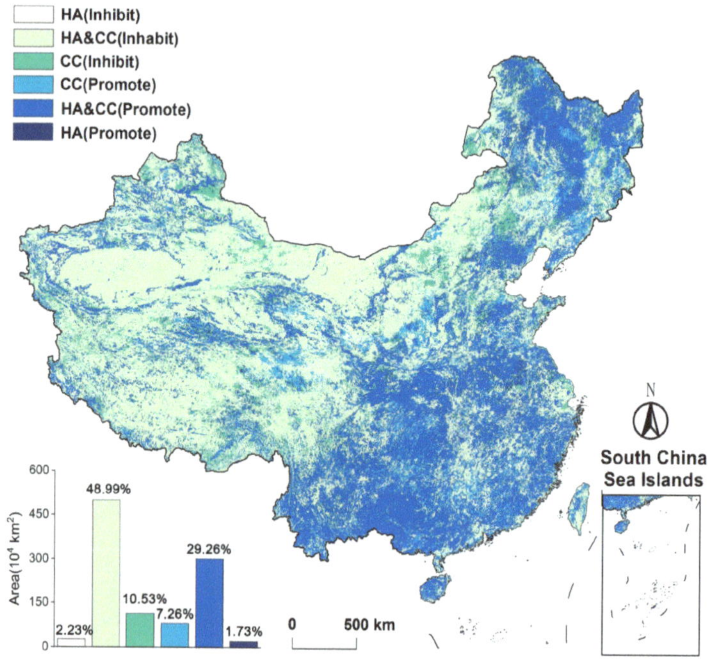

Figure 7. Trend distribution of natural factors and human activities on EEQ changes in China during 2002–2019.

As delineated in Figure 7, approximately 78.25% (750.60 $\times$ 10^4 km^2) of the EEQ alterations in China over the past 18 years can be attributed to the joint influence of HA&CC. Among these changes, HA&CC synergistically promoted 29.26% (280.41 $\times$ 10^4 km^2) of the area, primarily concentrated in the Yunnan–Guizhou Plateau, Loess Plateau, and the northeastern, central, eastern, and southern regions of China. The Loess Plateau and Yungui Plateau, characterized by intricate topography and diverse climates, have witnessed the implementation of extensive soil and water conservation measures, including terracing, row-break planting, and protective forest construction. These efforts have substantially improved the local ecological environment and, in turn, the quality of EEQ [39]. Similarly, several nature reserves and scenic areas have been established in Northeast, Central, East, and South China to mitigate the disruption of the natural environment by human activities. Furthermore, a series of ecological restoration strategies have been undertaken to safeguard and sustain ecosystem stability [7,40]. Moreover, with the progression of urbanization, these regions have begun to recognize the vital role of the ecological environment in urban development. Consequently, they have introduced ecological protection policies that encompass restricting industrial pollutant emissions, intensifying urban greening and ecological development, promoting the utilization of energy-efficient and low-emission clean energy sources, and deploying intelligent detection equipment and pollution control technologies. These measures have notably contributed to enhancing local EEQ [41]. Conversely, the area jointly influenced by HA&CC and experiencing suppression encompasses 48.99% (470.07 $\times$ 10^4 km^2) of the total area, mainly situated in the northern regions of the Inner Mongolia Plateau, Tarim Basin, and Sichuan Basin within China. These areas contend with severe natural climatic conditions, including drought and rising temperatures, which exacerbate grassland degradation. Additionally, water scarcity, over-exploitation, and irrational resource allocation have contributed to environmental challenges in these regions. Human activities have further compounded these issues, with overgrazing leading to excessive land use and grassland degradation, impairing vegetation growth and recovery. Frequent construction projects have disrupted land integrity, further unsettling the local ecological balance [12].

The EEQ change area influenced solely by climatic factors accounts for 17.79% (170.55 $\times$ 10^4 km^2) of the total, with these areas generally characterized by lower population densities and, thus, fewer human activities. Among them, 7.26% (69.65 $\times$ 10^4 km^2) of the area is predominantly found in the Shanxi, Shaanxi, and Qinghai regions. These areas enjoy ample sunshine, a variety of suitable land types for diverse plant categories, four distinct seasons, and moderate rainfall, all contributing to the water requirements of vegetation [42]. In contrast, the area experiencing inhibition amounts to 10.53% (100.90 $\times$ 10^4 km^2) and is primarily distributed along the Tibetan Himalayan border and in the Xinjiang Altay region. These regions grapple with harsh climates, aridity, soil fragility, stoniness, and nutrient deficiencies, rendering plant establishment difficult. Rapid water evaporation compounds these challenges, further hindering local vegetation growth [43].

Finally, EEQ alterations driven solely by human activities encompass 3.96% (38.02$\times$ 10^4 km^2) of the total area. Within this subset, 1.73% (16.6 $\times$ 10^4 km^2) is facilitated by human activities, while 2.23% (21.42 $\times$ 10^4 km^2) is inhibited. On the whole, climatic factors and human activities together have significantly shaped EEQ changes across the majority of China, with climatic factors exerting a more substantial influence on EEQ compared to human activities.

The contributions of climatic factors and human activities to the changes in EEQ across China from 2002 to 2019 are illustrated in Figure 8. Specifically, Figure 8a portrays the spatial distribution map of the influence of climatic factors on EEQ evolution, while Figure 8b delineates the spatial distribution map of the impact of human activities on EEQ evolution. As discernible in the figures, climatic factors emerge as the principal drivers of EEQ alterations across the majority of China. Regions where climatic factors account for nearly 100% of the influence are predominantly concentrated in the northern part of Inner Mongolia, the Loess Plateau, and Tibet. These areas exhibit lower population densities and reduced human activity, yet they benefit from favorable climates and ample precipitation, providing conducive conditions for local plant growth and ecosystem rehabilitation [39–44]. In contrast, the impact of human activities on EEQ changes in China is relatively modest, echoing the findings in Figure 8. The regions with more pronounced contributions from human activities to EEQ are primarily situated in central China, the northern Sichuan Basin, and the Junggar Basin. These areas are closely linked to a succession of ERPs initiated across China, including the Natural Forest Resources Protection Project and the construction of the Three-North Protective Forest System [45]. Additionally, these regions have progressively adopted advanced agricultural technologies such as intelligent agricultural machinery and drip irrigation, further supporting ecological development [46]. Overall, EEQ changes in China over the past 18 years have been jointly shaped by climatic factors and human activities. Climatic factors have accounted for approximately 79.19% of the contribution, whereas human activities have contributed approximately 20.81%. Notably, the influence of climatic factors has outweighed that of human activities in driving EEQ changes.

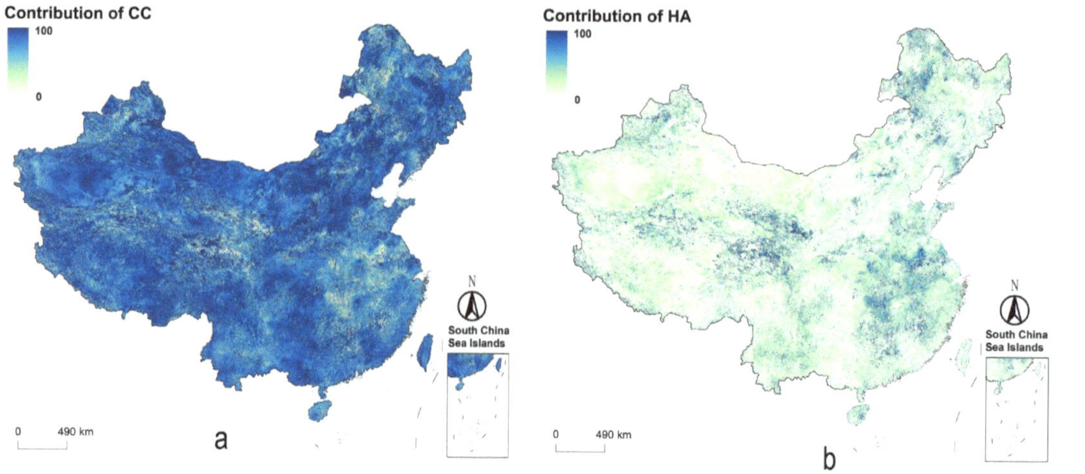

Figure 8. Spatial distribution of the contribution of natural factors and human activities to the trend of EEQ change in China during 2002–2019.

5. Discussion

5.1. Ecological and Environmental Benefits of ERPs

Since 1998, China has undertaken numerous ERPs (Table A1). Among them, there are nine ecological restoration projects (ERPs) with the highest investment and the most outstanding ecological benefits (Figure 9). These initiatives encompass a range of comprehensive undertakings: the Beijing–Tianjin Sand Source Comprehensive Control Project (BTSSCP), the Three-North Protective Forest Construction Project (TNSDP), the Sanjiangyuan Ecological Protection and Construction Project (SEPCP), the Natural Forest Resource Protection Project (NFRPP), the Returning Ploughland to Forestry Project (RPFP), the Returning Pasture to Grassland Project (RPGP), the Southwestern Karst Desertification Comprehensive Treatment Project (SKRDCTP), the Yangtze River Basin Protection Forest System (YRBPFSCP), the Pearl River Basin Protection Forest System, and several other major projects (PRBPFSCP) [17]. Across the nation's expanse, these ecological restoration endeavors span all 31 provinces, collectively covering an extensive land area of approximately 924.8×10^4 km^2 [2].

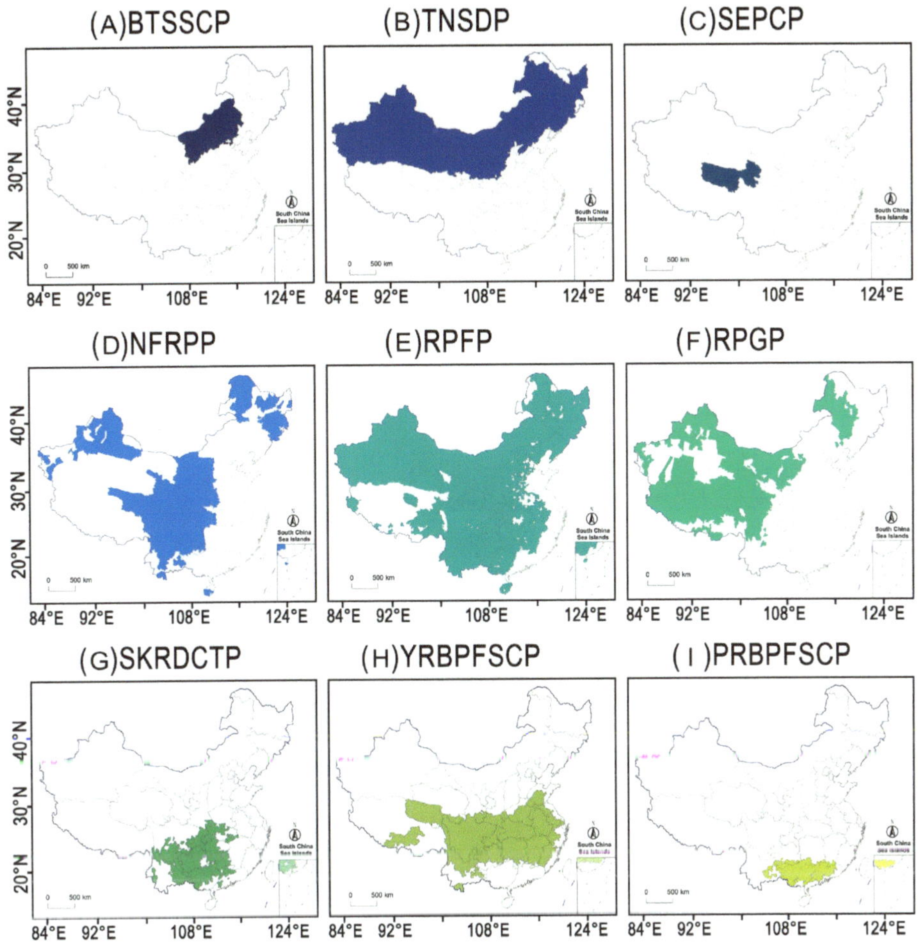

Figure 9. Spatial distribution of 9 ERPs in China. (**A**): the Beijing Tianjin Sand Source Comprehensive Control Project (BTSSCP), (**B**): the Three-North Protective Forest Construction Project (TNSDP), (**C**): the Sanjiangyuan Ecological Protection and Construction Project (SEPCP), (**D**): the Natural Forest Resource Protection Project (NFRPP), (**E**): the Returning Ploughland to Forestry Project (RPFP), the Returning Pasture to Grassland Project (RPGP), (**F**): the Southwestern Karst Desertification Comprehensive Treatment Project (SKRDCTP), (**G**): the Yangtze River Basin Protection Forest System (YRBPFSCP), (**H**): the Pearl River Basin Protection Forest System, (**I**): several other major projects (PRBPFSCP).

Figure 10 provides a comprehensive view of the spatio-temporal dynamics of EEQ within various ecological restoration project areas across China over the past 18 years. The dotted line denotes the average EEQ value for all ERP areas in China. Overall, there has been a discernible upward trend in China's EEQ during the 2002–2019 period. Notably, significant EEQ improvements are observed in regions associated with SKRDCTP, SEPCP, and PRBPFSCP. These areas are predominantly located in Hubei, Jiangxi, Guangxi, Guangdong, and Qinghai within China. They boast high vegetation coverage, particularly Qinghai, which has implemented a suite of ecological protection and restoration measures in recent years. These measures encompass initiatives like returning farmland to forests and grasslands, ecological compensation policies, soil and water conservation projects, and the encouragement of land restoration and resource rationalization. As a result, Qinghai's comprehensive vegetation cover in grasslands has reached 57.9%, encompassing 57.67% of Qinghai Province's area and 14.92% of China's grassland area, directly attributable to China's ecological restoration endeavors [47]. Furthermore, local EEQ in Guangxi and Guangdong, bolstered by PRBPFSCP and the SKRDCTP, has also seen remarkable enhancement. These regions are dedicated to restoring the water-sourcing function of the karst areas, ensuring a stable water supply, and providing water security for local plant growth. Additionally, they have implemented afforestation and grassland restoration initiatives, augmenting plant species diversity and population numbers, enhancing vegetation coverage, and fostering a conducive growth environment for indigenous plants [48,49]. Conversely, the progress of TNSDP and RPGP has been relatively sluggish. These areas grapple with complex climatic and ecological dynamics, with multiple ecological elements interplaying. Moreover, their climates are characterized by aridity and low temperatures, imposing constraints on vegetation growth due to factors like temperature and precipitation. The region's infertile soil further hampers plant growth, resulting in slow progress in ecological recovery. Consequently, substantial improvements in the local ecological environment in these areas may require an extended time frame and ongoing maintenance efforts to yield noticeable effects [50].

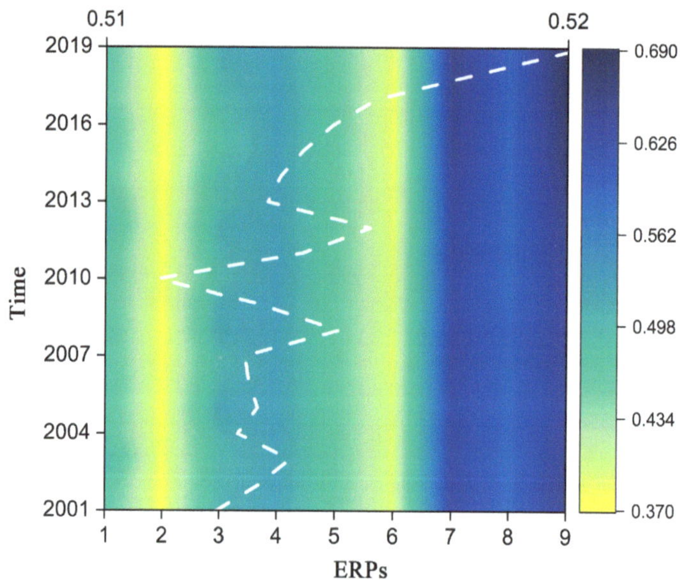

Figure 10. Spatio-temporal dynamics of EEQ in China during 2002–2019 (Numbers 1–9 denote 9 ERPs, of which 1. BTSSCP, 2. TNSDP, 3. SEPCP, 4. NFRPP, 5. RPFP, 6. RPGP, 7. SKRDCTP, 8. YRBPFSCP, 9. PRBPFSCP). The white dotted line represents the trend of the average annual EEQ of the nine ecological restoration projects in the last 20 years, and the corresponding values can be found on the upper axis.

5.2. *Trends in EEQ for ERPs*

Figure 11 provides insights into the extent of contribution from ERPs and the trends in EEQ changes across various districts and counties in China spanning from 2002 to 2019. Figure 11a delineates the number of ERPs implemented and the corresponding recovery of EEQ in each district and county across China. The horizontal axis represents the count of overlapping ERPs in China, while the vertical axis employs a five-tier grading system to facilitate the visualization of EEQ recovery in these regions throughout the study period. Examining the figure reveals a positive correlation between the number of ecological restoration project overlaps and EEQ recovery. Areas that experienced six instances of ERP coverage are predominantly situated within China's Yunnan–Guizhou Plateau, while regions with five ERP overlaps are primarily located in the central provinces of China, including Sichuan, Chongqing, Hubei, Henan, and others. Notably, these regions exhibit more pronounced EEQ improvements, consistent with the findings depicted in Figure 7. China's Yunnan–Guizhou region is characterized by its high-altitude plateau terrain. In recent years, the local government has actively pursued industrial restructuring efforts to curtail the overexploitation of natural resources and mitigate pollution. Moreover, they have initiated large-scale ERP implementations and established an ecological compensation mechanism. These measures have contributed significantly to the notable EEQ recovery in the region. On the other hand, the central region of China, located inland and marked by relatively scarce water resources, has intensified water resource management. This involves the robust construction of water conservancy projects and improved utilization of water resources, effectively safeguarding the local ecological equilibrium and promoting EEQ enhancements [25].

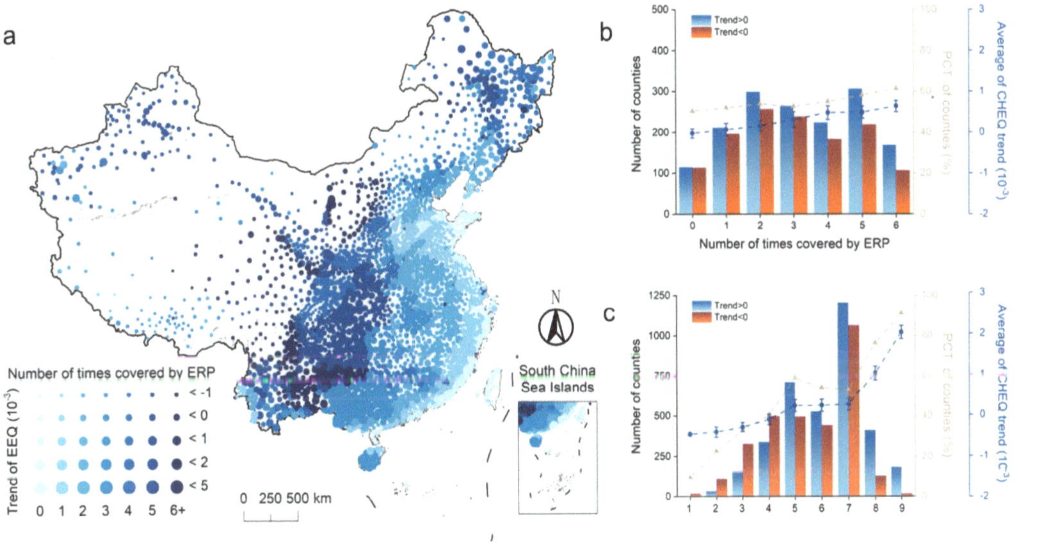

Figure 11. Degree of contribution of ERPs in China and the trend of EEQ by district and county, 2002–2019. (**a**): delineates the number of ERPs implemented and the corresponding recovery of EEQ in each district and county across China; (**b**): illustrates the relationship between the frequency of ERP coverage in specific districts and counties and their corresponding EEQ; (**c**): illustrates the temporal trends in the geographical coverage of nine major ERPs in China spanning from 2002 to 2019, along with the EEQ; ((**c**): Numbers 1–9 denote 9 ERPs, of which 1. BTSSCP, 2. TNSDP, 3. SEPCP, 4. NFRPP, 5. RPFP, 6. RPGP, 7. SKRDCTP, 8. YRBPFSCP, 9. PRBPFSCP).

Figure 11b illustrates the relationship between the frequency of ERP coverage in specific districts and counties and their corresponding EEQ. The Y-axis represents the number of districts covered by ERPs and the X-axis represents the number of times covered by ERPs. The blue bars represent districts and counties with EEQ greater than 0, while the red bars represent those with EEQ less than 0. The number of districts and counties with an ERP coverage of 0 is the lowest (115 districts and counties with EEQ greater than 0 and 113 districts and counties with EEQ less than 0), while the number of districts and counties with a coverage of 2 and 5 is the highest (299 and 306 districts and counties with EEQ greater than 0 and 256 and 219 districts and counties with EEQ less than 0). These regions are primarily situated in the central and eastern parts of China. The yellow dashed line illustrates the proportion of China's EEQ recovery trend exceeding 0, while the blue dashed line represents the average EEQ value associated with China's ERPs. It is evident that, as the number of ecological restoration project coverages increases, both the overall trend of EEQ recovery and the average value exhibit an upward trajectory. Consequently, it can be deduced that China's EEQ progressively improves with the ongoing advancement of ERPs.

Figure 11c illustrates the temporal trends in the geographical coverage of nine major ERPs in China spanning from 2002 to 2019, along with the EEQ. Notably, the reforestation initiative on retired farmland boasts the broadest presence, encompassing a noteworthy 1205 districts and counties with an EEQ greater than 0, while 1063 districts and counties exhibit an EEQ lower than 0. In contrast, SEPCP exhibits the most restricted geographical span, with a mere 2 districts and 19 counties displaying an EEQ greater and lower than 0, respectively. The yellow dashed line delineates the percentage of China experiencing an EEQ recovery trend surpassing 0, with the highest proportion evident in PRBPFSCP and the lowest proportion observed in SEPCP and Construction Project. Concurrently, the blue folded line traces the average EEQ value across ERPs in China. It is evident that PRBPFSCP demonstrates a more pronounced increase, while the SEPCP exhibits a comparatively slower rate of enhancement. Located primarily in the coastal regions of Guangdong and Guangxi provinces in southern China, PRBPFSCP benefits from favorable climatic conditions, essential for local vegetation restoration. Additionally, the relevant authorities have promulgated a series of favorable policies and regulations to facilitate a comprehensive assessment of the local climate and natural conditions. This has enabled the determination of optimal vegetation planting types, modes, and densities through the utilization of modern scientific and technological tools for the real-time monitoring and analysis of protection forest distribution and growth status. Timely measures are then implemented to contribute to the restoration of local EEQ [7,9]. Conversely, SEPCP is predominantly situated in China's Qinghai region, characterized by a complex ecosystem encompassing climate, soil, vegetation, and wildlife dimensions. Frequent natural disasters and climate fluctuations in the region further impede the recovery of local EEQ [42]. In summary, the development of ERPs in China necessitates a scientifically grounded approach based on the specific geographical attributes of each locality. This approach should involve the formulation of rational policies, the prioritization of long-term sustainability, and a commitment to achieving substantial outcomes.

5.3. China's Future EEQ Forecast

The prediction of China's future ecological effectiveness quotient (EEQ) development holds paramount significance for pertinent governmental agencies tasked with formulating environmental protection policies and sustainable development strategies. Such forecasts are instrumental in steering the trajectory of future development while facilitating a scientifically rigorous evaluation of ecological and environmental impacts. In this investigation, we employ the Hurst index rescaled polarity method (R/S) to project forthcoming EEQ changes in China, building upon the observed trends from 2002 to 2019 [29,30]. As illustrated in Figure 12, the projected area earmarked for EEQ improvement amounts to 544.00×10^4 square kilometers, constituting 56.70% of China's total land area. This implies that over half of China's territory is anticipated to witness enhanced EEQ in the future. Concurrently, the region projected to experience EEQ deterioration encompasses approximately 409.45×10^4 square kilometers, representing 42.72% of China's landmass. The most substantial change, spanning 311.18×10^4 square kilometers and comprising 32.43% of China's total land area, is primarily situated in the Loess Plateau, Tarim Basin, Inner Mongolia Plateau, and Tibet. These regions are poised to gradually restore their ecological quality due to the progression of ERPs. Conversely, regions expected to transition from high to low EEQ cover an area of roughly 132.81×10^4 square kilometers, making up 13.88% of China's total land area. These areas are primarily concentrated in the central and eastern parts of China and may impede future EEQ development owing to the rapid pace of urbanization.

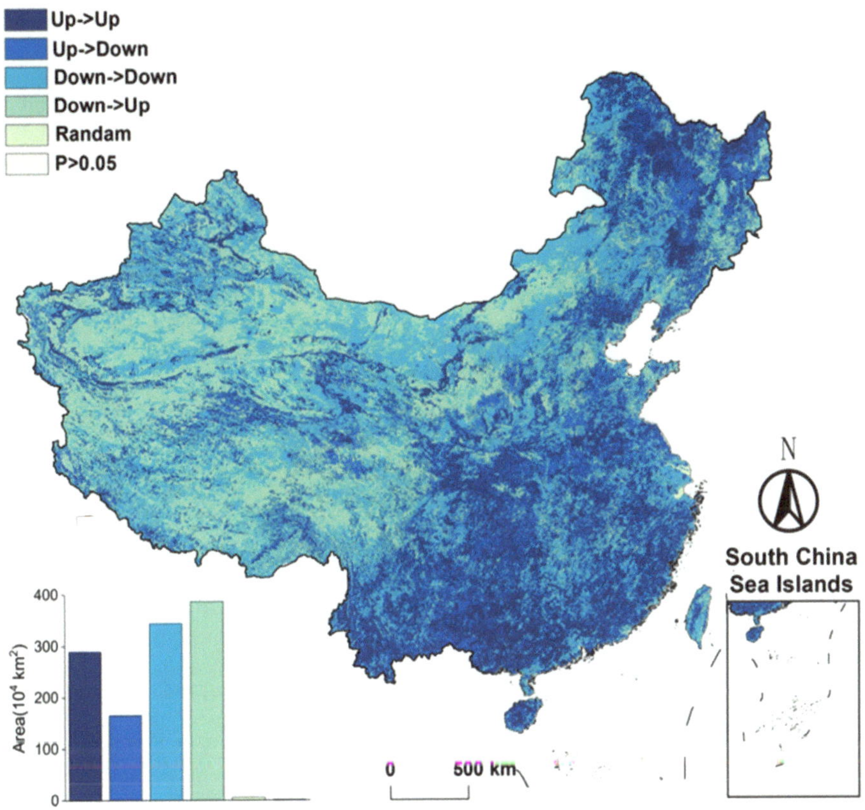

Figure 12. Spatial distribution of future EEQ in China.

Based on the aforementioned analysis, this paper presents the following recommendations: (1) Regions experiencing a decline in EEQ within China should consider adjusting their development strategies. Emphasis should be placed on achieving sustainable development by establishing a robust environmental monitoring system that enables real-time assessment of environmental quality. Concurrently, fostering collaboration among regions to collectively address challenges and facilitate resource sharing and environmental protection synergies is essential. (2) The ecological restoration strategy has played a pivotal role in enhancing China's EEQ since its inception. China should continue to implement this strategy and institute an ecological compensation mechanism to facilitate the necessary conditions for ecological restoration efforts. (3) The establishment of an ecological red line is imperative in order to rigorously regulate the exploitation and overgrazing of China's Inner Mongolian Plateau, Loess Plateau, and Tibetan Plateau. In regions characterized by delicate climatic and natural environments, tailored recovery strategies must be developed to ameliorate the local ecological landscape.

5.4. Contribution of the Study

RSEI was first proposed by Xu et al. in 2013 [11], who aimed to remedy the cumbersome calculation process of the EI index (proposed by the Ministry of Ecology and Environment of China) by creating a simple index that can be equivalently substituted with the EI index. Since its introduction, the RSEI has been widely used in China and other parts of the world to monitor the quality of ecosystems. However, we have not yet seen any article exploring the applicability of the RSEI index in China. For this reason, with the help of the Ministry of Ecology and Environment of China, we conducted the first evaluation study on the applicability of the RSEI index in China and found that the RSEI index is not suitable for regions with poor ecological environments, which is mainly due to the lack of indicators that can characterize regional ecosystem services in the evaluation system of the RSEI index. Therefore, we combined the RSEI and EI indices and introduced the land-use abundance index to create the CHEQ, a universal ecological quality index for China, which compensates for the low applicability of the RSEI index and the cumbersome calculation of the EI. In addition, we analyzed the drivers of spatial and temporal changes in EEQ in China in the last two decades from the perspectives of both climate change and human activities. Combined with the analyses in Sections 5.1–5.3, our study not only provides reference value for the implementation of future ecological restoration projects in China, but is also expected to provide certain valuable suggestions for scholars to carry out research on the mechanism of the impact of ecological restoration projects on EEQ.

6. Recommendation and Conclusions

6.1. Recommendation

Concerning the prospective development of China's EEQ, it is noteworthy that the projected EEQ improvement encompasses 56.70% of China's total land area. However, there remain regions exhibiting a declining trend in EEQ. These areas are primarily situated within the Loess Plateau, Tarim Basin, Inner Mongolia Plateau, and Tibet region of China, characterized by fragile ecosystems influenced by intricate climatic factors and geographic conditions. Consequently, pertinent authorities should establish more scientifically informed and rational ecological planning approaches. This should be coupled with a steadfast commitment to advance ERPs, emphasizing large-scale land conversion from farming to afforestation and grassland rejuvenation, averting soil degradation, and striking a harmonious balance between economic development and ecological preservation [25,39]. Furthermore, as urbanization continues to advance in China, regions of high economic development such as the Yangtze River Delta and the Pearl River Delta may encounter future declines in EEQ. These areas are especially vulnerable to EEQ degradation due to high population density and the gradual expansion of urban areas. To ameliorate the quality of the local ecological environment in these urban settings, the incorporation of resilient green spaces, including parks, gardens, and green infrastructure, is recommended in order to increase urban green space coverage, elevate the city's greening ratio, and mitigate the urban heat island effect.

6.2. Conclusions

This study focuses on unveiling the spatial and temporal characteristics of EEQ in China, exploring its response to both climate change and human activities spanning from 2002 to 2019. Our findings reveal a substantial improvement in EEQ across most regions of China over the past 18 years, with a particularly pronounced recovery observed in the northeastern, Loess Plateau, and southern regions. This improvement can be attributed to the successive launch of ERPs, including TNSDP, NFRPP, and wetland protection and restoration initiatives. Together, climatic factors and human activities account for 78.82% of the EEQ variation in China, significantly contributing to its overall enhancement. Notably, climatic factors exert a greater influence, representing approximately 79.19% of the total impact, while human activities contribute to the remaining 20.81%.

The outcomes of this research offer valuable insights into the dynamics of China's ecological quality and the factors influencing it. Our quantitative analysis, assessing the respective contributions of ecological restoration efforts and climate change, furnishes policymakers and stakeholders with actionable recommendations. The knowledge derived from this study can effectively guide efforts to promote ecological well-being and sustainable development in China, as the nation navigates the delicate balance between economic expansion and ecological preservation.

Author Contributions: Conceptualization, J.L., D.X. and X.P.; methodology, J.L. and X.P.; resources, J.L. and R.T.; data curation, J.G., X.P., T.B. and J.L.; writing—original draft preparation, J.L., Z.Z. and X.P.; writing—review and editing, X.P. and J.L.; visualization, X.P., D.X, Z.Z., J.G., T.B. and J.L.; supervision, Z.Z., J.L., R.T. and X.P.; funding acquisition, J.G. All authors have read and agreed to the published version of the manuscript.

Funding: This research was funded by the Innovation Research Team Project of Natural Science Foundation of Hainan Province (Grant No. 422CXTD515) and the Key Research and Development Plan of Hainan Province (No: ZDYF2022SHFZ062).

Institutional Review Board Statement: Not applicable.

Informed Consent Statement: Not applicable.

Data Availability Statement: The original contributions presented in the study are included in the article, further inquiries can be directed to the corresponding author.

Conflicts of Interest: The authors declare no conflict of interest.

Abbreviations

EEQ	Ecological environment quality
ERPs	Ecological restoration projects
HA	Human activities
CC	Climatic factors
CHEQ	Chinese High-Resolution Ecological Quality Dataset
EI	Ecological Environment Status Index
RSEI	Remote Sensing Ecological Index
NDVI	Normalized Vegetation Index data
NPP	Net primary productivity
PRE	Precipitation
AET	Actual evapotranspiration
PET	Potential evapotranspiration
SRAD	Solar radiation
VAP	Atmospheric pressure
VPD	Saturated water vapor pressure difference
VS	Wind speed
RO	Surface runoff
DI	Drought index
SOIL	Soil moisture
PDSI	Drought index
DEF	Water deficit
TEMP	Surface temperature
TMMN	Maximum temperature
TMMX	Minimum temperature
PCA	Principal component analysis
TSM	Theil–Sen median
TEMP	Temperature
NDBSI	Normalized Difference Built-Up Index
LST	Land surface temperature
WET	Humidity
AI	Abundance index
H	Hurst index
Root Mean Square Error	RMSE
BTSSCP	The Beijing–Tianjin Sand Source Comprehensive Control Project
TNSDP	The Three-North Protective Forest Construction Project
SEPCP	The Sanjiangyuan Ecological Protection and Construction Project
NFRPP	The Natural Forest Resource Protection Project
RPFP	The Returning Ploughland to Forestry Project
RPGP	The Returning Pasture to Grassland Project

SKRDCTP	The Southwestern Karst Desertification Comprehensive Treatment Project
YRBPFSCP	The Yangtze River Basin Protection Forest System
PRBPFSCP	The Pearl River Basin Protection Forest System, and several other major projects

Appendix A

Table A1. Ecological restoration projects in China in the past two decades.

Restore Object	ERPs	Period	Pilot Area	Investment (Billion CNY)
Forest	Natural Forest Protection project	1998–2010	Yunnan, Sichuan, Chongqing, Guizhou, Hunan, Hubei, Jiangxi, Shanxi, Shannxi, Gansu, Qinghai, Ningxia, Xinjiang, Inner Mongolia, Jilin, Heilongjiang, Hainan, Henan	962.02
	Reclaimed Farmland to Forest project	1999–2021	Gansu, Inner Mongolia, Guizhou, Shanxi, Shannxi, Hunan, Hubei, Sichuan, Chongqing, Yunnan	4311.30
	Three-North Shelter Forest Program	2001–2010	Xinjiang, Qinghai, Gansu, Ningxia, Inner Mongolia, Shannxi, Shanxi, Hebei, Liaoning, Jilin, Heilongjiang, Beijing, Tianjin	354.12
	Shelter Forest System in the Yangtze River Basin	2001–2010	Jiangxi, Hubei, Hunan, Sichuan, Guizhou, Yunnan, Shannxi, Gansu, Qinghai	205.61
	Beijing–Tianjin Sandstorm Source Control Project	2001–2010	Beijing, Tianjin, Hebei, Shanxi, Inner Mongolia	558.65
	Coastal Shelter Forest System Project	2001–2010	Liaoning, Heibei, Zhejiang, Fujian, Guangdong, Hainan, Guangxi	39.09
	Taihang Mountain Greening Project	2001–2010	Shanxi, Hebei, Henan, Beijing	35.97
	Plain Greening Project	2001–2010	More than 900 plains in China	12.47
	Coastal Shelter Forest Project	2006–2015	Liaoning, Heibei, Zhejiang, Fujian, Guangdong, Hainan, Guangxi	99.84
Wetland	National Wetland Protection Project	2005–2010	473 wetlands in China	90.04
	Reclaimed Farmland to Lake	1998–2005	Cover the whole country	—
Grassland	Reclaimed Pasture to Grass	2003–2007	Xinjiang, Xizang, Inner Mongolia, Qianghai, Gansu, Ningxia	143.00
Important ecological functions	Ecological Protection and Construction of Sanjiangyuan Nature Reserve	2005–2010	Qinghai	75.00
	National Nature Reserve	1999–2010	Cover the whole country	4.80

References

1. Yang, W.; Zhao, J. Study on China's economic development from the perspective of strong sustainability. *Singap. Econ. Rev.* **2020**, *65*, 161–192. [CrossRef]
2. Shao, Q.; Liu, S.; Ning, J.; Liu, G.; Yang, F.; Zhang, X.; Niu, L.; Huang, H.; Fan, J.; Liu, J. Remote sensing assessment of ecological benefits of major ecological projects in China from 2000 to 2019. *J. Geogr.* **2022**, *77*, 2133–2153.
3. Chen, C.; Park, T.; Wang, X.H.; Piao, S.L.; Xu, B.D.; Chaturvedi, R.K.; Fuchs, R.; Brovkin, V.; Ciais, P.; Fensholt, R.; et al. China and India lead in greening of the world through land-use management. *Nat. Sustain.* **2019**, *2*, 122–129. [CrossRef] [PubMed]
4. Fu, B.; Liu, Y.; Meadows, M.E. Ecological restoration for sustainable development in China. *Natl. Sci. Rev.* **2023**, *10*, nwad033. [CrossRef] [PubMed]
5. Lu, F.; Hu, H.; Sun, W.; Zhu, J.; Liu, G.; Zhou, W.; Zhang, Q.; Shi, P.; Liu, X.; Wu, X.; et al. Effects of national ERPs on carbon sequestration in China from 2001 to 2010. *Proc. Natl. Acad. Sci. USA* **2018**, *115*, 4039–4044. [CrossRef] [PubMed]
6. China National Forestry and Grass Administration. Available online: http://www.forestry.gov.cn (accessed on 2 May 2023).
7. Liu, T.; Yu, L.; Chen, X.; Wu, H.; Lin, H.; Li, C.; Hou, J. Environmental laws and ERPs enhancing ecosystem services in China: A meta-analysis. *J. Environ. Manag.* **2023**, *327*, 116810. [CrossRef]
8. Shan, W.; Jin, X.; Ren, J.; Wang, Y.; Xu, Z.; Fan, Y.; Gu, Z.; Hong, C.; Lin, J.; Zhou, Y. EEQ assessment based on remote sensing data for land consolidation. *J. Clean. Prod.* **2019**, *239*, 118126. [CrossRef]
9. Bai, T.; Cheng, J.; Zheng, Z.; Zhang, Q.; Li, Z.; Xu, D. Drivers of eco-environmental quality in China from 2000 to 2017. *J. Clean. Prod.* **2023**, *396*, 136408. [CrossRef]
10. Wang, Z.; Bai, T.; Xu, D.; Kang, J.; Shi, J.; Fang, H.; Wang, D. Temporal and Spatial Changes in Vegetation Ecological Quality and Driving Mechanism in Kökyar Project Area from 2000 to 2021. *Sustainability* **2022**, *14*, 7668. [CrossRef]
11. Hanqiu, X. Creation of urban remote sensing ecological index and its application. *J. Ecol.* **2013**, *33*, 7853–7862.
12. Xu, D.; Yang, F.; Yu, L.; Zhou, Y.; Li, H.; Ma, J.; Huang, J.; Wei, J.; Xu, Y.; Zhang, C.; et al. Quantization of the coupling mechanism between eco-environmental quality and urbanization from multisource remote sensing data. *J. Clean. Prod.* **2021**, *321*, 128948. [CrossRef]
13. Cai, Y.; Zhang, F.; Duan, P.; Jim, C.Y.; Chan, N.W.; Shi, J.; Liu, C.; Wang, J.; Bahtebay, J.; Ma, X. Vegetation cover changes in China induced by ecological restoration-protection projects and land-use changes from 2000 to 2020. *Catena* **2022**, *217*, 106530. [CrossRef]
14. National Earth System Science Data Center. National Science & Technology Infrastructure of China. Available online: http://www.geodata.cn (accessed on 2 May 2023).
15. Didan, K. *MOD13A2 MODIS/Terra Vegetation Indices 16-Day L3 Global 1 km SIN Grid v006*; NASA EOSDIS Land Processes DAAC. Sioux Falls, SD, USA, 2015. [CrossRef]
16. Running, S.; Mu, Q.; Zhao, M. *MOD17A3H MODIS/Terra Net Primary Production Yearly L4 Global 500 m SIN Grid v006*; NASA EOSDIS Land Processes DAAC: Sioux Falls, SD, USA, 2015. [CrossRef]
17. Yin, R.; Rothstein, D.; Qi, J.; Liu, S. *Methodology for an Integrative Assessment of China's Ecological Restoration Programs*; Springer: Dordrecht, The Netherlands, 2009.
18. Sen, P.K. Estimates of the regression coefficient based on Kendall's tau. *J. Am. Stat. Assoc.* **1968**, *63*, 1379–1389. [CrossRef]
19. Theil, H. A rank-invariant method of linear and polynomial regression analysis. *Indag. Math.* **1950**, *12*, 173.
20. Mann, H.B. Nonparametric tests against trend. *J. Econom. Soc.* **1945**, 245–259. [CrossRef]
21. Kendall, M.G. *Rank Correlation Methods*; American Psychological Association: Washington, DC, USA, 1948; Available online: https://psycnet.apa.org/record/1948-15040-000 (accessed on 2 May 2023).
22. Zhou, Y.; Fu, D.; Lu, C.; Xu, X.; Tang, Q. Positive effects of ecological restoration policies on the vegetation dynamics in a typical ecologically vulnerable area of China. *Ecol. Eng.* **2021**, *159*, 106087. [CrossRef]
23. Du, H.; Liu, X.; Jia, X.; Li, S.; Fan, Y. Assessment of the effects of ERPs on soil wind erosion in northern China in the past two decades. *Catena* **2022**, *215*, 106360. [CrossRef]
24. Li, Z.; Ning, K.; Chen, J.; Liu, C.; Wang, D.; Nie, X.; Hu, X.; Wang, L.; Wang, T. Soil and water conservation effects driven by the implementation of ERPs: Evidence from the red soil hilly region of China in the last three decades. *J. Clean. Prod.* **2020**, *260*, 121109. [CrossRef]
25. Sun, W.; Song, X.; Mu, X.; Gao, P.; Wang, F.; Zhao, G. Spatiotemporal vegetation cover variations associated with climate change and ecological restoration in the Loess Plateau. *Agric. For. Meteorol.* **2015**, *209*, 87–99. [CrossRef]
26. Burrell, A.L.; Evans, J.P.; Liu, Y. Detecting dryland degradation using time series segmentation and residual trend analysis (TSS-RESTREND). *Remote Sens. Environ.* **2017**, *197*, 43–57. [CrossRef]
27. Gao, A.; Yang, S.L.; Li, G.; Li, P.; Chen, S.L. Long-term morphological evolution of a tidal island as affected by natural factors and human activities, the Yangtze estuary. *J. Coast. Res.* **2010**, *26*, 123–131. [CrossRef]
28. Campbell, B.M.; Hansen, J.; Rioux, J.; Stirling, C.M.; Twomlow, S. Urgent action to combat climate change and its impacts (SDG 13): Transforming agriculture and food systems. *Curr. Opin. Environ. Sustain.* **2018**, *34*, 13–20. [CrossRef]
29. Davies, R.B.; Harte, D.S. Tests for Hurst effect. *Biometrika* **1987**, *74*, 95–101. [CrossRef]
30. Ndayisaba, F.; Guo, H.; Bao, A.; Guo, H.; Karamage, F.; Kayiranga, A. Understanding the spatial temporal vegetation dynamics in Rwanda. *Remote Sens.* **2016**, *8*, 129. [CrossRef]

31. Liao, W.; Jiang, W. Evaluation of the spatiotemporal variations in the eco-environmental quality in China based on the remote sensing ecological index. *Remote Sens.* **2020**, *12*, 2462. [CrossRef]
32. Li, S.; Liang, W.; Fu, B.; Lü, Y.; Fu, S.; Wang, S.; Su, H. Vegetation changes in recent large-scale ERPs and subsequent impact on water resources in China's Loess Plateau. *Sci. Total Environ.* **2016**, *569*, 1032–1039. [CrossRef]
33. Zhao, J.; Yang, Y.; Zhao, Q.; Zhao, Z. Effects of ERPs on changes in land cover: A case study on the Loess Plateau in China. *Sci. Rep.* **2017**, *7*, 44496. [CrossRef]
34. Qu, S.; Wang, L.; Lin, A.; Zhu, H.; Yuan, M. What drives the vegetation restoration in Yangtze River basin, China: Climate change or anthropogenic factors? *Ecol. Indic.* **2018**, *90*, 438–450. [CrossRef]
35. Wu, M.; Liu, Y.; Xu, Z.; Yan, G.; Ma, M.; Zhou, S.; Qian, Y. Spatio-temporal dynamics of China's ecological civilization progress after implementing national conservation strategy. *J. Clean. Prod.* **2021**, *285*, 124886. [CrossRef]
36. Zhang, J.; Gangopadhyay, P. Dynamics of environmental quality and economic development: The regional experience from Yangtze River Delta of China. *Appl. Econ.* **2015**, *47*, 3113–3123. [CrossRef]
37. Su, X.; Zhou, Y.; Li, Q. Designing ecological security patterns based on the framework of ecological quality and ecological sensitivity: A case study of Jianghan Plain, China. *Int. J. Environ. Res. Public Health* **2021**, *18*, 8383. [CrossRef]
38. Yin, R.; Yin, G. China's primary programs of terrestrial ecosystem restoration: Initiation, implementation, and challenges. *Environ. Manag.* **2010**, *45*, 429–441. [CrossRef]
39. Lü, Y.; Fu, B.; Feng, X.; Zeng, Y.; Liu, Y.; Chang, R.; Sun, G.; Wu, B. A policy-driven large scale ecological restoration: Quantifying ecosystem services changes in the Loess Plateau of China. *PLoS ONE* **2012**, *7*, e31782. [CrossRef] [PubMed]
40. Wang, L.J.; Ma, S.; Zhao, Y.G.; Zhang, J.C. ERPs did not increase the value of all ecosystem services in Northeast China. *For. Ecol. Manag.* **2021**, *495*, 119340. [CrossRef]
41. Zhang, J.; Mengting, L.; Hui, Y.; Xiyun, C.; Chong, F. Critical thresholds in ecological restoration to achieve optimal ecosystem services: An analysis based on forest ecosystem restoration projects in China. *Land Use Policy* **2018**, *76*, 675–678. [CrossRef]
42. Zhang, S.; Zhao, K.; Ji, S.; Guo, Y.; Wu, F.; Liu, J.; Xie, F. Evolution characteristics, eco-environmental response and influencing factors of production-living-ecological space in the Qinghai–Tibet Plateau. *Land* **2022**, *11*, 1020. [CrossRef]
43. He, L.; Shen, J.; Zhang, Y. Ecological vulnerability assessment for ecological conservation and environmental management. *J. Environ. Manag.* **2018**, *206*, 1115–1125. [CrossRef]
44. Li, C.; Zhang, S.; Cui, M.; Wan, J.; Rao, T.; Li, W.; Wang, X. Improved Vegetation Ecological Quality of the Three-North Shelterbelt Project Region of China during 2000–2020 as Evidenced from Multiple Remotely Sensed Indicators. *Remote Sens.* **2022**, *14*, 5708. [CrossRef]
45. Wang, X.M.; Zhang, C.X.; Hasi, E.; Dong, Z.B. Has the Three Norths Forest Shelterbelt Program solved the desertification and dust storm problems in arid and semiarid China? *J. Arid Environ.* **2010**, *74*, 13–22. [CrossRef]
46. Hasan, S.S.; Zhang, Y.; Chu, X.; Teng, Y. The role of big data in China's sustainable forest management. *For. Econ. Rev.* **2019**, *1*, 96–105. [CrossRef]
47. Qinghai Forestry and Grassland Bureau. Available online: http://www.qinghai.gov.cn (accessed on 2 May 2023).
48. Li, Y.; Bai, X.; Wang, S.; Tian, Y. Integrating mitigation measures for karst rocky desertification land in the Southwest mountains of China. *Carbonates Evaporites* **2019**, *34*, 1095–1106. [CrossRef]
49. Jiang, Z.; Lian, Y.; Qin, X. Rocky desertification in Southwest China: Impacts, causes, and restoration. *Earth Sci. Rev.* **2014**, *132*, 1–12. [CrossRef]
50. Liu, S.; Zhang, J.; Zhang, J.; Guo, Y. Simultaneously tackling ecological degradation and poverty challenges: Evidence from desertified areas in northern China. *Sci. Total Environ.* **2022**, *815*, 152927. [CrossRef] [PubMed]

land

MDPI

Article

Feature-Differencing-Based Self-Supervised Pre-Training for Land-Use/Land-Cover Change Detection in High-Resolution Remote Sensing Images

Wenqing Feng [1,*], Fangli Guan [1], Chenhao Sun [2] and Wei Xu [3]

[1] Computer & Software School, Hangzhou Dianzi University, Hangzhou 310018, China; fangli.guan@hdu.edu.cn
[2] Electrical & Information Engineering School, Changsha University of Science & Technology, Changsha 410114, China; chenhaosun@csust.edu.cn
[3] Information System and Management College, National University of Defense Technology, Changsha 410015, China; weixu@nudt.edu.cn
* Correspondence: wq_feng@hdu.edu.cn

Abstract: Land-use and land-cover (LULC) change detection (CD) is a pivotal research area in remote sensing applications, posing a significant challenge due to variations in illumination, radiation, and image noise between bi-temporal images. Currently, deep learning solutions, particularly convolutional neural networks (CNNs), represent the state of the art (SOTA) for CD. However, CNN-based models require substantial amounts of annotated data, which can be both expensive and time-consuming. Conversely, acquiring a large volume of unannotated images is relatively easy. Recently, self-supervised contrastive learning has emerged as a promising method for learning from unannotated images, thereby reducing the need for annotation. However, most existing methods employ random values or ImageNet pre-trained models to initialize their encoders and lack prior knowledge tailored to the demands of CD tasks, thus constraining the performance of CD models. To address these challenges, we introduce a novel feature-differencing-based framework called Barlow Twins for self-supervised pre-training and fine-tuning in CD (BTCD). The proposed approach employs absolute feature differences to directly learn unique representations associated with regions that have changed from unlabeled bi-temporal remote sensing images in a self-supervised manner. Moreover, we introduce invariant prediction loss and change consistency regularization loss to enhance image alignment between bi-temporal images in both the decision and feature space during network training, thereby mitigating the impact of variation in radiation conditions, noise, and imaging viewpoints. We select the improved UNet++ model for fine-tuning self-supervised pre-training models and conduct experiments using two publicly available LULC CD datasets. The experimental results demonstrate that our proposed approach outperforms existing SOTA methods in terms of competitive quantitative and qualitative performance metrics.

Keywords: Barlow Twins loss function; land use/land cover change detection; pre-training; self-supervised contrastive learning

Citation: Feng, W.; Guan, F.; Sun, C.; Xu, W. Feature-Differencing-Based Self-Supervised Pre-Training for Land-Use/Land-Cover Change Detection in High-Resolution Remote Sensing Images. *Land* **2024**, *13*, 927. https://doi.org/10.3390/land13070927

Academic Editor: Jūratė Sužiedelytė-Visockienė

Received: 13 May 2024
Revised: 16 June 2024
Accepted: 24 June 2024
Published: 26 June 2024

1. Introduction

Land-use and land-cover (LULC) change detection (CD) is applied to identify surface-related changes using bi-temporal remote sensing images, making it crucial for earth observation. The generated information can aid in the monitoring of urban development, natural resources, minerals, and assessments of military damage [1–4]. To date, many theoretical models and technical methods for diverse LULC CD applications have been proposed, including traditional algebraic comparison, change vector analysis, post-classification comparison, object-oriented image analysis, time series image analysis, and machine-learning-based methods [2]. Similar to other remote sensing image interpretation tasks, LULC CD involves dealing with multi-scene remote sensing images that cover an area at different times. Data sources can include both homologous and heterologous situations. Data processing involves radiometric and geometric preprocessing, CD algorithms, threshold segmentation, and accuracy evaluation [1,2].

In recent years, the emergence of remote sensing big data and the rapid development of artificial intelligence have led to a new academic trend in LULC CD research [5–12]. Deep learning has revolutionized traditional methods, as it adopts an end-to-end learning mode that directly extracts change area information from bi-temporal remote sensing images, thus avoiding the dependence on difference images. Furthermore, the features extracted by deep networks have strong noise robustness, making them suitable for handling bi-temporal images from the same sensor or different sources. A representative method is the fully convolutional Siamese network [10]. The key to its effectiveness lies in a large number of manually labelled variation samples, which are time-consuming to collect and annotate from bi-temporal remote sensing images and require professional domain knowledge. To overcome this challenge, many researchers have attempted transfer learning strategies that leverage the knowledge learned from a large-scale natural image dataset, ImageNet [13]. However, while pre-trained ImageNet models achieve better results than those trained from scratch, they still face significant limitations due to the domain gap between natural images and remote sensing images. To tackle these issues, the idea of utilizing a small set of manually labelled samples or even completely unlabeled samples has gained increasing popularity in LULC CD applications, and there are two main research directions. The first involves the construction of massive remote sensing image datasets (such as Million AID [14], fMoW [15], and BigEarthNet [16]), which usually come with sparsely noisy or clean labelled data. These datasets are leveraged for supervised remote sensing pre-training to acquire transferable image representations, which can then be tailored to downstream LULC CD tasks with only a few labelled samples. The other main direction involves utilizing unsupervised contrastive learning to pre-train an encoder on unlabeled massive remote sensing image datasets (such as SeCo [17] and SSL4EO-S12 [18]) to obtain well-initialized parameters for downstream CD tasks. As a unique form of unsupervised learning, self-supervised contrastive learning (SSCL) trains models through self-designed proxy tasks and sometimes produces better CD results than related supervised methods. However, gathering remote sensing imagery data on a scale of millions is costly, and carrying out unsupervised contrastive learning on such a dataset demands substantial GPU computing resources.

Recently, SSCL methods have made significant improvements in the field of computer vision, allowing for the learning of useful representations from unlabeled data [19–25]. This approach has become increasingly popular, particularly in situations where labelling is costly, such as in medical and satellite imaging. Self-supervised contrastive pre-training can learn meaningful feature representations by utilizing a large amount of unlabeled remote sensing image data. These meaningful feature representations can improve the performance of various downstream CD tasks, which has drawn the attention of many researchers [26,27]. Although self-supervised contrastive pre-training holds great potential, many advanced SSCL methods are currently difficult to implement in practice. They require a large number of data samples and computational resources for effective application, limiting their practicality. For instance, while SimCLR [21] and its improved versions achieve state-of-the-art (SOTA) performance on ImageNet classification tasks, they necessitate a batch size of 4096. Similarly, BYOL [24] achieves cutting-edge performance through positive contrastive learning, symmetric networks, and stop-gradient methods, but demands 64 graphical processing units (GPUs). MoCo [22] tackles negative samples by introducing momentum encoders and queue sampling strategies, yet its computation speed is hindered by the necessity of sampling negative samples across the entire queue. SimSiam [20], foregoing the need for negative samples, employs asymmetric network structures and cross-gradient updates to counter trivial solutions. However, it requires more computational resources and exhibits sensitivity to hyperparameters. SwAV [23], leveraging online clustering, circumvents the need for negative samples via multi-view prediction encoding, presenting advantages in high-label-cost scenarios. Nevertheless, it imposes high computational resource requirements and a relatively complex training process due to its utilization of online clustering algorithms. In contrast to the aforementioned methods, Barlow Twin (BT) [25] introduces innovative thinking into SSCL. Unconstrained by batch sample size limitations and devoid of the need for negative samples, BT focuses on the embedding itself to avoid asymmetric structural design. By computing the cross-correlation matrix of augmented samples and utilizing a loss function to reduce redundancy, BT ensures that the cross-correlation matrix closely approximates the identity matrix. This ensures that different augmented versions of the same sample possess similar feature vectors and minimize redundancy across different dimensional components, thereby enhancing the efficiency of feature representation.

Due to the advantages of BT, we adopted it as an SSCL objective in our proposed LULC CD framework, as it does not require large negative samples to avoid trivial solutions [25]. Here, we introduce the feature-differencing-based BT self-supervised pre-training and fine-tuning CD (BTCD) framework in this paper. Our proposed method utilizes feature difference to learn discriminatory representations that correspond to areas of change, which is highly beneficial for CD tasks. In addition, we introduce the invariant prediction (IP) loss and change consistency regularization (CCR) loss to enhance image alignment between bi-temporal images in the decision and feature space during network training, thereby reducing the effects of variation in radiation conditions, noise, and imaging viewpoints. In the proposed framework, an SSCL method is used to pre-train the encoder of the improved UNet++ model [4]. This encoder has excellent parameter initialization and can effectively solve downstream CD tasks. The SSCL algorithm does not require CD label images. Instead, bi-temporal images are used to construct sample pairs for comparison, and a powerful encoder can be pre-trained for downstream CD tasks. Our contributions can be summarized as follows:

(1) We propose a novel framework called BTCD, which consists of two cascaded stages: self-supervised contrastive pre-training and fine-tuning. In the first stage, the algorithm leverages an absolute feature difference self-supervised pre-training method to learn task-specific change representation for CD. In the second stage, the pre-trained encoder undergoes fine-tuning, which effectively enhances the downstream CD network by taking advantage of favorable parameter initialization.
(2) To mitigate the impact of radiation differences, noise variation, and imaging perspective differences caused by bi-temporal images, we incorporate an IP loss and a CCR loss into the self-supervised contrastive pre-training of the original BT loss function. This approach improves the performance of the pre-training process.
(3) We evaluated the proposed approach on two publicly available LULC CD datasets and demonstrated that fine-tuning using only the pre-trained encoder surpasses ImageNet supervised pre-training methods and several recently proposed SSCL pre-training methods without requiring additional data.
(4) We verified the effectiveness of self-supervised pre-training under insufficiently labelled sample data. When labelled data were insufficient, the proposed pre-training method significantly improved CD model performance.

In the following sections, we present our proposed BTCD method and experimental details (Section 2), present comparison results (Section 3), discuss our findings (Section 4), and conclude our work (Section 5).

2. Materials and Methods

2.1. BTCD Framework

Figure 1 provides an overview of our proposed BTCD framework. We assume that X^1_{train} and X^2_{train} are bi-temporal images of the same geographic location captured at times t^1 and t^2, respectively. We pre-train a model on the unlabeled training set $\{U = (X^1_{\text{train}}, X^2_{\text{train}})_i\}_{i=1}^{N}$ using a self-supervised algorithm, so that the pre-trained encoder can be easily fine-tuned on the labelled training set $\{L = (X^1_{\text{train}}, X^2_{\text{train}}, Y_{\text{train}})_i\}_{i=1}^{N}$. The proposed framework consists of two distinct stages. During the SSCL stage, only bi-temporal images X^1_{train} and X^2_{train} are used for training. In the subsequent CD stage, the framework is trained using both bi-temporal images X^1_{train} and X^2_{train}, as well as labelled images Y_{train} from the training set. The overall framework is enhanced in two key steps: the first CNN is trained on a pretext task, the encoder component that provides the best features for CD is selected, and the selected encoder is utilized by the second CNN to perform CD.

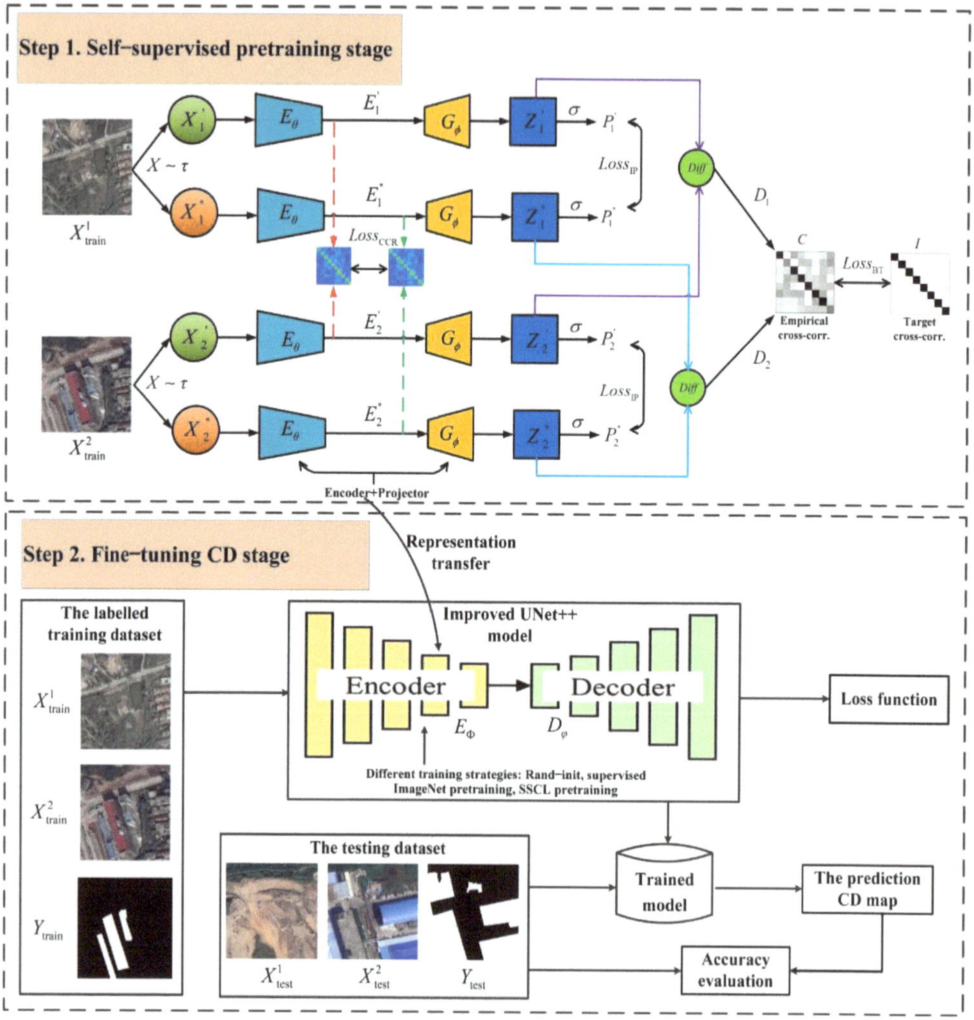

Figure 1. Pipeline of the proposed LULC CD Framework.

(1) Self-supervised Pre-training Stage. In this stage, we initialize our proposed self-supervised pre-training models with the supervised ImageNet weights instead of random initialization. An encoder E_θ is trained, where θ is a learnable parameter used to solve a pretext task. The BTCD algorithm uses three objective functions during the training process: BT loss, IP loss and CCR loss. Self-supervised by the hybrid loss

function *Loss* of a pretext task "contrasting", the parameters θ^i of the encoder E_θ for the i-th iteration are updated as follows:

$$\theta^{i+1} = \theta^i - \kappa\frac{\partial Loss}{\partial \theta^i} \quad (1)$$

where κ represents the learning rate. Ultimately, we obtain a pre-training encoder E_Φ with well-initialized parameters.

(2) Fine-tuning CD Stage. In this stage, we utilize the pre-trained encoder E_Φ to train the improved UNet++ model end-to-end. The improved UNet++ model extracts features from the given bi-temporal images X^1_{train} and X^2_{train} and classifies them as either "changed" or "unchanged" to predict the change map $\hat{Y}_{\text{train}}$.

$$\hat{Y}_{\text{train}} = D_\varphi(E_\Phi(X^1_{\text{train}}), E_\Phi(X^2_{\text{train}})) \quad (2)$$

where D_φ represents the decoder and the CD head, and φ is a learnable parameter. With parameters Φ and φ, the improved UNet++ model is updated as follows:

$$\Phi^{i+1} = \Phi^i - \kappa\frac{\partial Loss_{\text{CD}}(\hat{Y}_{\text{train}}, Y_{\text{train}})}{\partial \Phi^i} \quad (3)$$

$$\varphi^{i+1} = \varphi^i - \kappa\frac{\partial Loss_{\text{CD}}(\hat{Y}_{\text{train}}, Y_{\text{train}})}{\partial \varphi^i} \quad (4)$$

where $Loss_{\text{CD}}$ represents the loss function and Y_{train} represents the ground truth.

2.2. BTCD Pretraining Method

2.2.1. BTCD

The primary objective of CD is to detect and identify changes in surface objects between bi-temporal images. Therefore, the radiometric space of the bi-temporal images must be aligned with the low-level features of the CD network. This alignment poses a significant challenge because regions of change are highly susceptible to variation in seasons and noise. To facilitate low-level feature alignment between bi-temporal images, we employed a self-supervised pre-training algorithm as the baseline model utilizing the BT loss function [25]. By maximizing the cross-correlation between unchanged regions, this function implicitly minimizes the differences between the bi-temporal images in the feature space. The BT loss function was selected because it intersects with other competitive self-supervised contrastive learning algorithms while requiring only a few negative samples to learn robust representations. To strengthen the correlation between bi-temporal images, we developed a self-supervised pre-training algorithm that uses the absolute feature difference to learn essential representations for CD. Figure 2 provides an overview of the BTCD algorithm.

The developed BTCD algorithm takes bi-temporal images (X_1, X_2) as inputs and applies color distortion and Gaussian blur transformation to obtain augmented image pairs $(X_1 \rightarrow X_1', X_1'', X_2 \rightarrow X_2', X_2'')$. Random clipping was excluded from preprocessing of the training set, as areas of change are significantly smaller than unchanged areas. The BTCD algorithm has a Siamese architecture that uses ResNet50 (excluding the final classification layer) as a feature extractor, followed by a projector network. The input image, sized at 256 × 256 × 3, undergoes several stages of processing through ResNet50. Initially, it passes through multiple layers of convolution and pooling operations. Additionally, dilated convolutions are applied to enhance the receptive field without increasing the number of parameters significantly. The stride of 16 is then used to reduce the dimensionality of the input image, resulting in a more compact representation. This series of operations ultimately produces a feature vector with dimensions of 16 × 16 × 2048, effectively capturing essential features while maintaining computational efficiency and a wide context range. Then, we applied 2D adaptive average pooling to the feature vector and passed it to the projection head. The projector network comprises two linear layers, each with a hidden layer size of 512 output units. Due to high computational requirements, the output of the projection network was modified to achieve an embedding size of 256, while the BT network had an embedding size of 8192 [25]. The first layer of the projector was followed by a batch normalization layer and rectified linear unit. Afterward, we applied the absolute difference to the output embeddings to obtain the difference embeddings. We used the BT loss function on the difference embeddings of size 1×256 to generate a cross-correlation matrix of size 256×256. The Siamese encoder (E_θ), which has shared parameters θ, was employed to extract the feature vectors E_1' and E_2' from bi-temporal images. Next, nonlinear projection G_ϕ was applied to the encoded feature vectors E_1' and E_2' to produce the representations Z_1' and Z_2'. Then, the absolute feature difference was applied to the projection outputs to learn representations of changed features between input images and is described as follows:

$$\begin{aligned} D_1 &= |Z_1' - Z_2'| = |G(E(X_1')) - G(E(X_2'))| \\ D_2 &= |Z_1'' - Z_2''| = |G(E(X_1'')) - G(E(X_2''))| \end{aligned} \tag{5}$$

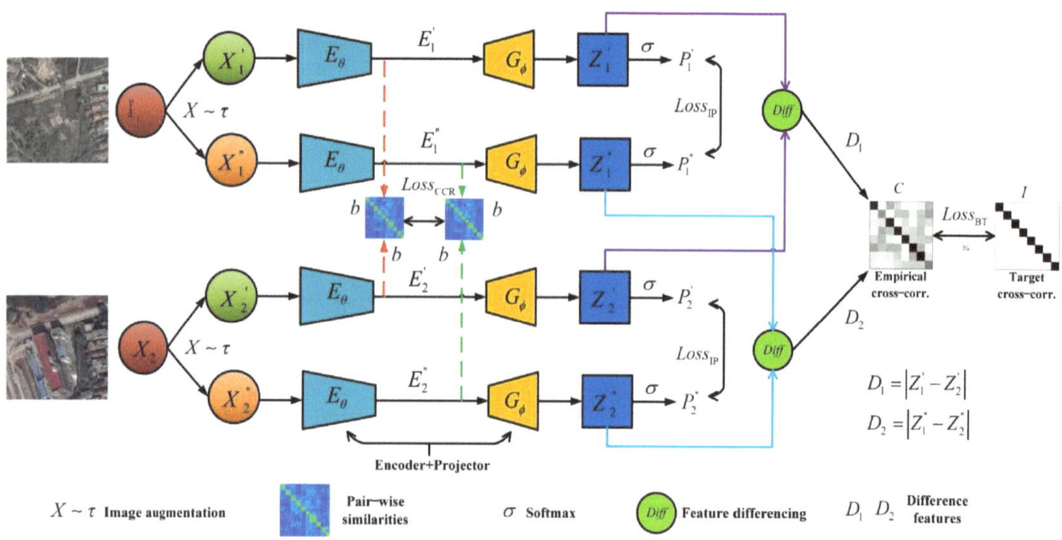

Figure 2. Flow chart of the BTCD algorithm.

We assumed that the semantic changes between the bi-temporal image features (X_1', X_2') and (X_1'', X_2'') should remain consistent. Therefore, we constrained our model to ensure $D_1 \simeq D_2$, regardless of the image enhancement method applied. To achieve this, we applied the BT objective function (Section 2.2.2) on the difference feature representations D_1 and D_2 to maximize the cross-correlation of the change features. This encouraged the model to learn specific information about the changes that occur between the bi-temporal images. After the self-supervised pre-training stage was complete, the encoder (E_Φ) parameters were passed on to the downstream CD task.

2.2.2. Loss Function

BT Loss. The algorithm uses the BT loss function proposed by Jure et al. [25] for self-supervised training. However, unlike those authors, who maximized the cross-correlation of the augmented views of the same input image to be closer to the identity matrix, we maximized the information of the difference representations (D_1, D_2) between the corresponding bitemporal images to be similar in the feature space. Afterwards, the model was trained in a self-supervised manner using the modified BT loss function $Loss_{\mathrm{BT}}$, which is defined as follows:

$$Loss_{\mathrm{BT}} \triangleq \underbrace{\sum_i (1 - C_{ii})^2}_{\text{invariance term}} + \lambda \underbrace{\sum_i \sum_{j \neq i} C_{ij}^2}_{\text{redundancy reduction term}} \tag{6}$$

$$C_{ij} \triangleq \frac{\sum_b (D_1)_{b,i} (D_2)_{b,j}}{\sqrt{\sum_b \left((D_1)_{b,i}\right)^2} \sqrt{\sum_b \left((D_2)_{b,j}\right)^2}} \tag{7}$$

where λ is the trade-off constant, b indexes batch samples, i, j index the vector dimensions of the network outputs, and C is the cross-correlation matrix calculated between the difference representations D_1 and D_2.

IP Loss. When performing CD using bi-temporal imagery, we ensured that samples with semantic similarity produced similar predictions, regardless of different radiation conditions. To achieve IP, we enhanced the features between the augmented views (X_1', X_1'') and (X_2', X_2'') of the bi-temporal imagery. To accomplish this, we used Jensen-Shannon divergence D_{JS} as the objective to achieve IP [28].

$$D_{\mathrm{JS}}^{X_1} = \frac{1}{2} \times \left(D_{\mathrm{KL}}(\sigma(Z_1'))||AP_1 + D_{\mathrm{KL}}(\sigma(Z_1''))||AP_1\right) = \frac{1}{2} \times \left(D_{\mathrm{KL}}(P_1')||AP_1 + D_{\mathrm{KL}}(P_1'')||AP_1\right) \tag{8}$$

$$D_{\mathrm{JS}}^{X_2} = \frac{1}{2} \times \left(D_{\mathrm{KL}}(\sigma(Z_2'))||AP_2 + D_{\mathrm{KL}}(\sigma(Z_2''))||AP_2\right) = \frac{1}{2} \times \left(D_{\mathrm{KL}}(P_2')||AP_2 + D_{\mathrm{KL}}(P_2'')||AP_2\right) \tag{9}$$

$$AP_1 = \frac{P_1' + P_1''}{2} \tag{10}$$

$$AP_2 = \frac{P_2' + P_2''}{2} \tag{11}$$

$$Loss_{\mathrm{IP}} = D_{\mathrm{JS}}^{X_1} + D_{\mathrm{JS}}^{X_2} \tag{12}$$

where σ is the softmax function, D_{KL} is the Kullback–Leibler divergence [29], and AP_1 and AP_2 are the average probabilities belonging to the corresponding features. We utilized the loss function $Loss_{\mathrm{IP}}$ during network training to compel the model to learn features that are resilient to radiometric differences between bi-temporal images.

CCR Loss. CD is challenging because bi-temporal images can be heavily influenced by noise, seasonal changes, and differences in imaging angles. Therefore, it is critical to strengthen the similarity between images in the feature space, because the probability of change between images is typically lower than the probability of no change. To strengthen the similarity between images, we developed a CCR that preserved the semantic similarity between the temporal image pairs in the feature space. Specifically, the network encoder in the BTCD algorithm was coerced into producing similar activations for the bi-temporal images, thus incorporating temporal invariance into the network model. This, in turn, implicitly enhanced the model's robustness to noise variation during the pre-training stage. The BTCD network encoder generated activation maps from input images X_1 that underwent augmented transformations. These activation maps are represented by $E_1' \in \mathbb{R}^{b \times c \times h \times w}$ and $E_1'' \in \mathbb{R}^{b \times c \times h \times w}$, where b denotes the batch size, c denotes the number of output channels, and h and w denote the spatial dimensions. Similarly, the network encoder generated activation maps for transformed X_2 images, represented by E_2' and E_2''. To maintain temporal consistency between the input images X_1 and X_2 in the feature space, we constructed a pairwise similarity matrix $G' = \langle E_1', E_2' \rangle$ between the feature maps E_1'

and E'_2. Similarly, we construct a pairwise similarity matrix $G'' = \langle E''_1, E''_2 \rangle$ between the feature maps E''_1 and E''_2. We defined the CCR loss function as follows:

$$Loss_{\text{CCR}} = \frac{\|G' - G''\|_F^2}{b^2} \tag{13}$$

where $\|.\|_F$ is the Frobenius norm and $\langle .,. \rangle$ is the dot product [30].

Overall Loss. As shown in Figure 2, the output of the projector was utilized to calculate the IP loss before the differences are computed, while the CCR loss was applied to the output of the feature extractor. We developed a weighted combination of three loss functions (BT, IP, and CCR) to derive the final network loss function:

$$Loss_{\text{Total}} = Loss_{\text{BT}} + \alpha Loss_{\text{IP}} + \beta Loss_{\text{CCR}} \tag{14}$$

where α and β are the loss balancing weights.

2.3. CD Algorithm

We evaluated the performance of a self-supervised pre-trained model through fine-tuning using an existing CNN-based CD algorithm. We selected the improved UNet++ model [4] due to its ability to achieve SOTA results with CD datasets. The improved UNet++ model can capture multi-scale features using skip connections and integrate them via feature superposition. It utilizes an encoder–decoder architecture that includes a spatial and channel squeeze and excitation (scSE) module [31]. To train the model, the bands of the bi-temporal images were superimposed to create a new image with twice the number of bands, which was then fed into the improved UNet++ model. This approach has the advantage of allowing the model to learn both low- and high-level features of the input images, which is not possible with Siamese networks. It also simplifies the CD problem into a semantic segmentation task.

2.4. Experimental Datasets

To assess the BTCD algorithm, we utilized two publicly available datasets, LEVIR-CD [12] and SYSU-CD [32]. LEVIR-CD consists of 637 pairs of highly detailed very high-resolution (0.5 m/pixel) Google Earth image patches, each measuring 1024 × 1024 pixels. These bi-temporal images were collected from 20 distinct areas across Texas, USA, spanning the years 2002 to 2018. Images from different regions may have been captured at different times, ranging from 5 to 14 years apart, and exhibit notable changes in land use, particularly in building structures. LEVIR-CD encompasses various types of constructions, including villa residences, high-rise apartments, small-scale garages, and large warehouses, with a specific focus on changes related to buildings, such as growth and demolition [12]. To prepare for training, we randomly partitioned the images into three non-overlapping sets: a training set of 7120 image pairs, a validation set of 1024 image pairs, and a testing set of 2048 image pairs, all of which were further divided into patches of size 256 × 256. Examples of the LEVIR-CD dataset are shown in Figure 3. The SYSU-CD dataset comprises 20,000 pairs of aerial images, each measuring 256 × 256 pixels, captured in Hong Kong between 2007 and 2014, with a resolution of 0.5 m per pixel. The primary types of changes documented in the dataset include (a) newly erected urban structures; (b) the expansion of suburban areas; (c) preliminary groundwork prior to construction; (d) alterations in vegetation cover; (e) widening of roads; and (f) construction activities in marine environments [32]. We used 12,000 image pairs for training, 4000 for validation, and 4000 for testing. Examples of the SYSU-CD dataset are shown in Figure 4.

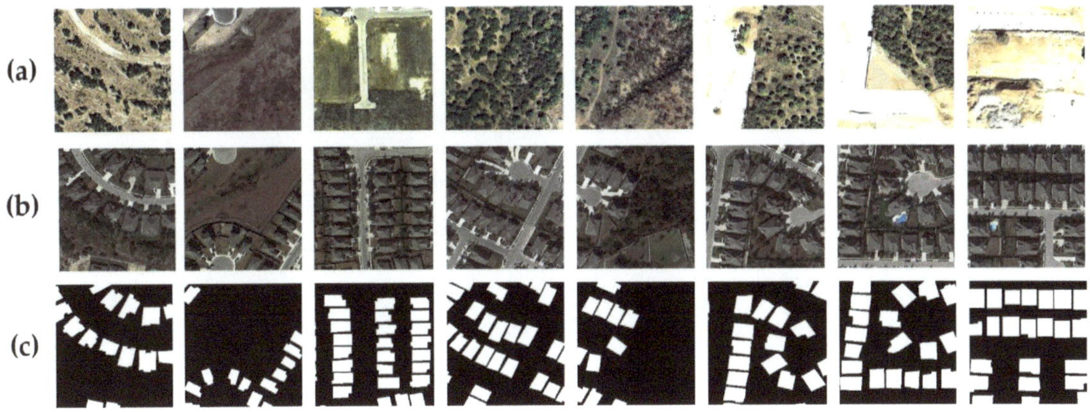

Figure 3. Examples of the LEVIR-CD dataset. (a) Remote sensing images at time T1. (b) Remote sensing images at time T2. (c) Ground truth.

Figure 4. Examples of the SYSU-CD dataset. (a) Remote sensing images at time T1. (b) Remote sensing images at time T2. (c) Ground truth.

2.5. Implementation Details and Evaluation Metrics

We utilized the PyTorch framework to implement the BTCD algorithm and conducted experiments on a workstation featuring a 12th Gen Intel Core i9–12900K @ 3.19 GHz processor, 64 GB RAM, and a NVIDIA GeForce RTX 3090 (24 GB) graphics card. To optimize the model, we followed the BT protocol of two previous studies [25,33]. During the SSCL stage, we trained the ResNet50 model using image pairs from the training and validation sets without labels. We set the batch size to 40 and used the LARS optimizer to train for 400 epochs. The initial learning rate was 0.003, which was adjusted through multiplication with the batch size and division by 256. We incorporated a learning rate warm-up period of 10 epochs, followed by a reduction in the learning rate by a factor of 1000 using a cosine decay schedule following Ramkumar et al. [33]. The trade-off parameter λ of the loss function was set to $\lambda = 5 \times 10^{-3}$, and we used a weight decay parameter value of 1.5×10^{-6}. We set the hyperparameters α and β to 100 and 2000, respectively. The sensitivity analysis results for these hyperparameters are presented in the Section 4. Further experimental details can be found in Section 4.2.

We assessed the performance of the BTCD algorithm on downstream CD tasks through fine-tuning. For consistency, we used ResNet50 as the feature extractor and fine-tuned with the SOTA UNet++ model. We applied data augmentation techniques, including random flipping, clipping, and Gaussian blurring. We employed the AdamW optimizer with an initial learning rate of 0.003, batch size of 30, and 150 epochs. We employed the hybrid loss function developed in a previous study [34], which combines weighted cross-entropy and dice loss with equal weights. In the CD stage, we also applied the cosine decay schedule. We compared the performance of our algorithm to SOTA methods using four metrics: precision, recall, F1 score, and intersection over union (IoU). These metrics are defined as follows:

$$\text{precision} = \frac{TP}{TP + FP} \tag{15}$$

$$\text{recall} = \frac{TP}{TP + FN} \tag{16}$$

$$\text{F1} = \frac{2 \times \text{precision} \times \text{recall}}{\text{precision} + \text{recall}} \tag{17}$$

$$\text{IoU} = \frac{TP}{TP + FP + FN} \tag{18}$$

where TP denotes true-positive values, TN denotes true-negative values, FP denotes false-positive values, and FN denotes false-negative values.

2.6. Comparison to SOTA Approaches

We compared our proposed method for CD with several SOTA approaches, including FC-EF [10], FC-Siam-Diff [10], FC-Siam-Conc [10], DASNet [11], STANet [12], SNUNet [34], DSIFN [35], DSAMNet [32], SRCDNet [36], BiDateNet [37], and the latest transformer-based methods BIT [38], ChangeFormer [39], MSTDSNet-CD [40], and Hybrid-TransCD [41]. We fine-tuned the improved UNet++ model on the LEVIR-CD and SYSU-CD datasets using three strategies: random initialization (Rand-init), supervised ImageNet pre-training (ImageNet-sup), and pre-training with BTCD. To evaluate the effectiveness of the proposed pre-training method, we conducted a detailed comparative analysis with several other self-supervised pre-training approaches, namely CMC [19], MoCo v2 [22], SimCLR [21], and BT [25].

3. Experimental Results

3.1. Performance Comparison for the LEVIR-CD Dataset

Our method achieved the highest F1 score (91.30%) and IoU (83.81%) among all tested methods (Table 1). Hybrid-TransCD and ChangeFormer had the second- and third-highest F1 score (90.06% and 90.04%, respectively) and the third- and second-highest IoU (81.92% and 82.48%, respectively). Both methods use self-attention mechanisms, which can effectively capture global information and map it to multiple spaces using Multi-Head, thereby enhancing the models' expressive power. Thus, Hybrid-TransCD and ChangeFormer achieved better CD results than CNN-based methods.

Our developed method does not employ transformer-based global self-attention mechanisms but instead uses the improved UNet++ model framework to address the problem of varying change area scales. In the encoder, we used an efficient ResNet 50, while the decoder embedded the scSE module. The dense skip connections in the improved UNet++ model effectively captured multi-scale and multi-level feature information, ensuring the efficient utilization of fine-grained local features. This enhanced the model's ability to perceive change areas. This approach was more suitable for handling local variations in images compared to the global self-attention mechanism of transformers, which may fall short in capturing subtle local features. Furthermore, the scSE module applied attention weighting to the spatial and channel dimensions of the feature map, highlighting important features while suppressing irrelevant information. This module enhanced the utilization of more useful and discriminative features, enabling the generation of more accurate feature maps of the change area. This demonstrates that the combination of dense skip connections in the improved UNet++ and the scSE module can effectively improve the accuracy of CD results. For the LEVIR-CD dataset, we found that ImageNet pre-training did not significantly improve the accuracy of CD results compared to random initialization. By contrast, our SSCL method, which involved BTCD, outperformed ImageNet pre-training.

Table 1. Performance comparison for the LEVIR-CD dataset.

Methods	LEVIR-CD			
	Precision	**Recall**	**F1**	**IoU**
FC-EF	86.91	80.17	83.40	71.53
FC-Siam-Diff	89.53	83.31	86.31	75.92
FC-Siam-Conc	91.99	76.77	83.69	71.96
DASNet	80.76	79.53	79.91	74.65
STANet	83.81	**91.00**	87.26	77.40
DSIFN	**94.02**	82.93	88.13	78.77
SNUNet	89.18	87.17	88.16	78.83
BiDateNet	85.65	89.98	87.76	78.19
BIT	89.24	89.37	89.31	80.68
ChangeFormer	**92.05**	88.80	**90.04**	**82.48**
MSTDSNet-CD	85.52	**90.84**	88.10	78.73
Hybrid-TransCD	91.45	88.72	**90.06**	**81.92**
Ours (UNet++ with Rand-init)	88.89	86.74	87.80	78.25
Ours (UNet++ with ImageNet-sup)	89.99	85.92	87.91	78.43
Ours (UNet++ with BTCD)	**92.03**	**90.59**	**91.30**	**83.81**

Developed as part of the present study. All values are percentages. Bold red text indicates highest, bold blue text indicates second-highest, and bold black text indicates third-highest performances.

To evaluate the effectiveness of different pre-training methods, we fine-tuned downstream CD network (the improved UNet++) using the pre-trained models CMC, MoCo v2, SimCLR, and BT. To ensure a fair comparison, we only used the pre-trained model parameters to initialize the backbone part of the improved UNet++ model, i.e., the ResNet 50 encoder's parameters. Furthermore, we compared our BTCD method with several mainstream self-supervised pre-training methods, including random initialization, and ImageNet pre-training. These mainstream self-supervised pre-training methods were pre-trained using the ResNet-50 model on the ImageNet dataset and achieved good parameter initialization. We fine-tuned the encoder part of the ResNet-50 model obtained by these methods and used it for downstream CD tasks. Our pre-training method achieved the highest F1 score (91.30%) and IoU (83.81%) (Table 2). MoCo v2 and BT had the second- and third-highest F1 scores (89.19% and 89.06%, respectively) and IoU (80.49% and 80.28%, respectively) (Table 2). Visual comparison results for the LEVIR-CD dataset are shown in Figure 5. These scenes reflect the growth of buildings, mainly including changes from soil, grassland, and paved ground to new building areas. Because CMC, MoCo v2, SimCLR, and BT are all self-supervised pre-training methods based on the ImageNet dataset, they lack prior knowledge about CD tasks, which, to some extent, limits the performance of CD models. Considering the domain gap between ImageNet and remote sensing images, our BTCD method directly trained on the LEVIR-CD training dataset for self-supervised contrastive pre-training to learn representations within the domain. The results show that this pre-training strategy is more suitable for downstream CD tasks.

Table 2. Performance of the improved UNet++ model on the LEVIR-CD dataset using different pre-training methods.

Methods			LEVIR-CD			
			Precision	Recall	F1	IoU
Backbone ResNet50	Reliant on the ImageNet dataset	CMC	**89.92**	86.32	88.08	78.71
		MoCo v2	**90.32**	88.09	**89.19**	**80.49**
		SimCLR	89.23	**88.49**	88.86	79.96
		BT	89.86	**88.28**	**89.06**	**80.28**
	Not reliant on the ImageNet dataset	BTCD	**92.03**	**90.59**	**91.30**	**83.81**

All values are percentages. Bold red text indicates highest, bold blue text indicates second-highest, and bold black text indicates third-highest performances.

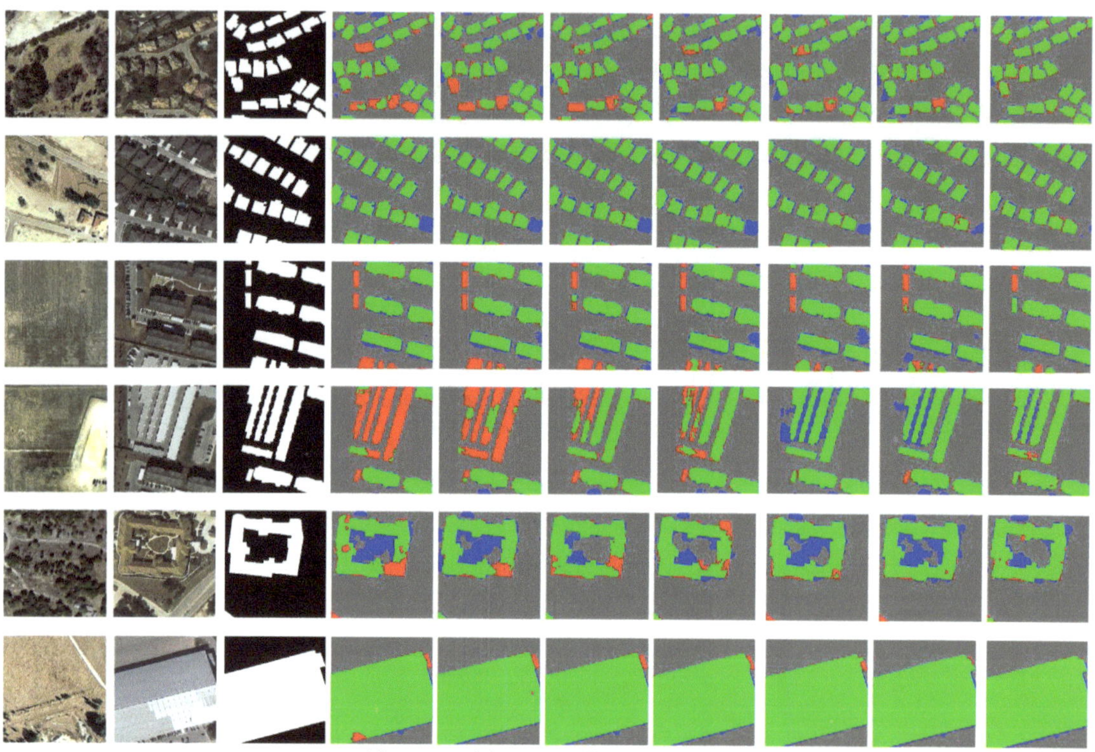

Figure 5. Visual comparisons of the UNet++ model applied to the LEVIR-CD dataset using different pre-training methods. (**a**) Images at time T1, (**b**) images at time T2; (**c**) Ground Truth; (**d**) Rand-init; (**e**) ImageNet-sup; (**f**) CMC; (**g**) SimCLR; (**h**) BT; (**i**) MoCo v2; and (**j**) BTCD. Gray: *TN* pixels; Green: *TP* pixels; Blue: *FP* pixels; Red: *FN* pixels.

3.2. Performance Comparison for the SYSU-CD Dataset

Our method achieved the highest F1 score (81.69%) and IoU value (69.04%). MSTDSNet-CD ranked second with an F1 score of 80.33% and an IoU of 67.13%, while Hybrid-TransCD ranked third with an F1 score of 80.13% and an IoU of 66.84%. MSTDSNet-CD combines a multi-scale Swin Transformer and a deep supervision network for CD tasks, while the Hybrid-TransCD method adopts a hybrid multi-scale transformer module that simulates the multi-scale mixed attention representations of bi-temporal images using fine-grained self-attention mechanisms. Both methods can enhance the models' expressive ability, thus achieving better CD results compared to CNN-based methods.

For the SYSU-CD dataset, the F1 score of the ImageNet pre-training method was slightly higher than that of the random initialization method, but the difference was not significant. Due to the significant distribution difference between the ImageNet dataset and the SYSU-CD dataset, domain shift occurs. Therefore, training a universal model in an unsupervised manner is more effective and robust for downstream CD tasks. Obtaining an ImageNet pre-trained model requires millions of labelled samples for supervised learning. The results in Table 3 further demonstrate that our BTCD method achieved the best detection performance and higher accuracy compared to other methods.

Table 3. Performance comparison for the SYSU-CD dataset.

Methods	SYSU-CD			
	Precision	**Recall**	**F1**	**IoU**
FC-EF	74.32	75.84	75.07	60.09
FC-Siam-Diff	**89.13**	61.21	72.57	56.96
FC-Siam-Conc	82.54	71.03	76.35	61.75
DASNet	68.14	70.01	69.14	60.65
STANet	70.76	**85.33**	77.37	63.09
DSAMNet	74.81	**81.86**	78.18	64.18
SRCDNet	75.54	**81.06**	78.20	64.21
BiDateNet	81.84	72.60	76.94	62.52
BIT	82.18	74.49	78.15	64.13
MSTDSNet-CD	79.91	80.76	**80.33**	**67.13**
Hybrid-TransCD	**83.05**	77.40	**80.13**	**66.84**
Ours (UNet++ with Rand-init)	80.67	75.82	78.17	64.16
Ours (UNet++ with ImageNet-sup)	80.92	77.94	79.40	65.84
Ours (UNet++ with BTCD)	**85.80**	77.95	**81.69**	**69.04**

Developed as part of the present study. All values are percentages. Bold red text indicates highest, bold blue text indicates second-highest, and bold black text indicates third-highest performances.

For the SYSU-CD dataset, we fine-tuned downstream CD networks (the improved UNet++) using the pre-trained models CMC, MoCo v2, SimCLR, and BT. Our BTCD pre-training method achieved the highest F1 score (81.69%) and IoU (69.04%) (Table 4). MoCo v2 method ranked second with an F1 score of 81.31% and an IoU of 68.50%, while BT ranked third with an F1 score of 80.35%) and an IoU of 67.19%. The visual comparison results for the SYSU-CD dataset are shown in Figure 6. These scenes depict various types of changes, including newly constructed urban buildings, preliminary groundwork before construction, and offshore construction activities. The CD results obtained using methods based on random initialization, ImageNet-sup, and CMC had more omissions (columns 4–6 in Figure 6). By contrast, MoCo v2 and BTCD performed better and provided more complete detection results. Rows 1–3 in Figure 6 demonstrate that for complex CD scenarios, our BTCD method was better able to detect the contour information than the other tested methods. Overall, our BTCD method had fewer errors and greater robustness than all other tested methods, providing superior performance for CD tasks.

Table 4. Performance of the improved UNet++ model on the SYSU-CD dataset using different pre-training methods.

Methods			SYSU-CD			
			Precision	Recall	F1	IoU
Backbone ResNet50	Reliant on the ImageNet dataset	CMC	83.69	76.26	79.80	66.39
		MoCo v2	**82.45**	80.19	81.31	68.50
		SimCLR	82.33	76.84	79.49	65.96
		BT	82.03	78.71	**80.35**	**67.19**
	Not reliant on the ImageNet dataset	BTCD	85.80	**77.95**	81.69	69.04

All values are percentages. Bold red text indicates highest, bold blue text indicates second-highest, and bold black text indicates third-highest performances.

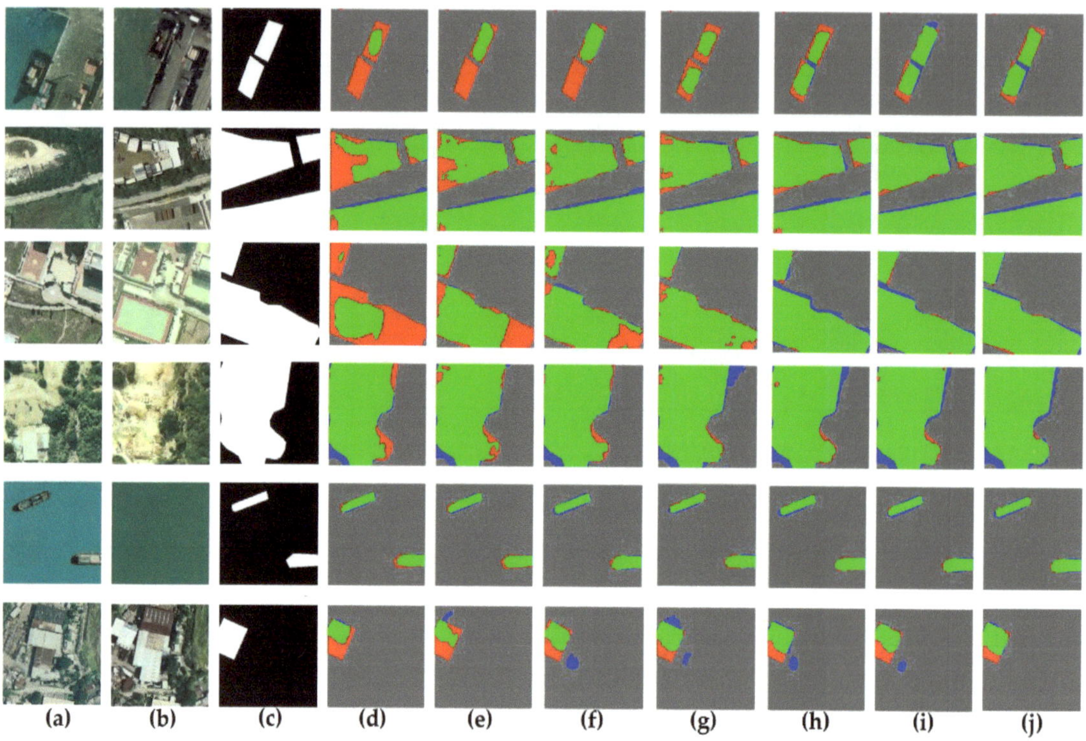

Figure 6. Visual comparisons of the UNet++ model applied to the SYSU-CD dataset using different pre-training methods. (**a**) Images at time T1, (**b**) Images at time T2; (**c**) ground truth; (**d**) Rand-init; (**e**) ImageNet-sup; (**f**) CMC; (**g**) SimCLR; (**h**) BT; (**i**) MoCo v2; and (**j**) BTCD. Gray: *TN* pixels; green: *TP* pixels; blue: *FP* pixels; red: *FN* pixels.

4. Discussion

4.1. Ablation Experiment

We fine-tuned the UNet++ model on the LEVIR-CD and SYSU-CD datasets through three strategies: Rand-init, ImageNet-sup, and pretraining with BTCD. To gauge the efficacy of our proposed pretraining method, we conducted a comprehensive comparative analysis against several other self-supervised pretraining approaches, including CMC, MoCo v2, SimCLR, and BT. According to the results in Table 5, our proposed pretraining method (BTCD) yields the best performance. Compared to Rand-init, applying our BTCD pretraining process increases the F1 score of LEVIR-CD by nearly 3.50%, and the IoU by approximately 5.56%; for SYSU-CD, the F1 score rises by around 3.52%, with IoU increasing by approximately 4.88%. On the ImageNet dataset, utilizing CMC, MoCo v2, SimCLR, and BT for self-supervised contrastive pretraining outperforms ImageNet-sup when transferred to LEVIR-CD and SYSU-CD datasets in terms of F1 score and IoU. Specifically, on the LEVIR-CD dataset, our proposed pretraining method surpasses the best self-supervised contrastive pretraining method, MoCo v2, by 2.11% in performance and increases IoU by nearly 3.32%. For the SYSU-CD dataset, our proposed pretraining method is roughly equivalent to MoCo v2 in terms of F1 score and IoU, with slight improvement. In summary, ablation experiments demonstrate that our proposed self-supervised contrastive pretraining method for unlabeled images can achieve or even surpass widely used pretraining methods based on the ImageNet dataset. Furthermore, our method mitigates the domain shift problem caused by transferring knowledge from ImageNet weights obtained through pretraining with datasets vastly different from CD datasets. Qualitative and quantitative comparisons indicate that our proposed BTCD exhibits significant effectiveness and superiority for CD tasks.

Table 5. Results of ablation experiments on LEVIR-CD and SYSU-CD datasets.

Methods		**Pretraining**	**ImageNet Dataset**	**LEVIR-CD**		**SYSU-CD**	
				F1	**IoU**	**F1**	**IoU**
Backbone ResNet50	Rand-init	×	×	87.80	78.25	78.17	64.16
	ImageNet-sup	✓	✓	87.91	78.43	79.40	65.84
	CMC	✓	✓	88.08	78.71	79.80	66.39
	MoCo v2	✓	✓	89.19	80.49	81.31	68.50
	SimCLR	✓	✓	88.86	79.96	79.49	65.96
	BT	✓	✓	89.06	80.28	80.35	67.19
	BTCD	✓	×	91.30	83.81	81.69	69.04

All values are percentages. × signifies steps excluded during the training process, while ✓ denotes their inclusion.

4.2. Hyperparameter Sensitivity Analysis

Inspired by the observation that the predictions of semantically similar remote sensing images should be similar irrespective of different lighting conditions, to achieve invariant prediction, we utilized the IP loss function to compel the BTCD model to learn features unaffected by imaging lighting conditions. Additionally, in order to mitigate the impact of lighting, noise, seasonal variations, and differences in remote sensing imaging angles on CD results, we employed the CCR Loss function. This dual approach not only enhances the alignment of bi-temporal images in both decision and feature spaces but also diminishes the influence of these confounding factors on CD results. To further analyze the impact of different loss balance hyperparameter settings on the final results, akin to the approach by Tung and Mori [30], we determined the CD accuracy results of different hyperparameter settings for the SYSU-CD dataset (Table 6). To evaluate the performance of the IP loss, we kept the coefficient β of the CCR loss constant and varied the α value. Increasing the α value had a limited impact on the downstream CD task performance of the BTCD method (Table 6). Consequently, we fixed the α value at 100 and further analyzed the final impact on CD task accuracy by changing the coefficient of the CCR loss, i.e., by modifying β. Because the scale of the CCR loss is smaller compared to the BT loss and IP loss, we increased β in increments of 1000 to balance losses on the same scale. Increasing β initially slightly improved the performance (Table 6). When $\alpha = 100$ and $\beta = 2000$, the BTCD method achieved the highest F1 score and IoU for the SYSU-CD dataset. After further increasing β, the effect reached saturation. Therefore, the optimal settings for α and β for the SYSU-CD dataset were 100 and 2000, respectively. Subsequently, we also applied the same loss function hyperparameter values to the LEVIR-CD dataset and obtained good CD results. This suggests that the performance of the BTCD algorithm developed in the present paper is not too sensitive to the hyperparameters of the final loss function. Furthermore, the parameters can be further optimized by cross-validating the validation sets of the two CD datasets.

Table 6. Table of sensitivity analysis for different loss balance hyperparameter settings for the SYSU-CD dataset.

Method	α	β	F1 (%)	IoU (%)	Method	α	β	F1 (%)	IoU (%)
	100	0	80.28	67.06		100	1000	79.42	65.86
BTCD	500	0	79.95	66.60	BTCD	100	2000	81.69	69.04
	1000	0	79.68	66.23		100	3000	80.35	67.19

4.3. *Efficiency under Limited Labels*

Manually marking areas of change is an expensive process that requires a significant amount of manpower and resources. Therefore, the lack of access to large-scale annotated data remains a major challenge in using bi-temporal remote sensing images for CD. We tested multiple proportions of training data (20%, 40%, 60%, and 100%) and attempted various pre-training methods for the LEVIR-CD and SYSU-CD datasets. Compared to the pre-training methods Rand-init, ImageNet-sup, CMC, MoCO v2, SimCLR, and BT, our BTCD pre-training method performed well on both datasets, even with limited annotated data (Figure 7). The F1 score gradually improved as the percentage of training data increased. When using 60% of the training data, the BTCD method achieved F1 scores of 0.8973 and 0.788 when applied to the LEVIR-CD and SYSU-CD datasets, respectively, both of which were higher than the F1 scores of CNN-based methods using all training data. When the percentage of training data was further reduced to 40%, BTCD achieved an F1 score of 0.8943 when applied to the LEVIR-CD dataset, which was only slightly lower than those achieved by ChangeFormer and Hybrid-TransCD, and an F1 score of 0.7787 when applied to the SYSU-CD dataset, which was comparable to the performance metrics of DSAMNet, SRCDNet, and BIT. When the percentage of training data was reduced to 20%, BTCD achieved F1 scores of 0.8486 and 0.7504 when applied to the LEVIR-CD and SYSU-CD datasets, respectively, which were slightly higher than F1 scores of FC-EF, FC-Siam-Diff, FC-Siam-Conc, and DASNet. We demonstrate that our developed BTCD pre-training method can significantly improve the generalization ability and data efficiency of downstream CD models. It is worth noting that we achieved competitive results using only 20% of the training samples for model training. In summary, when annotated data for CD are limited, our pre-training method can significantly enhance the robustness and general performance of CD models.

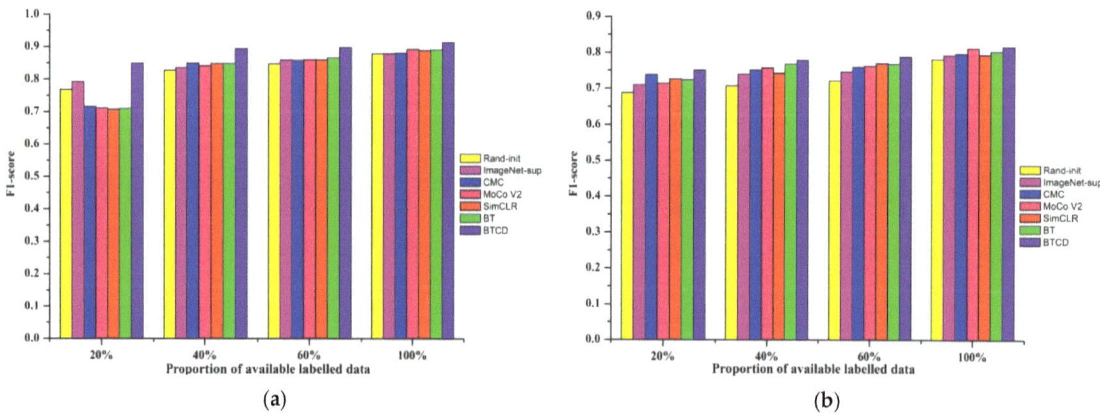

Figure 7. Performance (F1-score) of different pre-training methods evaluated using the improved UNet++ model under varying label availability. (**a**) LEVIR-CD. (**b**) SYSU-CD.

4.4. Evaluation of Model Generalization Performance

The LULC CD mandates that models exhibit genuine generalization capabilities. To validate the generalization performance of the proposed BTCD framework, we conducted experiments on the extensive WHU Building CD (WHU-BCD) dataset [42]. This dataset encompasses an area impacted by a 6.3-magnitude earthquake in February 2011, which was subsequently reconstructed over the following years. Our research focuses on Christchurch, New Zealand, a city that has seen significant new construction since the earthquake. The dataset includes aerial images captured in April 2012, covering 20.5 km^2 and featuring 12,796 buildings (16,077 buildings in the 2016 dataset within the same area). The images have a ground sampling distance of 0.2 m/pixel. Additionally, the dataset provides reference change masks and a pair of co-registered aerial images with a combined size of 15,354 × 32,507 × 3 pixels. The 2012 and 2016 remote sensing images are depicted in Figure 8a and 8b, respectively, while the ground truth for building changes is shown in Figure 8c. In this figure, black areas indicate unchanged regions, and white areas denote changed building regions. We cropped the bi-temporal images into smaller patches of 256 × 256 × 3 pixels, resizing edge patches to 256 × 256 × 3 pixels, resulting in a total of 7620 pairs of image patches.

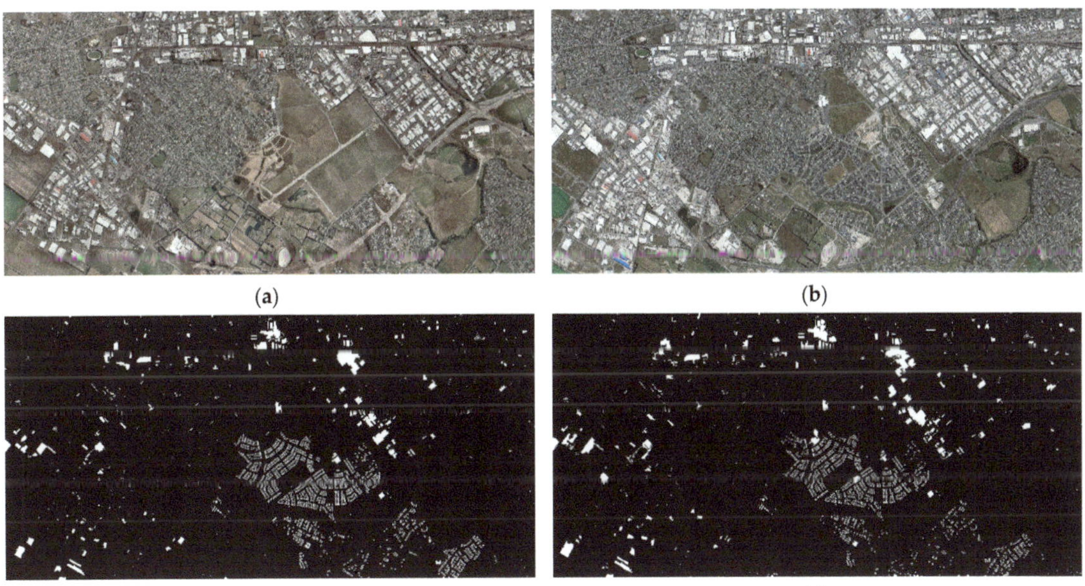

Figure 8. The WHU-BCD dataset. (**a**) An old temporal image from 2012; (**b**) a new temporal image from 2016; (**c**) ground truth; (**d**) the prediction CD map.

During the self-supervised pre-training stage, we utilized these 7620 pairs of unlabeled image patches for SSCL. SSCL involves designing an instance discrimination task to identify the most similar images to the input features within the dataset. It maximizes the similarity between two positive instances while increasing the distance between positive and negative instances. In the fine-tuning CD stage, the trained encoder, equipped with robust feature representations, is transferred to the downstream CD task. We conducted experimental analysis on the WHU-BCD dataset using the improved UNet++ model. The results were stitched together based on the relative positions of the cropped image patches, and the edge patches were resized back to their original size. The final experimental results on the WHU-BCD dataset are shown in Figure 8d. We evaluated the accuracy of the results, yielding a precision of 0.7674, a recall of 0.7771, an F1 score of 0.7722, and an IoU of 0.6289.

Both quantitative and qualitative experimental outcomes demonstrate that the proposed BTCD framework possesses strong generalization performance in practical applications.

Furthermore, it is noteworthy that our method was pre-trained on RGB images due to the nature of the two public datasets employed, which only include RGB imagery. Consequently, we had to exclude information beyond the visible light spectrum. However, in practical scenarios, most remote sensing images encompass not only RGB bands but also information beyond the visible light spectrum. This limitation affects the application of our method to multispectral satellite image data such as that from Sentinel-2 and Landsat. In future work, we plan to extend the proposed BTCD algorithm and investigate various strategies for adapting from RGB images to multispectral images, thereby enhancing the method's broad applicability.

5. Conclusions

When performing LULC CD in bi-temporal remote sensing images, deep learning models based on supervised paradigms require a large amount of annotated data. Unfortunately, collecting and annotating samples that contain the desired areas of change is both time-consuming and labor-intensive. Transfer learning of pre-trained models is an effective method that can alleviate this problem. In the present study, we developed the BTCD framework, which is composed of a self-supervised contrastive pre-training stage and a fine-tuning stage and can learn the inherent differential features of original bi-temporal remote sensing images in a self-supervised manner. Our method can be easily integrated with existing SOTA CD methods. We conducted extensive analyses using two public remote sensing CD datasets and demonstrated the superiority of BTCD over other commonly used ImageNet pre-training and self-supervised pre-training methods. In addition, BTCD does not require additional image data. Our proposed self-supervised pre-training strategy is effective even when using a limited sample size of labelled data, which is particularly valuable for CD applications for which obtaining labelled data for changed regions is difficult and expensive. In the future, we plan to replace the original ResNet50 encoder component of BTCD with a vision transformer to further improve the accuracy of the CD results.

Author Contributions: Conceptualization, W.F. and W.X.; methodology, W.F.; validation, F.G. and C.S.; writing—original draft preparation, W.F. and W.X.; writing—review and editing, F.G. and C.S.; supervision, F.G. and W.X. All authors have read and agreed to the published version of the manuscript.

Funding: This work was supported by the National Natural Science Foundation of China under Grant No. 42101358.

Data Availability Statement: Data associated with this research are available online. The LEVIR-CD dataset is available for download at https://justchenhao.github.io/LEVIR/, accessed on 22 May 2020. The SYSU-CD dataset is available for download at https://github.com/liumency/DSAMNet/, accessed on 4 October 2022. The WHU-BCD dataset is available for download at http://gpcv.whu.edu.cn/data/building_dataset.html, accessed on 14 April 2023.

Acknowledgments: We sincerely thank the anonymous reviewers for their critical comments and suggestions for improving the manuscript.

Conflicts of Interest: The authors declare no conflicts of interest.

References

1. Zhang, L.; Wu, C. Advance and future development of change detection for multi-temporal remote sensing imagery. *Acta Geod. Cartogr. Sin.* **2017**, *46*, 1447–1459.
2. Sui, H.; Feng, W.; Li, W.; Sun, K.; Xu, C. Review of change detection technology for multi-temporal remote sensing imagery. *Geomat. Inf. Sci. Wuhan Univ.* **2018**, *43*, 1885–1891.
3. Hussain, M.; Chen, D.; Cheng, A.; Wei, H.; Stanley, D. Change detection from remotely sensed images: From pixel-based to object-based approaches. *ISPRS J. Photogramm. Remote Sens.* **2013**, *80*, 91–106. [CrossRef]

4. Peng, D.; Zhang, Y.; Guan, H. End-to-End Change Detection for High Resolution Satellite Images Using Improved UNet++. *Remote Sens.* **2019**, *11*, 1382. [CrossRef]
5. Geng, J.; Wang, H.; Fan, J.; Ma, X. Change detection of SAR images based on supervised contractive auto-encoders and fuzzy clustering. In Proceedings of the IEEE International Workshop on Remote Sensing with Intelligent Processing (RSIP), Shanghai, China, 19–21 May 2017.
6. Su, L.; Gong, M.; Zhang, P.; Zhang, M.; Liu, J.; Yang, H. Deep learning and mapping based ternary change detection for information unbalanced images. *Pattern Recognit.* **2017**, *66*, 213–228. [CrossRef]
7. Gong, M.; Zhang, P.; Zhan, T.; Miao, Q. Superpixel-based difference representation learning for change detection in multispectral remote sensing images. *IEEE Trans. Geosci. Remote Sens.* **2017**, *55*, 2658–2673. [CrossRef]
8. Feng, G.; Dong, J.; Bo, L.; Xu, Q. Automatic Change Detection in Synthetic Aperture Radar Images Based on PCANet. *IEEE Geoence Remote Sens. Lett.* **2017**, *13*, 1792–1796.
9. Chen, H.; Wu, C.; Du, B.; Zhang, L.; Wang, L. Change detection in multisource VHR images via deep siamese convolutional multiple-layers recurrent neural network. *IEEE Trans. Geosci. Remote Sens.* **2019**, *58*, 2848–2864. [CrossRef]
10. Daudt, R.C.; Saux, B.L.; Boulch, A. Fully convolutional Siamese networks for change detection. In Proceedings of the 25th IEEE International Conference on Image Processing (ICIP), Athens, Greece, 7–10 October 2018; pp. 4063–4067.
11. Chen, J.; Yuan, Z.; Peng, J.; Chen, L.; Huang, H.; Zhu, J.; Liu, Y.; Li, H. DASNet: Dual attentive fully convolutional Siamese networks for change detection in high-resolution satellite images. *IEEE J. Sel. Topics Appl. Earth Observ. Remote Sens.* **2020**, *14*, 1194–1206. [CrossRef]
12. Chen, H.; Shi, Z. A spatial-temporal attention-based method and a new dataset for remote sensing image change detection. *Remote Sens.* **2020**, *12*, 1662. [CrossRef]
13. Deng, J.; Dong, W.; Socher, R.; Li, L.J.; Li, K.; Li, F.F. ImageNet: A large-scale hierarchical image database. In Proceedings of the IEEE Conference on Computer Vision and Pattern Recognition, Miami, FL, USA, 20–25 June 2009; pp. 248–255.
14. Long, Y.; Xia, G.; Li, S.; Yang, W.; Yang, M.Y.; Zhu, X.; Zhang, L.; Li, D. On Creating Benchmark Dataset for Aerial Image Interpretation: Reviews, Guidances, and Million-AID. *IEEE J. Sel. Top. Appl. Earth Obs. Remote Sens.* **2021**, *14*, 4205–4230. [CrossRef]
15. Minetto, R.; Segundo, M.P.; Sarkar, S. Hydra: An Ensemble of Convolutional Neural Networks for Geospatial Land Classification. *IEEE Trans. Geosci. Remote Sens.* **2019**, *57*, 6530–6541. [CrossRef]
16. Sumbul, G.; Charfuelan, M.; Demir, B.; Markl, V. BigEarthNet: A Large-Scale Benchmark Archive for Remote Sensing Image Understanding. In Proceedings of the IEEE International Geoscience and Remote Sensing Symposium, Yokohama, Japan, 28 July–2 August 2019; pp. 5901–5904.
17. Manas, O.; Lacoste, A.; Giro-i-Nieto, X.; Vazquez, D.; Rodriguez, P. Seasonal contrast: Unsupervised pretraining from uncurated remote sensing data. In Proceedings of the IEEE/CVF International Conference on Computer Vision (ICCV), Los Alamitos, CA, USA, 10–17 October 2021; pp. 9394–9403.
18. Wang, Y.; Albrecht, C.M.; Braham, N.A.A.; Mou, L.; Zhu, X. Self-Supervised Learning in Remote Sensing: A review. *IEEE Geosci. Remote Sens. Mag.* **2022**, *10*, 213–247. [CrossRef]
19. Tian, Y.; Krishnan, D.; Isola, P. Contrastive Multiview Coding. *arXiv* **2019**, arXiv:1906.05849.
20. Chen, X.; He, K. Exploring simple siamese representation learning. In Proceedings of the IEEE/CVF Conference on Computer Vision and Pattern Recognition, Nashville, TN, USA, 20–25 June 2021; pp. 15750–15758.
21. Chen, T.; Kornblith, S.; Norouzi, M.; Hinton, G. A simple framework for contrastive learning of visual representations. *arXiv* **2020**, arXiv:2002.05709.
22. He, K.; Fan, H.; Wu, Y.; Xie, S.; Girshick, R. Momentum contrast for unsupervised visual representation learning. *arXiv* **2019**, arXiv:1911.05722.
23. Caron, M.; Misra, I.; Mairal, J.; Goyal, P.; Bojanowski, P.; Joulin, A. Unsupervised Learning of Visual Features by Contrasting Cluster Assignments. *arXiv* **2021**, arXiv:2006.09882.
24. Grill, J.B.; Strub, F.; Altché, F.; Tallec, C.; Richemond, P.H.; Buchatskaya, E.; Doersch, C.; Pires, B.A.; Guo, Z.; Azar, M.G.; et al. Bootstrap your own latent: A new approach to self-supervised learning. *arXiv* **2020**, arXiv:2006.07733.
25. Jure, Z.; Li, J.; Ishan, M.; Yann, L.C.; Stéphane, D. Barlow twins: Self-supervised learning via redundancy reduction. *arXiv* **2021**, arXiv:2103.03230.
26. Jiang, F.; Gong, M.; Zheng, H.; Liu, T.; Zhang, M.; Liu, J. Self-Supervised Global–Local Contrastive Learning for Fine-Grained Change Detection in VHR Images. *IEEE Trans. Geosci. Remote Sens.* **2023**, *61*, 4400613. [CrossRef]
27. Wang, J.; Zhong, Y.; Zhang, L. Change Detection Based on Supervised Contrastive Learning for High-Resolution Remote Sensing Imagery. *IEEE Trans. Geosci. Remote Sens.* **2023**, *61*, 5601816. [CrossRef]
28. Englesson, E.; Azizpour, H. Generalized jensen-shannon divergence loss for learning with noisy labels. *Adv. Neural Inf. Process. Syst.* **2021**, *34*, 30284–30297.
29. Kim, T.; Oh, J.; Kim, N.Y.; Cho, S.W.; Yun, S.Y. Comparing kullback-leibler divergence and mean squared error loss in knowledge distillation. *arXiv* **2021**, arXiv:2105.08919.
30. Tung, F.; Mori, G. Similarity-preserving knowledge distillation. In Proceedings of the IEEE/CVF International Conference on Computer Vision, Seoul, Republic of Korea, 27 October–2 November 2019; pp. 1365–1374.

31. Abhijit, G.R.; Nassir, N.; Christian, W. Concurrent spatial and channel squeeze & excitation in fully convolutional networks *arXiv* **2018**, arXiv:1803.02579.
32. Shi, Q.; Liu, M.; Li, S.; Liu, X.; Wang, F.; Zhang, L. A Deeply Supervised Attention Metric-Based Network and an Open Aerial Image Dataset for Remote Sensing Change Detection. *IEEE Trans. Geosci. Remote Sens.* **2022**, *60*, 1–16. [CrossRef]
33. Ramkumar, V.R.T.; Arani, E.; Zonooz, B. Differencing based self-supervised pretraining for scene change detection. *arXiv* **2022**, arXiv:2208.05838.
34. Fang, S.; Li, K.; Shao, J.; Li, Z. SNUNet-CD: A Densely Connected Siamese Network for Change Detection of VHR Images. *IEEE Geosci. Remote Sens. Lett.* **2022**, *19*, 8007805. [CrossRef]
35. Zhang, C.; Yue, P.; Tapete, D.; Jiang, L.; Shangguan, B.; Huang, L.; Liu, G. A deeply supervised image fusion network for change detection in high resolution bi-temporal remote sensing images. *ISPRS J. Photogramm. Remote Sens.* **2020**, *166*, 183–200. [CrossRef]
36. Liu, M.; Shi, Q.; Marinoni, A.; He, D.; Liu, X.; Zhang, L. Super-resolution-based change detection network with stacked attention module for images with different resolutions. *IEEE Trans. Geosci. Remote Sens.* **2021**, *60*, 4403718. [CrossRef]
37. Papadomanolaki, M.; Verma, S.; Vakalopoulou, M.; Gupta, S.; Karantzalos, K. Detecting urban changes with recurrent neural networks from multitemporal Sentinel-2 data. In Proceedings of the 2019 IEEE International Geoscience and Remote Sensing Symposium, Yokohama, Japan, 28 July–2 August 2019; IEEE: Piscataway, NJ, USA, 2019; pp. 214–217.
38. Chen, H.; Qi, Z.; Shi, Z. Remote Sensing Image Change Detection with Transformers. *IEEE Trans. Geosci. Remote Sens.* **2022**, *60*, 5920416. [CrossRef]
39. Bandara, W.G.C.; Patel, V.M. A transformer-based siamese network for change detection. *arXiv* **2022**, arXiv:2201.01293.
40. Song, F.; Zhang, S.; Lei, T.; Song, Y.; Peng, Z. MSTDSNet-CD: Multiscale swin transformer and deeply supervised network for change detection of the fast-growing urban regions. *IEEE Geosci. Remote Sens. Lett.* **2022**, *19*, 6508505. [CrossRef]
41. Ke, Q.; Zhang, P. Hybrid-transCD: A hybrid transformer remote sensing image change detection network via token aggregation. *ISPRS Int. J. Geo-Inf.* **2022**, *11*, 263. [CrossRef]
42. Ji, S.; Shen, Y.; Lu, M.; Zhang, Y. Building Instance Change Detection from Large-Scale Aerial Images using Convolutional Neural Networks and Simulated Samples. *Remote Sens.* **2019**, *11*, 1343. [CrossRef]

land

Article

A Spatio-Temporal Examination of Land Use and Land Cover Changes in Smart Cities of the Delhi–Mumbai Industrial Corridor

Arun Kanchan [1,*], Vilas Nitivattananon [1], Nitin K. Tripathi [2], Ekbordin Winijkul [3] and Ranadheer Reddy Mandadi [2]

1 Urban Innovation and Sustainability, Asian Institute of Technology, Khlong Luang 12120, Thailand; vilasn@ait.asia
2 Remote Sensing and Geographic Information Systems, Asian Institute of Technology, Khlong Luang 12120, Thailand; nitinkt@ait.asia (N.K.T.); ranadheer@ait.asia (R.R.M.)
3 Environmental Engineering and Management, Khlong Luang 12120, Thailand; ekbordinw@ait.asia
* Correspondence: st122541@ait.asia

Abstract: This study provides a detailed analysis of land use and land cover (LULC) changes at the district level within the Delhi–Mumbai Industrial Corridor (DMIC) from 2001 to 2021. Using the Indian Meteorological Department's sub-divisional framework and MODIS data across seven primary LULC classes, the analysis is instrumental in informing infrastructure planning for existing and future smart cities and industrial clusters within the DMIC. The key findings reveal a yearly increase of 3031.40 sq. km. per year in agricultural land, with decreases in shrubland, grassland, and bareland of −1774.72 sq. km. per year, −1119.62 sq. km. per year, and −203.76 sq. km. per year, respectively. On the other hand, forests grew by a modest 148.14 sq. km. per year, while waterbodies and built-up lands saw minor increases of 55.73 sq. km. and 21.48 sq. km. per year. Ecologically Sensitive Areas (ESAs) were evaluated for LULC changes. The smart cities of Pune and Thane serve as excellent examples of balanced urban development and natural growth management. However, the study also highlights the need for further research to investigate LULC impacts on climatic variables, advocating for a regional planning approach in the DMIC.

Keywords: Delhi–Mumbai industrial corridor; land use and land cover; remote sensing; smart cities mission; Indian Meteorological Department

Citation: Kanchan, A.; Nitivattananon, V.; Tripathi, N.K.; Winijkul, E.; Mandadi, R.R. A Spatio-Temporal Examination of Land Use and Land Cover Changes in Smart Cities of the Delhi–Mumbai Industrial Corridor. *Land* **2024**, *13*, 957. https://doi.org/10.3390/land13070957

Academic Editor: Jūratė Sužiedelytė-Visockienė

Received: 15 May 2024
Revised: 21 June 2024
Accepted: 24 June 2024
Published: 29 June 2024

1. Introduction

The Delhi–Mumbai Industrial Corridor (DMIC) stands as a testament to India's commitment to economic growth and sustainable development. It is a monumental infrastructure project, one of the country's largest and most ambitious, to foster economic growth and sustainable development in the region (NICDC, 2021) [1]. Spanning over 1483 km and encompassing several states and Union Territories (UTs), this mega-regional project holds immense potential to impact existing land use and land cover significantly (LULC), as well as climatology in the affected areas (Jain, 2014; Jain, 2021) [2,3].

The zone of influence (ZoI) encompasses eight states and three Union Territories (UTs), covering fourteen Indian Meteorological Department (IMD) sub-divisions, as shown in Table 1. It also encompasses 24 smart cities under the Smart Cities Mission (SCM) program [4] and 121 other districts. The plan for the ZoI includes investment regions (minimum 200 sq. km.) and industrial areas (minimum 100 sq. km.), integrated with road and rail for freight transit. These investment regions and investment areas are estimated to generate 3500 sq. km. of built-up land, excluding supporting feeder roads and infrastructure.

Table 1. IMD sub-divisions, zone of influence, and smart cities within the DMIC.

IMD Sub-Division/Union Territory	Total Area under the DMIC ZoI in sq. km.	Total Area Under the DMIC ZoI in %	Total # of Districts	Smart Cities [1]
Chandigarh, Haryana, and Delhi	61,339.00	9.47%	30	Faridabad, Karnal, and New Delhi
Dadra and Nagar Haveli [2]	622.25	0.10%	3	Diu and Silvassa [2]
East Madhya Pradesh	3630.25	0.56%	1	
East Rajasthan	156,423.75	24.15%	23	Ajmer
East Uttar Pradesh	285.50	0.04%	1	Agra
Gujarat Region	91,741.00	14.17%	18	Ahmedabad, Dohad, Gandhinagar, Surat, and Vadodara
Konkan and Goa	23,403.25	3.61%	5	Thane
Madhya Maharashtra	66,206.00	10.22%	8	Nashik
Marathwada	464.00	0.07%	1	Pune
Punjab	1293.75	0.20%	3	
Saurashtra and Kachh	52,533.25	8.11%	7	Rajkot
Uttarakhand	11,374.50	1.76%	7	Dehradun
West Madhya Pradesh	17,077.75	2.64%	8	
West Rajasthan	92,522.00	14.29%	7	Jaipur, Kota, and Udaipur
West Uttar Pradesh	68,711.50	10.61%	23	Aligarh, Bareilly, Moradabad, and Saharanpur
15 IMD sub-divisions	647,627.75	100.00%	145	24 Smart Cities

[1]. Smart cities under the Smart Cities Mission (https://smartcities.gov.in/cities-profiles, accessed on 1 March 2024) [2]. Diu and Silvassa are smart cities in the Dadra and Nagar Haveli Union Territory, which is within the DMIC zone of influence.

The DMIC aims to foster regional economic growth by maximizing existing capacities and improving the investment climate by developing supportive, long-term infrastructure. The Government of India (GoI, 2007) [5] has planned a project area with a potential zone of influence (ZoI) of 150 to 200 km on either side of the Western Dedicated Freight Corridor (WDFC).

The DMIC's topographical and climatic diversity necessitates a methodological approach that transcends state or UT administrative boundaries for analysis. The IMD sub-divisions offer a suitable framework for this purpose, providing a uniform basis for facilitating a more accurate assessment of local ecosystems and climate changes in the 14 sub-divisions and the UT of Dadra and Nagar Haveli within the DMIC. Identifying sub-division-specific impacts of LULC changes enables policymakers to craft policies addressing the impact of environmental dynamics and infrastructural development on the region's ecological and socio-economic systems.

Under the SCM in India, a city qualifies as "smart" by meeting several critical indicators and criteria focused on sustainable and inclusive development. Launched in June 2015, the SCM aims to leverage technology to enhance citizens' quality of life and economic prospects, emphasizing sustainable environment practices such as waste management and energy efficiency. The integration of comprehensive urban planning, focusing on land use and spatial planning, ensures balanced regional development. These criteria collectively contribute to creating technologically advanced, environmentally sustainable, and socially inclusive urban spaces. Therefore, evaluating existing smart cities within the DMIC for baseline LULC conditions and changes is critical.

Upon its full development, experts classify the DMIC as a mega-region [6], encompassing mega-cities like Delhi and Mumbai (with populations exceeding 10 million) and cities poised to become mega-cities by 2030, such as Ahmedabad, in addition to various smaller townships and rural agricultural zones. Experts anticipate that the DMIC's development will catalyze substantial economic growth, urbanization, and industrialization in the forthcoming decades.

1.1. Remote Sensing and GIS in LULC Studies

Previous studies have demonstrated the utility of remote sensing in analyzing urban sprawl, particularly in Ajmer City, revealing key metrics and relationships essential for environmental monitoring. These studies have used metrics to understand urban expansion and its environmental impact (Jat et al., 2008) [7]. MODIS has significantly contributed to environmental studies, with refinements in data collection and analysis methods enhancing our understanding of land surface dynamics. The improvements in MODIS data provide a clearer picture of land cover changes over time (Sulla-Menashe et al., 2019) [8]. Lal et al. (2022) [9] highlighted the effects of LULC changes on hydro-climatic variables, noting significant cooling impacts in major regions and variations in the atmosphere's lower boundary. The study emphasizes the importance of monitoring these changes for better climate predictions.

In an urban context, Mathan and Krishnaveni (2019) [10] observed substantial urban expansion in the Chennai Metropolitan Area, with marked decreases in green areas and waterbodies, as seen in other global urbanization trends. Urbanization has resulted in significant environmental changes, particularly affecting surface air temperature and contributing to the cooling impacts observed in Chennai. The study showcases how urban expansion modifies local climate conditions. Similarly, Sahana, Ahmed, and Sajjad (2016) [11] analyzed the Sundarban Biosphere Reserve, finding that land cover shifts led to increased surface temperatures, a pattern echoed by Nayak (2021) [12] in central India, where changes favored agricultural and dense forests over small vegetation lands and open forests.

While these urban studies provide critical insights into urban expansion and its environmental consequences, comparable studies are needed to understand similar impacts in a rapidly developing industrial corridor such as the DMIC.

1.2. LULC Change Dynamics in India

In recent years, research has focused on land use and land cover change (LULCC) across different regions of India, highlighting the spatio-temporal dynamics and environmental repercussions of such transformations. Duraisamy, Bendapudi, and Jadhav (2018) [13] investigated the semi-arid Mula Pravara river basin in Maharashtra, identifying significant LULCC driven by factors enhancing water resource access. Garai and Narayana (2018) [14] explored the Godavari basin's LULC changes, highlighting significant environmental impacts from industrial activities. The satellite data revealed how industrial activities have altered land use and cover, leading to environmental degradation. Moulds, Buytaert, and Mijic´ (2018) [15] developed high-resolution land use/cover maps for India from 1960 to 2010, utilizing a regional change model informed by district-level data, marking a methodological advance in tracking LULCC.

These studies collectively paint a complex picture of LULC in India, reflecting both natural ecosystem alterations and anthropogenic influences, with varying implications for local and regional climates, hydrology, and biodiversity. Further, these regional studies underscore the varied drivers of LULC changes across different ecological zones in

India. However, a gap remains in understanding these drivers within the context of mega-infrastructure projects like the DMIC.

1.3. Focused Studies on DMIC

Previous studies on the DMIC have utilized remote sensing techniques and geospatial analysis to evaluate impacts, emphasizing the importance of these tools for comprehensive assessments. This review is helpful and important for monitoring regional planning and development in the DMIC (Jain, 2014) [2]. Mukhopadhyay (2017) [6] evaluated the DMIC as a mega-region, emphasizing the use of various tools to assess the impacts of this extensive industrial corridor on regional development and environmental sustainability. The study underscores the DMIC's role in restructuring spatial development and accelerating industrialization. Regional planning plays a crucial role in reducing spatial disparities, particularly in large cities where inadequate urban infrastructure often exacerbates inequalities. Evaluation tools have highlighted these disparities, providing insights for better urban planning and infrastructure development (Jain, 2019) [16].

Ramachandran (2019) [17] reviewed issues with intrastate corridor development, focusing on the utilization of evaluation tools to address challenges in planning and implementation. These tools are critical for overcoming obstacles and achieving effective development outcomes in intrastate corridors. Williams et al. (2019) [18] analyzed the implementation of the DMIC in Gujarat, restructuring spatial development to accelerate urbanization and industrialization. The study details significant changes in land use patterns, which are indicative of broader regional development strategies. The impact of urbanization on surface air temperature has been modeled using weather research and forecasting, which highlight the critical role of LULC changes in influencing temperature variability (Jain, 2021) [3].

Kumar and Sharma (2018) [19] investigated land transformation due to urbanization in the DMIC region, highlighting the rapid changes in land use driven by highway peripheral developments. The study emphasizes the need for effective land management to mitigate adverse environmental impacts. Biswas et al. (2019) [20] examined the LULC impact in the vicinity of industrial areas, employing spatio-temporal analysis to assess how industrialization has accelerated urban land cover changes globally and locally. The study provides a detailed temporal analysis of land cover changes, crucial for understanding urbanization trends.

Despite these prior studies, research focusing on LULC changes at the district level and implications for smart cities within the DMIC framework remains sparse.

1.4. Comparative Studies: DMIC and BRI

The DMIC in India benefits from land use research on international mega-projects such as China's Belt and Road Initiative (BRI). While the BRI is significantly larger and transboundary and the DMIC is within India, both projects alter LULC in their respective zones of influence, especially due to the expected increase in built-up land and urbanization.

Recent BRI-related studies have increasingly focused on its impacts on land use land cover change (LULCC), trade, and ecosystem services across Central Asia, highlighting the multifaceted consequences of large-scale infrastructure projects. Zhang et al. (2022) [21] demonstrated that the Belt and Road Initiative (BRI) significantly influences agricultural land use, confirming earlier findings about the increase in trade activities and economic development driving land use changes. The study highlights the BRI's role in reshaping agricultural landscapes and its broader economic implications. Complementing this perspective, Zhang et al. (2022) [22] explored the spatio-temporal dynamics of LULC in BRI regions, utilizing the Landscape Ecological Risk Index to assess how the initiative impacts land cover and ecological risks over time. The study provides a comprehensive analysis of ecological risks associated with the BRI, emphasizing the need for sustainable land management.

Adding a monitoring dimension, Naboureh et al. (2020) [23] reviewed the challenges associated with LULCC mapping in the China–Central Asia–West Asia Economic Corridor (a critical component of the BRI), emphasizing the need for improved methodologies to address issues in accuracy and consistency. The study highlights the importance of accurate mapping for effective land use planning and the development of robust environmental policies. Chen et al. (2018) [24] examined spatio-temporal changes in cultivated land in China, utilizing FAO statistics and GlobeLand30 data to highlight significant fluctuations in land use over recent decades. This analysis is crucial for understanding agricultural trends and planning for sustainable land management.

Dong et al. (2021) [25] analyzed the environmental and socioeconomic drivers of LUCC in the China–Mongolia–Russia Economic Corridor, identifying key factors influencing land use changes between 1992 and 2015. The study emphasizes the role of economic activities and policy decisions in shaping land use patterns. In a parallel vein, Zuo et al. (2020) [26] investigated the effects of land use changes on ecosystem services, quantifying significant changes and their implications for environmental sustainability. The study highlights the critical need to balance development and ecosystem conservation. Teo et al. (2019) [27] characterized how several types of infrastructure development affect other components of the Earth's systems, including the atmosphere, hydrosphere, geosphere, and biosphere.

These BRI findings underscore the importance of monitoring and understanding LULC changes in large infrastructure projects. Lessons from BRI studies can inform the approach to analyzing and mitigating similar impacts within the DMIC.

1.5. Research Gap

Many LULC-based studies have been conducted in India, primarily focused on specific regions or smaller research areas. However, a comprehensive understanding of LULC changes across the entire DMIC, particularly concerning the impact of large-scale infrastructure development, is still lacking. As a significant economic corridor in India, it is crucial to study the changes in land use and their ecological impacts within the DMIC.

The primary objective of this study is to provide a comprehensive analysis of the current LULC patterns within the DMIC and to understand how these patterns have evolved over the last two decades. Additionally, the study aims to investigate the specific changes in LULC within the IMD sub-divisions of the DMIC to identify regional variations. Another key objective is to compare the LULC changes in smart cities with other districts within the DMIC, highlighting the impacts of urbanization and infrastructure development. Furthermore, the study looks at the Ecologically Sensitive Areas (ESAs) within the DMIC, thereby providing valuable insights for sustainable land use planning and policy formulation.

Despite prior studies, as detailed above, research focusing on LULC changes at the district level and implications for smart cities within the DMIC framework remains sparse. This gap highlights a critical need for detailed examination, particularly in understanding the spatio-temporal dynamics of large-scale infrastructure projects. This paper aims to fill the research gap by analyzing the baseline conditions in LULC in the DMIC before its significant and accelerated expansion over the next few decades. Accordingly, our study discusses the following:

1. What are the current LULC patterns within the DMIC, and how have they changed in the last two decades?
2. How have the IMD sub-divisions within the DMIC experienced changes in LULC?
3. How do changes in LULC in smart cities compare with other districts within the DMIC?

This research provides a baseline reference for land use decision-makers in the DMIC and equips them with the necessary insights to formulate effective land use development and optimization policies. By ensuring sustainable land use development and protecting the ecological environment, this study has the potential to significantly impact the DMIC's future. Further, synthesizing the spatio-temporal dynamics documented in Indian contexts and comparative international projects like the BRI, this study contributes to understanding the relationship between large-scale infrastructure development and land use changes.

2. Materials and Methods

The inter-relationships between various LULC classes are complex, presenting challenges in identifying the most appropriate methods to analyze their connections over time (Hansen, 2013) [28]. Researchers have employed machine learning (ML) models to enhance the understanding of this relationship and the accuracy of LULC prediction. In the DMIC region, the study utilized various machine-learning models, notably the Support Vector Machine (SVM) (Cortes and Vapnik, 1995) [29], the Random Forest (Breiman, 2001) [30], and the Classification and Regression Tree (Breiman, 1984) [31]. This information can contribute to assessing the land allocation aspects of urban planning (Ouma, 2022) [32] and determining the priorities of different policy measures in retaining the overall ecological balance in the DMIC region.

The study employs these three advanced classification techniques to ensure the accuracy and reliability of the LULC classification results (Foody and Mathur, 2004; Congalton and Green, 2008) [33,34]. During ground truthing, multiple classifiers are tested separately to identify the most accurate individual classifier, thus improving overall prediction accuracy (Mountrakis et al., 2011) [35].

This study employs various frameworks and concepts to enhance the accuracy and reliability of LULC classification results. Key frameworks include using IMD sub-divisions as analytical units, which provide a uniform basis for assessing local ecosystems and climate changes at the district level. Integrating remote sensing and GIS techniques is central to this study, enabling detailed spatio-temporal analysis. Figure 1 displays the methodology flowchart, detailed in the following sections.

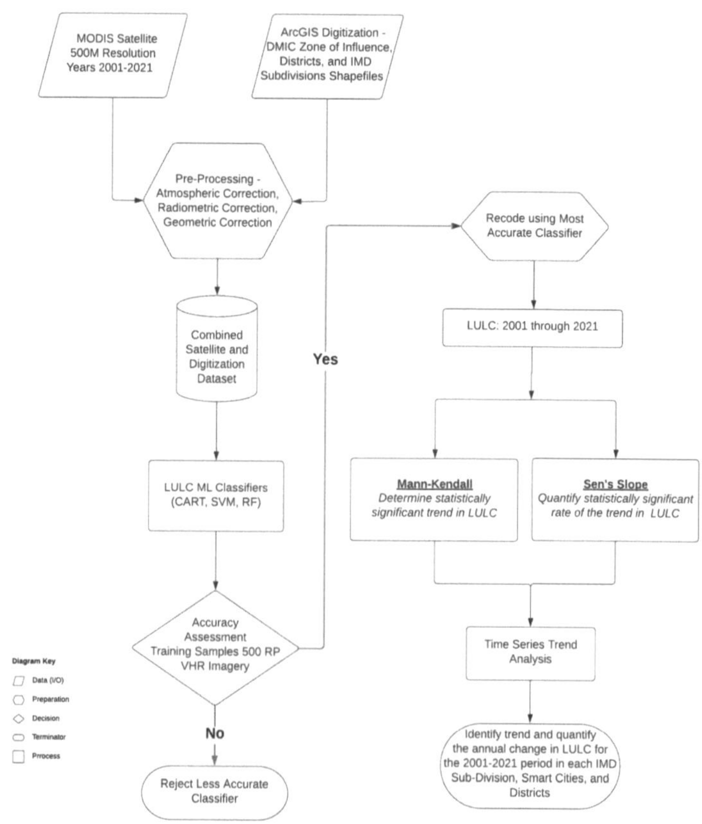

Figure 1. Methodology flowchart.

2.1. Study Area

The DMIC runs adjacent to the WDFC, connecting Dadri in the National Capital Region to Mumbai's Jawaharlal Nehru Port, and is approximately 1483 km in length. As shown in Figure 2, the corridor lies between 17°13′ N and 30°17′ N latitude and between 70°22′ E and 79°18′ E longitude.

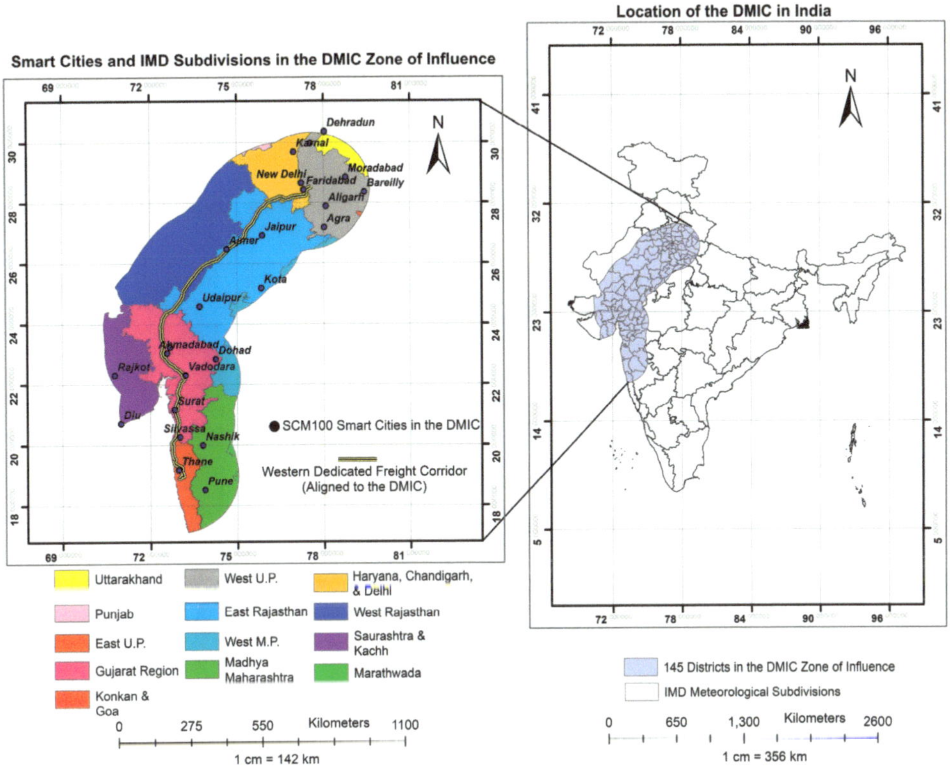

Figure 2. The study area (DMIC) with the smart cities and IMD sub-divisions.

Regarding land mass, the DMIC covers an extensive region passing through states such as Gujarat, Haryana, Madhya Pradesh, Maharashtra, Uttar Pradesh, and Rajasthan. The ZoI of 647,627.75 sq. km. also includes portions of the UT of Daman and Diu, Dadra and Nagar Haveli, the National Capital Territory (NCT of Delhi), and parts of Uttarakhand and Punjab states. In terms of IMD sub-divisions under the ZoI, East Rajasthan (24.15% and 23 districts), West Rajasthan (14.29% and 7 districts), Gujarat Region (14.17% and 18 districts), West Uttar Pradesh (10.61% and 23 districts), and Madhya Maharashtra (10.22% and 8 districts) have the largest shares.

The region's topography is diverse and includes parts of the Indo-Gangetic Plain, the Thar Desert, the Aravalli Range, and the Western Ghats. Major rivers, such as the Yamuna, Chambal, and Sabarmati, traverse the region. Additionally, a range of Ecologically Sensitive Areas, including national parks, wildlife sanctuaries, biosphere reserves, wetlands, rivers, grasslands, and coastal ecosystems, are bound by the DMIC. These areas span multiple districts and IMD sub-divisions, reflecting the region's rich biodiversity and ecological significance. Effective conservation strategies and sustainable development practices are essential to protect these valuable natural resources while promoting economic growth within the DMIC. As the DMIC extends towards the south, the elevation decreases, and the corridor approaches the coastline of the Arabian Sea. Coastal areas within the DMIC's ZoI include Mumbai and the Gulf of Khambhat in Gujarat.

2.2. Data Collection

We used Google Earth Engine (GEE) to acquire MODIS satellite imagery with a resolution of 500 m for the 20-year "lookback period" from 2001 to 2021. Concurrently, we sourced shapefiles delineating DMIC Buffer, District, and IMD sub-divisions through ArcGIS digitization.

2.3. Data Analysis Methods

The study employs machine learning models, notably the Support Vector Machine (SVM), the Random Forest (RF), and the Classification and Regression Tree (CART), to enhance the understanding and accuracy of LULC classification. Ground truthing was conducted using high-resolution base maps to validate the accuracy of the classifications, with a Kappa coefficient of 0.85 as the benchmark for acceptable agreement.

2.4. Study Process/Steps

2.4.1. Data Acquisition and Integration

We combined the processed satellite data with the digitized GIS data to create an integrated dataset, forming a comprehensive base for further analysis. GEE was used in this study as it offers a variety of tools and algorithms for image classification, change detection, time series analysis, and machine learning, making it an ideal platform for conducting comprehensive studies on land use and land cover changes, land surface temperature variations, and other environmental applications (Donchyts et al., 2016) [36]. The study used Land Cover Type 1 from the Annual International Geosphere-Biosphere Programme (Schneider et al., 2013; IGBP, 2022) [37,38]. LULC classes, measured in square kilometers, were derived from the MODIS data and aggregated from monthly to annual scales (Friedl et al., 2010) [39]. The observed seventeen LULC classes were compressed into seven classes critical for the baseline analysis (Congalton and Green, 2008) [34], as noted in Table 2.

Table 2. IGBP classifications of land cover and reclassification.

LC #	LC Description	Reclassification Description
1	Evergreen Needleleaf Forests: dominated by evergreen conifer trees (canopy > 2 m). Tree cover > 60%.	Combined into Forests
2	Evergreen Broadleaf Forests are dominated by evergreen broadleaf and palmate trees (canopy > 2 m). Tree cover > 60%.	Combined into Forests
3	Deciduous Needleleaf Forests: dominated by deciduous needleleaf (larch) trees (canopy > 2 m). Tree cover > 60%.	Combined into Forests
4	Deciduous Broadleaf Forests: dominated by deciduous broadleaf trees (canopy > 2 m). Tree cover > 60%.	Combined into Forests
5	Mixed Forests: dominated by neither deciduous nor evergreen (40–60% of each) tree type (canopy > 2 m). Tree cover > 60%.	Combined into Forests
6	Closed Shrublands: dominated by woody perennials (1–2 m height) >60% cover.	Reclassified and combined into Shrublands
7	Open Shrublands: dominated by woody perennials (1–2 m height) 10–60% cover.	Reclassified and combined into Shrublands
8	Woody Savannas: tree cover 30–60% (canopy > 2 m).	Reclassified and combined into Grasslands/Savannas
9	Savannas: tree cover 10–30% (canopy > 2 m).	Reclassified and combined into Grasslands/Savannas
10	Grasslands: dominated by herbaceous annuals (<2 m).	Reclassified and combined into Grasslands/Savannas

Table 2. Cont.

LC #	LC Description	Reclassification Description
11	Permanent Wetlands: permanently inundated lands with 30–60% water cover and >10% vegetated cover.	Combined into Waterbody
12	Croplands: at least 60% of area is cultivated cropland.	Combined into Agriculture
13	Urban and Built-up Lands: at least 30% impervious surface area including building materials, asphalt and vehicles.	Renamed as Built-up Land
14	Cropland/Natural Vegetation Mosaics: mosaics of small-scale cultivation 40–60% with natural tree, shrub, or herbaceous vegetation.	Combined into Agriculture
15	Permanent Snow and Ice: at least 60% of area is covered by snow and ice for at least 10 months of the year.	No Change
16	Barren: at least 60% of area is non-vegetated barren (sand, rock, soil) areas with less than 10% vegetation.	Renamed as Bareland
17	Water Bodies: at least 60% of area is covered by permanent waterbodies.	Combined into Waterbody

The research derived seven classes from the original seventeen by combining similar land covers, as noted in the "Reclassification Description" column. Table 3 shows the combined LULC classes and the associated area in the DMIC ZoI.

Table 3. Combined LULC classes and area in the DMIC ZoI.

LC #	LULC Class	Area in sq. km. and Percent of Area (%)
1	Agriculture	514,503.25 (79.44%)
2	Bareland	10,916.5 (1.69%)
3	Built-up land (including urban)	10,949.5 (1.69%)
4	Forest	37,446.25 (5.8%)
5	Grassland	55,838.5 (8.6%)
6	Shrubland	13,892.75 (2.15%)
7	Waterbody	4081.00 (0.63%)
		647,627.75 (100%)

2.4.2. Classifier Application

Three ML classifiers were applied to these seven classes to ensure the accuracy and reliability of the LULC classification results. These classifiers include SVM, RF, and CART. SVM is a supervised learning method for constructing optimal separating hyperplanes in multi-dimensional space. RF is an ensemble learning method that constructs multiple decision trees and combines their results to improve the overall prediction accuracy. CART is a decision tree-based classifier that recursively splits the dataset into subsets based on the most significant feature to minimize the Gini impurity.

2.4.3. Accuracy Assessment via Ground Truthing

Next, to validate the accuracy of the LULC classification, ground truthing was conducted using high-resolution base maps (Congalton and Green, 2008) [34]. This process involves comparing the classified satellite images with the reference data obtained from the ground or other reliable sources. Researchers evaluate the classification accuracy using measures like the Kappa coefficient (Jensen, 2005) [40], with a Kappa coefficient of 0.85 as the benchmark for acceptable agreement. Landis and Koch (1977) [41] and Congalton and Green (2008) [34] categorize a Kappa value above 0.80 as a high agreement and a value between 0.40 and 0.80 as moderate to substantial agreement. In this study, the Random Forest (RF) Kappa score of 0.88 surpassed those of SVM (0.86) and CART (0.84), leading to the selection of RF for further analysis.

2.5. Time-Series Trend Analysis

Using the LULC classified via RF, a time-series analysis [42] was run on the 2000–2021 period to detect trends in LULC changes. In the time-series trend analysis, we first applied the Mann–Kendall test to determine statistically significant trends in LULC, followed by Sen's slope method to quantify the magnitude of these trends. The study conducted trend analysis for LULC changes across the DMIC ZoI, its 14 IMD sub-divisions, and 145 districts, which include 24 smart cities and 121 other districts. The utilization of the Mann–Kendall trend test and Sen's slope estimator played a crucial role in the analysis of climatic trends, such as in identifying and measuring alterations in surface air temperatures, precipitation, and various ecological factors in diverse geographical areas. Notably, studies have shown their efficacy in investigating temperature and precipitation patterns in Iraq, uncovering statistically significant climatic fluctuations over time (Roboaa, 2015) [43]. Researchers utilized comparable approaches to assess temperature trends in Gombe State, addressing local climatic dynamics (Alhaji, 2018) [44]. The use of the Mann–Kendall test with innovative trend methodologies for water quality parameters in Turkey demonstrates the versatility and relevance of these techniques in environmental investigations (Kisi, 2014) [45].

Kendall's tau and Sen's slope (Sen, 1968) [46] are non-parametric statistical methods in time-series analysis, notably in environmental studies that handle non-normally distributed data or outliers. Kendall's tau assesses the strength and direction of the monotonic relationship between two variables. A significant p-value indicates a monotonic trend. The tau value suggests the strength and direction of the LULC change trend—a positive value indicates an increasing trend, and a negative value indicates a decreasing trend. The study further employs Sen's slope to estimate the magnitude of Kendall's tau trend. This value, representing the median of all slopes between pairs of time points, offers a robust estimate of the rate of change over time. Table 4 displays a p-value and associated probabilities for evaluating Kendall's tau significance. For each p-value noted in the table, a symbol (+, *, **, ***) has been assigned and used throughout Section 3.

Table 4. Alpha (p-value) symbols in Kendall's tau and associated probabilities used in the study.

Alpha (p-Value)	Symbols Used in Kendall's Tau Significance	Probability
0.10	For p-value <= 0.1, the symbol used is +	90.00%
0.05	For p-value <= 0.05, the symbol used is *	95.00%
0.01	For p-value <= 0.01, the symbol used is **	99.00%
0.001	For p-value <= 0.001, the symbol used is ***	99.90%

3. Results

3.1. Overall LULC Changes in the DMIC ZoI

Our analysis of land use and land cover changes over two decades has provided valuable insights into the dynamics of different land use classes. The findings in Table 5 highlight important trends in LULC changes. Graphs for Sen's slopes for the DMIC have been included in Supplementary Section A (Sen's slope graphs for the DMIC, smart cities, and other districts). This section summarizes the results and discusses the following questions:

Table 5. LULC changes during 2001–2021 in the DMIC ZoI.

LULC Classes	Current LULC Area in sq. km. as of 2021 (% of Total)	Kendall's tau (LULC Changes during 2001–2021)	p-Value (Significance Symbol) [1]	Sen's Slope in sq. km. per Year Change during 2001–2021
Agriculture	514,503.25 (79.44%)	0.676	<0.0001 (***)	3031.40
Bareland	10,916.5 (1.69)	−0.667	<0.0001 (***)	−203.76
Forest	10,949.5 (1.69%)	0.343	0.031 (*)	148.14

Table 5. *Cont.*

LULC Classes	Current LULC Area in sq. km. as of 2021 (% of Total)	Kendall's tau (LULC Changes during 2001–2021)	p-Value (Significance Symbol) [1]	Sen's Slope in sq. km. per Year Change during 2001–2021
Grassland	37,446.25 (5.8%)	−0.819	<0.0001 (***)	−1119.62
Shrubland	55,838.5 (8.6%)	−0.505	0.001 (***)	−1774.72
Built-up (including urban)	13,892.75 (2.15%)	1.000	<0.0001 (***)	21.48
Waterbody	4081 (0.63%)	0.619	<0.0001 (***)	55.73
	647,627.75 (100%)			

[1] For p-value <= 0.05, significance symbol = * (95% probability) and for p-value <= 0.001, significance symbol = *** (99.9% probability).

"What are the current LULC patterns within the DMIC, and how have they changed in the last two decades?"

3.1.1. Current LULC Patterns

In 2021, agricultural land is the largest LULC class with 514,503.25 sq. km, which is 79.44% of the total land mass of 647,627.75 sq. km. covered under the ZoI of the DMIC. Shrubland is the next-largest LULC class at 55,838.50 sq. km. (8.6% of total DMIC ZoI). Grassland is the third-largest LULC class with 37,446.25 sq. km. (5.78% of total DMIC ZoI). Built-up land with 13,892.75 sq. km. (2.15% of total DMIC ZoI), forests with 10,949.5 sq. km. (1.69% of total DMIC ZoI), and bareland with 10,916.5 sq. km. (1.69% of total DMIC ZoI) are the next-largest classes represented in the DMIC. The waterbody is the smallest LULC class with a total area of 4081.00 sq. km. (0.63% of total DMIC ZoI). The infrastructure developed under the DMIC program is estimated at a minimum of 3500 sq. km.—a significant addition of 25% over the current built-up area of 13,892.75 sq. km.

Therefore, agriculture dominates the current landscape of the DMIC. All other classes are significantly less and limited. Therefore, any greenfield infrastructural development must consider the paucity of natural land classes and the limited built-up land and waterbody areas.

3.1.2. Decadal Changes in LULC

Between 2001 and 2021, agriculture experienced a positive trend, with a Kendall's tau of 0.676***, indicating a moderate, statistically significant correlation with an increase of 3031.40 sq. km. per year in the DMIC ZoI. Built-up areas (including urban) were the only category to demonstrate a perfect positive trend, with a Kendall's tau of 1.0***, indicating a strong correlation that is statistically significant with an increase of 21.48 sq. km. per year. The waterbody exhibited a positive trend, with a Kendall's tau of 0.619***, indicating a moderate, statistically significant correlation with an increase of 55.73 sq. km. per year. Forest areas exhibited a positive trend, with a Kendall's tau of 0.343*, indicating a weak correlation that is statistically significant with an increase of 148.14 sq. km. per year.

During this period, bareland showed a negative trend, with a Kendall's tau of −0.667***, indicating a moderate correlation that is statistically significant with a reduction of −203.76 sq. km. per year. Grassland revealed a negative trend, with a Kendall's tau of −0.819***, indicating a strong correlation that is statistically significant with a reduction of −1119.62 sq. km. per year. Shrubland showed a negative trend, with a Kendall's tau of −0.505***, indicating a moderate correlation that is statistically significant with a reduction of −1774.72 sq. km. per year.

Notably, all observed changes except for the forest category had p-values well below 0.001, indicating highly significant trends. The considerable expansion in agricultural land highlights the substantial impact of human intervention on land use patterns. The statistically significant negative trend in grassland and shrubland suggests conversions to other land uses due to agricultural expansion and increased built-up areas.

The reduction in bareland, alongside the increase in waterbodies, may reflect changes in land management practices, climate change impacts, or both. These shifts are essential for understanding ecological balances, water resource management, and conservation efforts. The strong positive correlation observed in urban built-up areas, with a statistically significant *p*-value, underscores the pace of urbanization requiring careful urban planning and policy interventions to mitigate environmental impacts.

3.1.3. Land Use Transfer Matrix

In addition to the time-series analysis, a correlation test was run to understand how land cover types relate over time. This provides insights into land use patterns and land cover change dynamics.

Understanding the relationships between different LULC classes is crucial for sustainable land management and planning. This study examines the land use transfer matrix for various LULC classes within the DMIC to identify significant direct and inverse relationships. The LULC classes analyzed include forest, shrubland, grassland, waterbody, agriculture, built-up, and bareland, as shown in Table 6 for the entire DMIC at a 95% confidence level.

Table 6. Land use transfer matrix.

LULC Class	Forest	Shrubland	Grassland	Waterbody	Agriculture	Built-Up	Bareland
Forest	1	−0.095	−0.333	0.267	0.152	0.343	−0.390
Shrubland	−0.095	1	0.514	−0.619	−0.829	−0.505	0.495
Grassland	−0.333	0.514	1	−0.648	−0.686	−0.819	0.581
Waterbody	0.267	−0.619	−0.648	1	0.676	0.619	−0.610
Agriculture	0.152	−0.829	−0.686	0.676	1	0.676	−0.552
Built-up	0.343	−0.505	−0.819	0.619	0.676	1	−0.667
Bareland	−0.390	0.495	0.581	−0.610	−0.552	−0.667	1

The positive correlation between waterbody and agriculture (r = 0.676) indicates that regions with significant agricultural activities also have substantial waterbodies. This relationship highlights the importance of sustainable water management practices in agricultural areas to ensure water availability for crop production. The positive correlation between forests and built-up areas (r = 0.343) suggests a link between urban development in certain regions and forest conservation or afforestation efforts, albeit a weak association. This relationship might indicate a trend towards integrating green spaces within urban environments, focusing on sustainable urban planning practices.

Furthermore, the positive correlation between waterbody and built-up areas (r = 0.619) suggests that urban regions within the DMIC might be designed with waterbodies, possibly for aesthetic or recreational purposes. This trend may reflect efforts to incorporate natural water features into urban landscapes, contributing to improved urban livability and environmental quality. The positive correlation between grassland and bareland (r = 0.581) suggests that regions with significant grassland areas also tend to have considerable bareland. This relationship might indicate transitional zones where bareland is gradually converted to grassland or vice versa, underscoring the dynamic nature of land cover changes.

Conversely, the strong negative correlation between shrubland and agriculture (r = −0.829) indicates that an increase in agricultural activities is associated with a significant reduction in shrubland areas. This relationship underscores the pressure on shrubland ecosystems due to the expansion of agricultural lands, which can affect biodiversity and ecosystem services. Similarly, the strong negative correlation between grassland and built-up areas (r = −0.819) suggests that urban expansion is often at the expense of grassland. This highlights the competition between urban development and natural land covers, raising concerns about the loss of grassland habitats due to urbanization.

The negative correlation between grassland and waterbody (r = −0.648) indicates that increases in grassland areas are associated with decreases in waterbodies. This relationship might be due to land conversion practices affecting water retention or drainage patterns, emphasizing the need for integrated land and water management strategies. The negative correlation between bareland and agriculture (r = −0.552) suggests that areas with higher agricultural activities tend to have less bareland, reflecting the conversion of bareland to cultivated fields. This relationship indicates active land use changes aimed at increasing agricultural productivity but also points to potential land degradation issues if not managed sustainably.

Additionally, the negative correlation between shrubland and waterbody (r = −0.619) suggests that increased waterbodies are associated with reduced shrubland areas. This relationship could indicate the inundation of shrubland areas or changes in land use priorities that favor water retention and management over maintaining shrubland ecosystems.

3.2. LULC Changes in the IMD Sub-Divisions

The Sen's slope for each LULC class within various IMD sub-divisions reflects the same general growth or decline patterns as the overall DMIC ZoI. Table 7 documents the LULC changes at the IMD sub-division level, where Kendall's tau shows the statistical significance for each Sen's slope listed. This section provides a summary of the results and discusses the following question:

Table 7. IMD sub-divisions and Sen's slope for LULC changes between 2001 and 2021.

		Sen's Slope for LULC Changes in sq. km. per Year [1]						
No.	**IMD Sub-Division** [2]	**Agriculture**	**Bareland**	**Forest**	**Grassland**	**Shrub-Land**	**Built-Up**	**Waterbody**
1	Chandigarh, Haryana, and Delhi	64.33	-	1.07	(34.60)	(33.44)	7.50	0.08
2	East Madhya Pradesh	4.69	(0.45)	-	(4.50)	-	-	0.29
3	East Rajasthan	599.14	(1.64)	8.46	(288.24)	(317.25)	2.27	6.35
4	East Uttar Pradesh	0.29	(0.06)	-	(0.25)	-	-	-
5	Gujarat Region	423.85	(7.14)	35.48	(253.98)	(177.06)	3.08	15.37
6	Konkan and Goa	(151.05)	(0.64)	56.47	85.95	-	3.05	4.65
7	Madhya Maharashtra	123.84	(7.04)	41.12	(172.29)	(4.37)	2.19	5.71
8	Marathwada	0.14	-	-	(0.15)	-	-	-
9	Punjab	(0.03)	-	-	0.02	(0.03)	-	-
10	Saurashtra and Kachh	361.18	(48.44)	8.72	(187.55)	(117.66)	0.48	22.08
11	Uttarakhand	(25.28)	(0.06)	24.72	5.21	-	1.61	0.74
12	West Madhya Pradesh	38.50	(0.24)	0.67	(37.36)	(2.13)	0.07	0.17
13	West Rajasthan	1593.31	(137.90)	(28.57)	(232.98)	(1122.79)	0.93	0.05
14	West Uttar Pradesh	(3.29)	(2.80)	(3.42)	2.15	-	5.92	0.12

[1] Sen's slopes for LULC changes in sq. km. per year shown in the table are associated with p-values of lesser than 0.1 (90% to 99.90% probability), [2] Dadra and Nagar Haveli is a Union Territory within the DMIC zone of influence. This Union Territory was also modeled as it has two smart cities, Diu and Silvassa. The Sen's slope for LULC changes for Dadra and Nagar Haveli for agriculture is (1.51) sq. km. per year, for bareland is (0.15) sq. km. per year, for grassland is 1.11 sq. km. per year, for built-up land is 0.31 sq. km. per year, and for waterbody is 0.23 sq. km. per year.

"How have the IMD sub-divisions within the DMIC experienced changes in LULC?"

Highlights for each of the IMD sub-division LULC changes are noted below:

1. Chandigarh, Haryana, and Delhi: Noteworthy is the increase in agriculture with a Sen's slope of 64.33 sq. km. per year. Concurrently, grassland and shrubland decreased at a rate of −34.60 sq. km. per year and −33.44 sq. km. per year, respectively. The area also witnessed the highest increase in built-up land at 7.50 sq. km. per year.

2. East Madhya Pradesh: Noteworthy is the increase in agriculture, with a Sen's slope of 4.69 sq. km. per year, and a concurrent decrease in grassland, at a rate of −4.50 sq. km. per year.
3. East Rajasthan: This IMD sub-division witnessed the second-highest increase in agriculture, with a Sen's slope of 599.15 sq. km. per year. The data show the third-highest increase in waterbodies among all the IMD sub-divisions, at 6.35 sq. km. per year.
4. East Uttar Pradesh: There are minor increases in agriculture of 0.29 sq. km. per year, with decreases in bareland of −0.06 sq. km. per year and grassland of −0.25 sq. km. per year.
5. Gujarat Region: This IMD sub-division witnessed the third-highest increase in agriculture, with a Sen's slope of 423.85 sq. km. per year. The study notes significant decrease in grassland (253.98 sq. km. per year) and shrubland (−177.06 sq. km. per year). The Gujarat Region also has the second-highest increase in waterbodies among all the IMD sub-divisions, at 15.37 sq. km. per year.
6. Konkan and Goa: Amongst all IMD sub-divisions, Konkan and Goa exhibited the highest increase in forest with a Sen's slope of 56.47 sq. km. per year with the highest decrease in agriculture at −151.05 sq. km. per year.
7. Madhya Maharashtra: Agriculture has increased by 123.84 sq. km. per year, with a concurrent decrease in grassland of −172.49 sq. km. per year. It also witnessed the second-highest increase in forest, with a Sen's slope of 41.12 sq. km. per year
8. Marathwada: A minor increase in agriculture with a Sen's slope of 0.14 sq. km. per year was offset by a decrease of −0.15 in grassland for this IMD sub-division.
9. Punjab: There are minor increases in agriculture of 0.03 sq. km. per year, with a concurrent decrease in shrubland of −0.03 sq. km. per year.
10. Saurashtra and Kachh: The data show a considerable increase in agriculture of 361.18 sq. km. per year and significant decreases in grassland by −187.55 sq. km. and shrubland by −117.66 sq. km. per year. The data show the highest increase in waterbodies among all the IMD sub-divisions, at 22.08 sq. km. per year.
11. Uttarakhand: The data show the second-highest increase in forest among all the IMD sub-divisions at 24.72 sq. km. per year. Concurrently, the second-highest decrease in agriculture among all IMD sub-divisions is −25.28 sq. km. per year.
12. West Madhya Pradesh: Agriculture sees modest increases, at 38.50 sq. km. per year. Concurrently, grassland decreases by −37.36 sq. km. per year.
13. West Rajasthan: Amongst the IMD sub-divisions, West Rajasthan shows the highest increase in agriculture at 1593.31 sq. km. per year. The area also witnessed the highest decreases in shrubland at −1122.79 sq. km. per year, a decrease in grassland at −232.98 sq. km. per year, and a decrease in bareland at −137.90 sq. km. per year.
14. West Uttar Pradesh: There is a minor decrease in agriculture of −3.29 sq. km. per year and forests of 3.42 sq. km. per year. The area also witnessed the second-highest increase in built-up land at 5.92 sq. km. per year.

A substantial agricultural expansion was witnessed in West Rajasthan during 2001–2021, potentially reflecting successful agricultural policies or shifts in land use, but this comes with losses in shrubland. East Rajasthan, the Gujarat Region, and Saurashtra and Kachh are other IMD sub-divisions with notable higher-than-average increases in agricultural land. A decline in natural land covers (grassland and shrubland) with a higher-than-average growth in agricultural land in the Gujarat Region may reflect the impact of human activities and necessitate measures for ecological conservation. The Gujarat Region is also an area with the second-highest increase in waterbody due to effective water resource management or changing patterns in precipitation and hydrology.

Urban development remained fairly stable in most IMD sub-divisions, with the Chandigarh, Haryana, Delhi, and West Uttar Pradesh regions registering a higher-than-average increase in built-up land. The dominance of rural areas within the DMIC ZoI and the 500 m spatial resolution of MODIS data may necessitate further research with higher resolutions to accurately determine the extent of urban LULC changes.

The results suggest that while some IMD sub-divisions have experienced significant agricultural growth, others show a decline in natural land covers, indicating potential overuse or conversion to other land uses. There is limited evidence of planned conservation and ecological maintenance. These findings highlight the need for a regional strategy to balance economic development with environmental conservation within the DMIC.

3.3. LULC Changes in the Smart Cities and the District Level

The study area encompasses 121 districts and 24 smart cities. The study conducted further analysis at the district level for each of the seven LULC classes, identifying smart cities and districts with notable LULC changes. The smart cities and districts with notable changes in LULC are identified. The complete list of 121 districts and 24 smart cities, along with the associated Kendall's tau, *p*-value, and Sen's slope, can be found in Supplementary Section B (Summary of time series for smart cities and other districts-Table S1. Summary of Kendall's tau and Sen's slope for time series analysis of decadal LULC changes for smart cities in the DMIC, and Table S2. Summary of Kendall's tau and Sen's slope for time series analysis of decadal LULC changes for districts in the DMIC). This section summarizes the results and discusses the following questions:

"How do changes in LULC in smart cities compare with other districts within the DMIC?"

3.3.1. Changes in Agricultural LULC

Agricultural land is the largest class in the area (514,503.25 sq. km. or 79.44% of DMIC ZoI), and changes in this LULC can significantly impact regional economic development and ecological balance. In general, throughout the DMIC ZoI, the agricultural land increased in size, with some exceptional decreases in smart cities or other districts. The change was throughout the study area. Details of the findings on smart cities and other districts are stated below.

Smart Cities: A significant positive trend in agricultural land change was observed in the analysis of smart cities within the DMIC. The top two cities belonging to the IMD sub-division of the Gujarat Region with the most substantial positive trends were Ahmedabad, with a Kendall's tau of 0.92*** and a Sen's slope of 139.60 sq. km. per year, and Vadodara, with a Kendall's tau of 0.86*** and a Sen's slope of 51.45 sq. km. per year. These districts displayed strong positive correlations between time and agricultural land area increase, signaling the robust growth of farming from 2001 to 2021.

Conversely, Thane in the Konkan and Goa IMD sub-division, with a Kendall's tau of −0.80*** and a Sen's slope of −36.07 sq. km. per year, showed a significant decline in agricultural land.

Other Districts: Among other districts within the study area, the top districts exhibiting positive trends in agricultural land change were Churu, with a Kendall's tau of 0.60*** and a Sen's slope of 437.74 sq. km. per year; Jodhpur, with a Kendall's tau of 0.52*** and a Sen's slope of 403.42 sq. km. per year; and Nagaur, with a Kendall's tau of 0.62*** and a Sen's slope of 315.85 sq. km. per year. These figures indicate considerable increases in agricultural land, with strong statistical significance.

On the other end of the spectrum, the districts with negative trends were Raigarh, with a Kendall's tau of −0.88*** and Sen's slope of −69.33 sq. km. per year; Ratnagiri, with a Kendall's tau of −0.86*** and a Sen's slope of −44.58 sq. km. per year; and Valsad, with a Kendall's tau of −0.75*** and a Sen's slope of −12.00 sq. km. per year.

The overall trends suggest that while some districts within the DMIC are experiencing growth in agricultural land, others face challenges that may lead to stagnation or decline. Figure 3 shows the changes in agricultural land area in contiguous districts (for example, West Rajasthan) within the individual IMD sub-divisions, with some exceptions. The factors driving these changes may vary significantly across the regions, and further investigation is required to understand these trends' underlying causes and develop region-specific strategies for sustainable land use management.

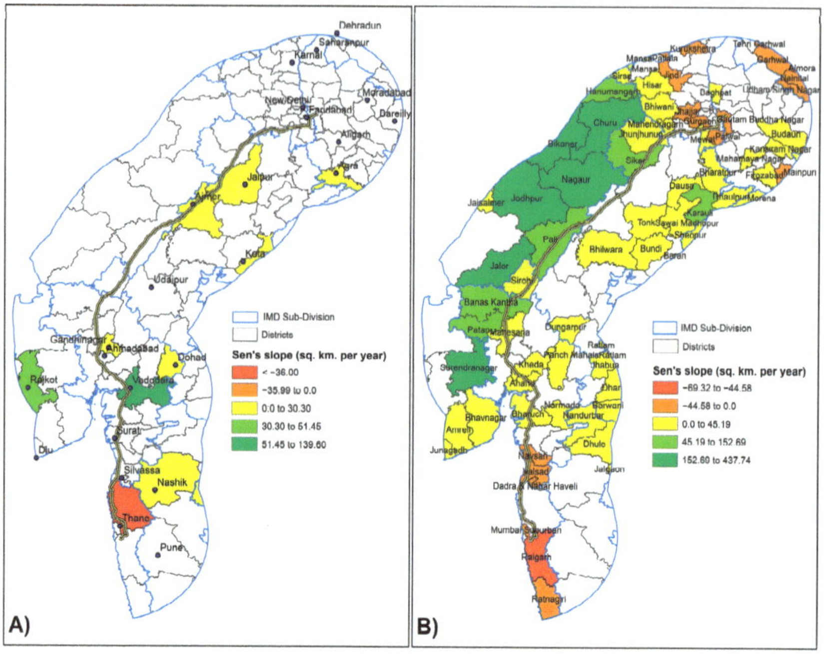

Figure 3. Agricultural LULC changes between 2000 and 2021 in the DMIC ZoI—(**A**) Sen's Slope—Agriculture LULC Changes in Smart Cities. (**B**) Sen's Slope—Agriculture LULC Changes in Other Districts.

3.3.2. Changes in Bareland LULC

Bareland is one of the smallest classes in area size (10,916.5 sq. km. or 1.69% of the DMIC ZoI), and changes in this LULC can indicate conversions to other LULCs, such as urban or agricultural land. Throughout the DMIC ZoI, the bareland areas decreased in size, with lesser decreases in smart cities and more decreases in other districts. Across the region, the changes in bareland were heterogenous, as noted in Figure 4. Details of the findings on smart cities and other districts are stated below.

Smart Cities: For the smart cities, the trend in the bareland LULC class showed a significant decline. The top three districts with the most pronounced negative trends, as indicated by Kendall's tau with corresponding *p*-values, were Ahmedabad, with a Kendall's tau of −0.60*** and a Sen's slope of −3.54 sq. km. per year; Pune, with a Kendall's tau of −0.66*** and a Sen's slope of −2.83 sq. km. per year; and Nashik, with a Kendall's tau of −0.58** and a Sen's slope of −2.48 sq. km. per year. All other smart cities showed reductions in bareland.

Other Districts. The analysis of other districts within the DMIC also showed a higher decline in bareland areas than in smart cities. The top three districts with significant negative trends were Jodhpur, with a Kendall's tau of −0.54*** and a Sen's slope of −96.97 sq. km. per year; Kachchh, with a Kendall's tau of −0.56*** and a Sen's slope of −44.10 sq. km. per year; and Bikaner with a Kendall's tau of −0.64*** and a Sen's slope of −37.63 sq. km. per year. Unlike the smart cities, which had only declined, some districts showed positive trends in bareland areas. These include Alirajpur, with a Kendall's tau of 0.44** and a Sen's slope of 0.20 sq. km. per year; Neemuch, with a Kendall's tau of 0.48** and a Sen's slope of 0.11 sq. km. per year; and Pali, with a Kendall's tau of 0.47** and a Sen's slope of 0.10 sq. km. per year.

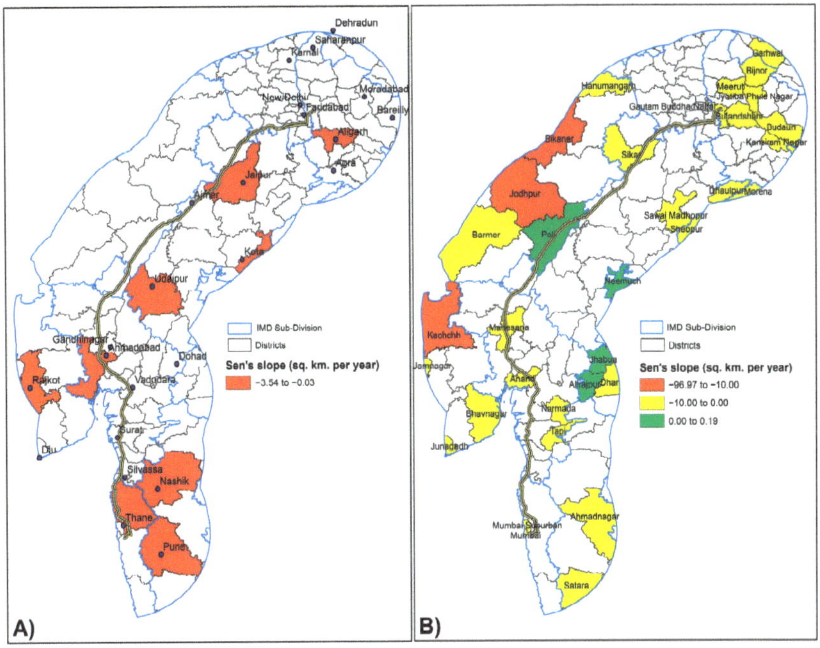

Figure 4. Bareland LULC changes between 2000 and 2021 in the DMIC ZoI—(**A**) Sen's Slope—Bareland LULC Changes in Smart Cities. (**B**) Sen's Slope—Bareland LULC Changes in Other Districts.

The decline in bareland in urban agglomerations such as Thane, Pune, and Nashik (all smart cities) is contiguous, as seen in Figure 4.

3.3.3. Changes in Forest LULC

Forests are one of the smallest classes in area size (37,446.25 sq. km. or 1.69% of the DMIC ZoI), and changes in this LULC can have critical implications on biodiversity and carbon sequestration. The forest areas increased in size in most of the smart cities but with substantial decreases in certain of the other districts of the DMIC ZoI. Figure 5 shows the distinction between smart cities and other districts regarding changes in forests. Details of the findings on smart cities and other districts are noted below.

Smart Cities: This loss of forest cover has implications for biodiversity and carbon sequestration. For the smart cities in the DMIC, forest area changes from 2001 to 2021 indicate increases but with varying trends. The top three districts with the greatest increases in forest area, as denoted by Kendall's tau and Sen's slope, are Pune, with a Kendall's tau of 0.600*** and Sen's slope of 28.46 sq. km. per year; Thane, with a Kendall's tau of 0.562*** and a Sen's slope of 15.60 sq. km. per year, suggesting notable afforestation or reforestation activities; and Surat, with a Kendall's tau of 0.752*** and a Sen's slope of 12.62 sq. km. per year, which could reflect successful environmental policies or natural forest regeneration.

Other Districts: Within the DMIC, other districts exhibit the largest increases in forest areas, notably Raigarh, with a Kendall's tau of 0.79*** and a Sen's slope of 21.25 sq. km. per year; Ratnagiri, with a Kendall's tau of 0.75*** and a Sen's slope of 18.97 sq. km. per year; and Garhwal, with a Kendall's tau of 0.84*** and a Sen's slope of 16.74 sq. km. per year.

These districts exhibit significant positive trends indicative of effective forest conservation strategies or natural reforestation.

On the contrary, the districts with the greatest decline in forest areas are Nagaur, with a Kendall's tau of −0.48** and a Sen's slope of −30.32 sq. km. per year; Anand, with a

Kendall's tau of −0.59*** and Sen's slope of −3.92 sq. km. per year; and Baghpat, with a Kendall's tau of −0.60*** and a Sen's slope of −3.45 sq. km. per year.

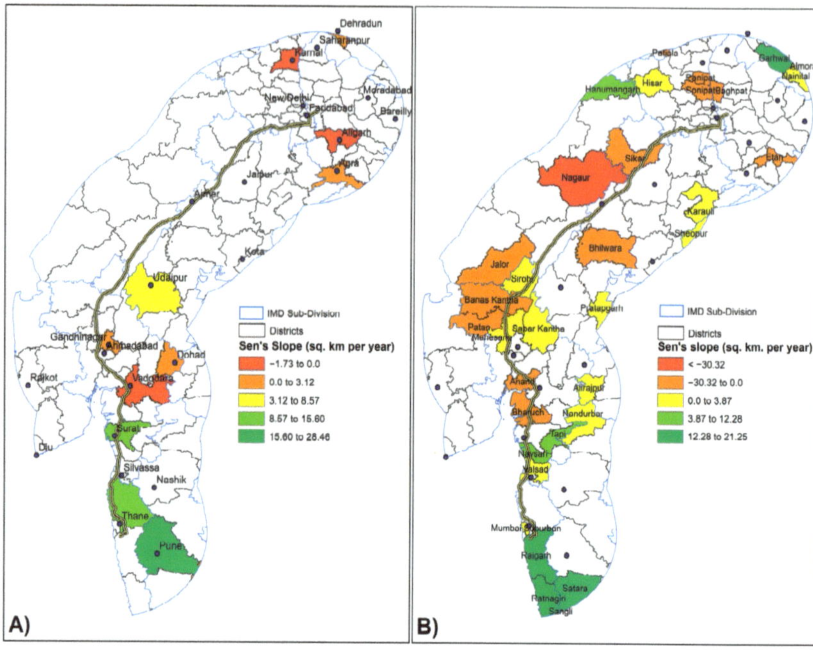

Figure 5. Forest LULC changes between 2000 and 2021 in the DMIC ZoI—(**A**) Sen's SlopeForest LULC Changes in Smart Cities. (**B**) Sen's Slope—Forest LULC Changes in Other Districts.

These reductions may be due to several factors, including land use change, agriculture-driven deforestation, or urban development pressures.

The reductions in forests are heterogeneous, as seen in Figure 5. However, the southern part of the DMIC, which includes the smart cities of Thane and Pune and the districts Raigarh, Ratnagiri, and Satara, has homogeneous increases in forests. Expanding forest areas in these districts demonstrates the potential for ecological recovery and the success of green initiatives. Conversely, the decline in other regions emphasizes the need for enhanced conservation efforts and the careful evaluation of land use policies to mitigate the loss of forested lands.

3.3.4. Changes in Grassland LULC

After agricultural land and shrubland, grassland is the third-largest class in the area (55,838.5 sq. km. or 5.78% of DMIC ZoI), and changes in this LULC can significantly impact the alteration of ecosystems. In general, throughout the DMIC ZoI, the grassland areas decreased with some exceptions. Figure 6 shows the heterogeneous spread of the grassland changes. Details of the findings on smart cities and other districts are noted below.

Smart Cities: For the smart cities, the greatest increases in grassland include Thane, with a Kendall's tau of 0.72*** and a Sen's slope of +20.80 sq. km. per year; Karnal, with a Kendall's tau of 0.58*** and a Sen's slope of +0.88 sq. km. per year; and Moradabad, with a Kendall's tau of 0.43** and a Sen's slope of +0.86 sq. km. per year. Notably, Sen's slope drops from Thane to Karnal by a significant margin.

The greatest declines in grassland are in Ahmadabad, with a Kendall's tau of −0.81*** and a Sen's slope of −96.91 sq. km. per year; Rajkot, with a Kendall's tau of −0.82*** and a Sen's slope of −51.15 sq. km. per year; and Vadodara, with a Kendall's tau of −0.87*** and a Sen's slope of −49.88 sq. km. per year.

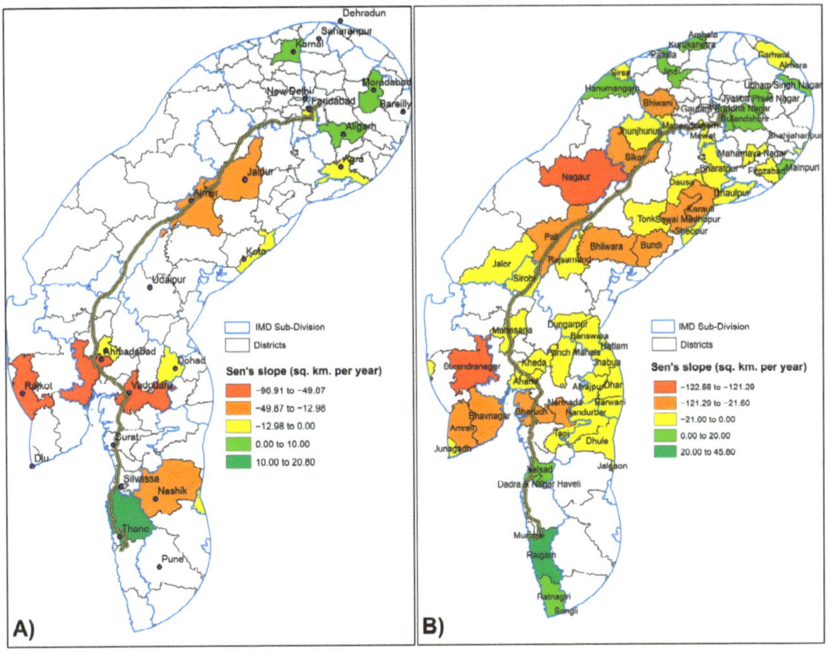

Figure 6. Grassland LULC changes between 2000 and 2021 in the DMIC ZoI—(**A**) Sen's Slope—Grassland LULC Changes in Smart Cities. (**B**) Sen's Slope—Grassland LULC Changes in Other Districts.

Other Districts: The different districts with the greatest increases in grassland include Raigarh, with a Kendall's tau of 0.86*** and Sen's slope of +45.80 sq. km. per year; Hanumangarh, with a Kendall's tau of 0.44** and a Sen's slope of +19.41 sq. km. per year; and Ratnagiri, with a Kendall's tau of 0.56*** and a Sen's slope of +19.38 sq. km. per year.

For the other districts, the greatest decreases in grassland include Nagaur, with a Kendall's tau of −0.71*** and a Sen's slope of −122.56 sq. km. per year; Surendranagar, with a Kendall's tau of −0.91*** and a Sen's slope of −121.29 sq. km. per year; and Bharuch, with a Kendall's tau of −0.93*** and a Sen's slope of −44.96 sq. km. per year.

The reduction in grassland areas and increases in agricultural land point to a significant transformation of natural and semi-natural landscapes, raising concerns about the loss of biodiversity and alteration of ecosystems.

3.3.5. Changes in Shrubland LULC

After agricultural land, shrubland is the largest class in the area (13,892.75 sq. km. or 8.62% of DMIC ZoI), and changes in this LULC can significantly impact local biodiversity, soil stability, and microclimate regulation. Throughout the DMIC ZoI, shrubland areas had the highest decreases in most districts. Figure 7 shows the homogeneous spread of the shrubland changes. Details of the findings on smart cities and other districts are noted below.

Smart Cities: In smart cities, Ahmadabad, Jaipur, and Ajmer exhibit decreases in shrubland, with a Kendall's tau of −0.67*** and a Sen's slope of −35.43 sq. km. per year for Ahmadabad; −0.67*** and −16.50 sq. km. per year for Jaipur; and −0.68*** and −8.27 sq. km. per year for Ajmer.

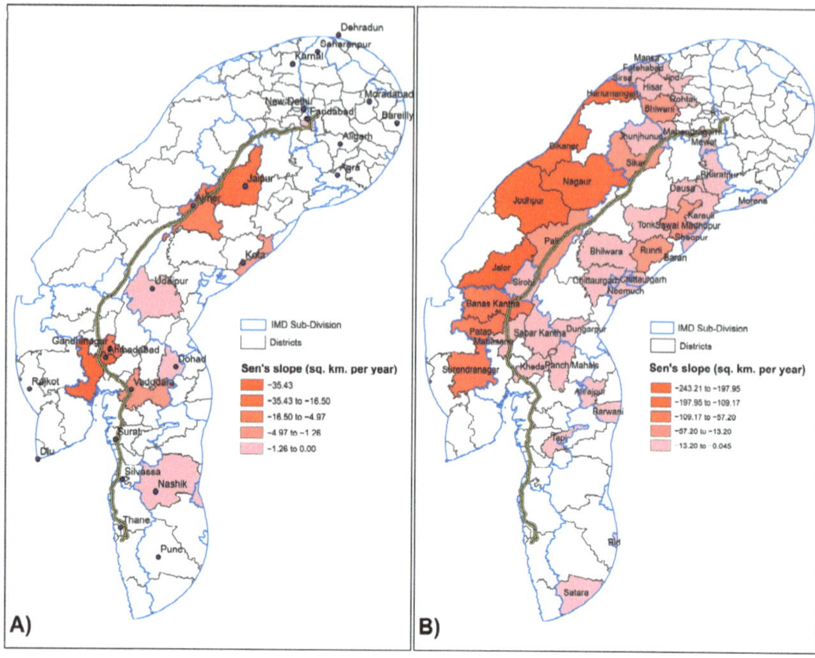

Figure 7. Shrubland LULC changes between 2000 and 2021 in the DMIC ZoI—(**A**) Sen's Slope—Shrubland LULC Changes in Smart Cities. (**B**) Sen's Slope—Shrubland LULC Changes in Other Districts.

Other Districts: Jind displayed the sole increase in shrubland, marked by a Kendall's tau of 0.70*** and a modest Sen's slope of 0.045 sq. km. per year. The districts of Mewat, Tapi, Morena, and Bid indicated no change in the Sen's slope for shrubland. The decreases in shrubland, on the other hand, were fairly substantial: Jodhpur had a Kendall's tau of −0.42** and Sen's slope of −243.21 sq. km. per year; Jalor had a Kendall's tau of −0.61*** and a Sen's slope of −197.94 sq. km. per year; and Nagaur had a Kendall's tau of −0.50** and a Sen's slope of −159.88 sq. km. per year. This rounded out the list of highest decreases in shrubland in other districts of the DMIC ZoI.

As seen in Figure 7, the shrubland reductions are homogeneous within the IMD subdivisions of West Rajasthan, East Rajasthan, and the Gujarat Region. These shrubland changes may affect local biodiversity, soil stability, and microclimate regulation. The transformation of shrublands raises concerns about the loss of these unique ecosystems, which often serve as important habitats for wildlife.

3.3.6. Changes in Built-Up LULC

Built-up land (including urban areas) is one of the smallest classes in area size (10,949.5 sq. km. or 2.15% of the DMIC ZoI). Throughout the DMIC ZoI, the urban areas increased, with no decreases in smart cities or other districts. Figure 8 illustrates the increase in homogeneity across built-up land, marked by green-shaded areas in small increments. The detailed findings on smart cities and other districts are noted below.

Smart Cities: The Sen's slope data for urban and built-up land LULC changes in the DMIC region show an increase in urbanization, particularly in districts near cities such as Thane, Pune, Surat, Ahmedabad, Faridabad, and Jaipur. Built-up areas for any smart cities within the DMIC ZoI did not decrease.

The highest increases were Thane, with a Kendall's tau of 1.00*** and Sen's slope of 2.11 sq. km. per year; Pune, with a Kendall's tau of 0.99*** and a Sen's slope of 1.91 sq. km. per year; and Surat, with a Kendall's tau of 0.99*** and a Sen's slope of 1.16 sq. km. per year.

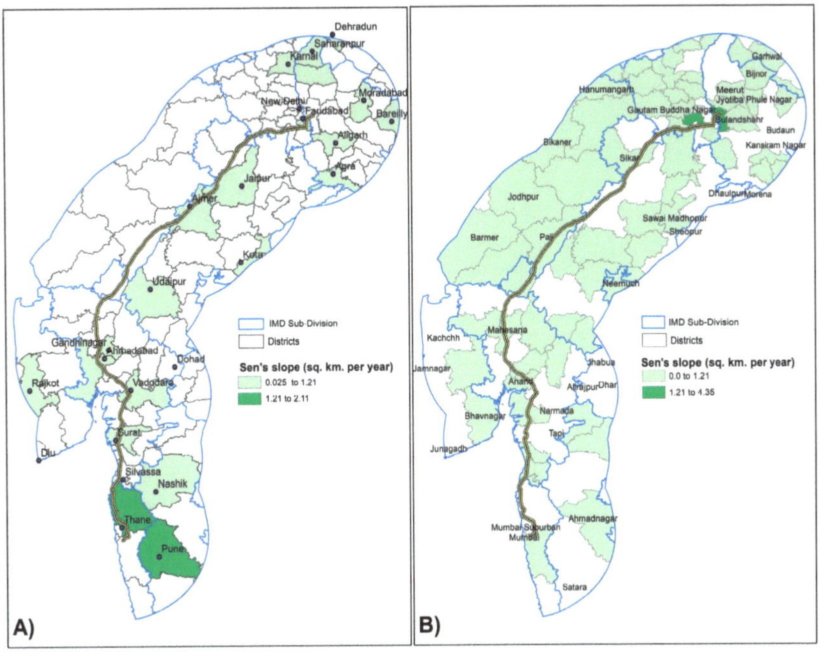

Figure 8. Built-up LULC changes between 2000 and 2021 in the DMIC ZoI—(**A**) Sen's Slope—Built-Up LULC Changes in Smart Cities. (**B**) Sen's Slope—Built-Up LULC Changes in Other Districts.

Other Districts: The other districts saw larger increases in built-up land than the smart cities, e.g., Gautam Buddha Nagar, with a Kendall's tau of 0.99*** and a Sen's slope of 4.35 sq. km. per year; Gurgaon, with a Kendall's tau of 0.99*** and a Sen's slope of 2.38 sq. km. per year; and Udham Singh Nagar, with a Kendall's tau of 0.97*** and a Sen's slope of 1.21 sq. km. per year. Districts like Gandhinagar, Udaipur, and Bareilly suggest a more moderated or planned urban growth due to their low Sen's slopes yet significant Kendall's tau results. The negligible changes in areas like North Delhi and South Delhi might reflect development saturation. Also, the expansion of built-up areas in urban settings is typically in brownfield areas and may be lesser than the minimum mapping unit for MODIS land cover products.

Hence, further research is recommended using the highest available imagery (30 m or higher resolution).

3.3.7. Changes in Waterbody LULC

Waterbodies are the smallest class in area size (4081 sq. km. or 0.63% of the DMIC ZoI). Throughout the DMIC ZoI, the waterbody LULC class increased, with no decreases in the smart cities or other districts. The increase can be seen in Figure 9 and is noteworthy as the upper half of the DMIC ZoI seems to have no districts with a statistically significant (p-value < 0.1) growth, except for Jhajjar. A notable observation is the correlation between the increase in urban and built-up areas and waterbody areas in several adjacent districts in the lower half of the DMIC ZoI. Details of the findings on smart cities and other districts are noted below.

Smart Cities: Among the smart cities were Pune, with a Kendall's tau of 0.47** and a Sen's slope of 2.56 sq. km. per year; Thane, with a Kendall's tau of 0.47** and a Sen's slope of 2.41 sq. km. per year; and Ahmadabad, with a Kendall's tau of 0.76*** and a Sen's slope of 2.31 sq. km. per year.

Other Districts: Other districts included Kachchh, with a Kendall's tau of 0.50** and a Sen's slope of 15.32 sq. km. per year; Bhavnagar, with a Kendall's tau of 0.94*** and a

Sen's slope of 4.80 sq. km. per year; and Panch Mahals, with a Kendall's tau of 0.70*** and a Sen's slope of 2.71 sq. km. per year.

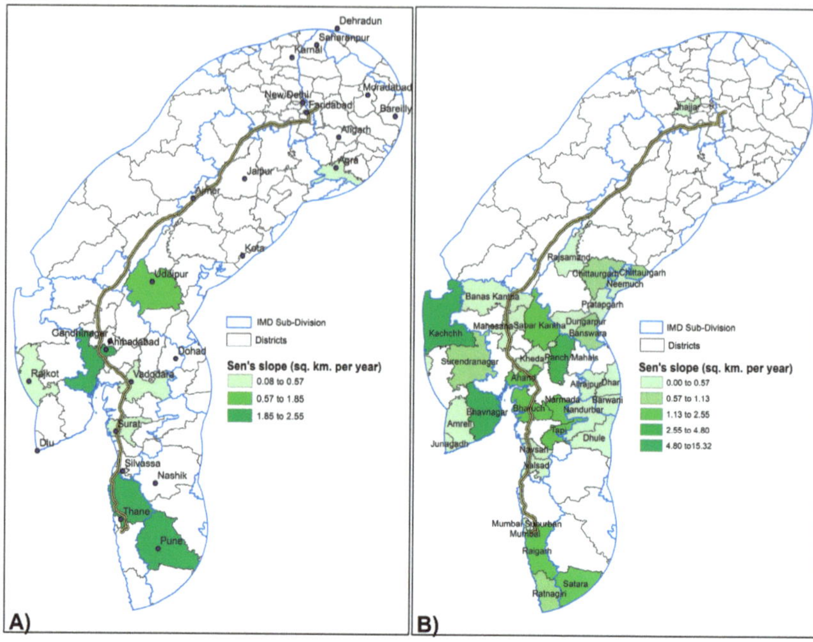

Figure 9. Waterbody LULC changes between 2000 and 2021 in the DMIC ZoI—(**A**) Sen's Slope—Waterbody LULC Changes in Smart Cities. (**B**) Sen's Slope—Waterbody LULC Changes in Other Districts.

The changes in waterbody areas across the smart cities and other districts within the DMIC from 2001 to 2021 reveal distinct patterns. Pune and Thane, with moderate increases in waterbody areas, might reflect efforts in water conservation or the result of seasonal variability captured over the years. Ahmadabad's higher Sen's slope suggests improved water management strategies.

Large-scale water conservation projects or changes in land use practices that favor waterbody expansion could explain the notable increases in waterbody areas in Kachchh and Bhavnagar.

4. Discussions

4.1. Implications for Smart Cities and Planned Industrial Clusters

4.1.1. Smart Cities

Smart cities, such as Ahmedabad (Gujarat Region), Jaipur (East Rajasthan), and Pune (Madhya Maharashtra), show significant positive trends in urban growth. This rapid urbanization, while driving economic development and infrastructure improvement, also poses challenges such as increased pollution, higher energy consumption, and pressure on existing urban services. For example, Pune's substantial urban sprawl often leads to encroachment on agricultural and natural lands, impacting local biodiversity and water resources. The rapid expansion of urban areas reduces green spaces, affecting the local climate and air quality and increasing the urban heat island effect.

Many cities, including Agra (West Uttar Pradesh), Ahmedabad (Gujarat Region), and Nashik (Madhya Maharashtra), show significant increases in agricultural land. While this boosts food production and supports local economies, it often comes at the cost of natural habitats. In Ahmedabad, agriculture has expanded significantly, converting natural ecosystems into farmlands. This expansion can lead to soil degradation, water scarcity due to increased irrigation demands, and biodiversity loss as native species are displaced by monocultures. Reducing natural habitats such as wetlands and forests can also disrupt local water cycles, exacerbating the effects of droughts and floods.

The decline in natural land cover, particularly forests and shrublands, is notable in cities like Faridabad (Chandigarh, Haryana, Delhi IMD sub-division), Gandhinagar (Gujarat Region), and Jaipur (East Rajasthan). Gandhinagar, for instance, has experienced a marked decrease in shrubland. This loss of natural habitats disrupts local flora and fauna, reducing biodiversity. Shrinking green spaces affect ecosystem services, climate regulation, and recreational spaces for urban populations.

4.1.2. Industrial Clusters and Industrial Areas Planned for the DMIC

The DMIC encompasses several projects that may significantly impact natural land cover. Here, we analyze five such projects based on current LULC changes and recommend whether they should proceed as greenfield or brownfield developments to minimize environmental impacts:

Faridabad-Palwal (Chandigarh, Haryana, Delhi IMD sub-division): The baseline conditions in this area indicate a substantial reduction in forest cover and shrubland, accompanied by increased urbanization and industrial activities. The decrease in natural habitats has fragmented ecosystems and increased human-wildlife conflict, reducing biodiversity and essential ecosystem services like air purification. Given the extensive urban and industrial infrastructure, the Faridabad-Palwal project, under the auspices of the DMIC, should proceed as a brownfield development, focusing on revitalizing degraded areas. Implementing green belts and reforestation projects will help mitigate the adverse environmental impacts and enhance local biodiversity.

Jaipur-Dausa (East Rajasthan IMD sub-division): The Jaipur-Dausa project area has experienced significant declines in shrubland and savanna, driven by expanding agriculture and urban areas since 2001, as indicated in the study. These changes have disrupted local flora and fauna, especially species dependent on these habitats. This project should be approached as a brownfield initiative to minimize further environmental degradation and repurposing existing agricultural lands and urban spaces. Establishing green corridors and incorporating sustainable urban planning can help maintain ecological balance and support biodiversity conservation efforts.

Vadodara-Ankleshwar (Gujarat Region IMD sub-division): The area has reduced natural waterbodies and forest cover due to increased industrial and agricultural activities. These changes have impacted aquatic ecosystems and terrestrial biodiversity, with industrial activities contributing to pollution. Given the existing industrial landscape, a brownfield development approach is recommended. This will limit the conversion of remaining natural habitats, and incorporating advanced water management systems and pollution control measures will be essential to mitigate environmental impacts and protect local ecosystems.

Nashik-Sinnar (Madhya Maharashtra IMD sub-division): Urban expansion and agricultural development in the Nashik-Sinnar project area have significantly declined forest and shrubland areas. This loss of natural habitats has led to deforestation, reduced biodiversity, and exacerbated the urban heat island effect. The project should be developed as a brownfield initiative, focusing on existing urban and agricultural lands to prevent further habitat loss. Integrating extensive green spaces and sustainable land management practices within the urban planning framework will mitigate environmental impacts and enhance the area's ecological resilience.

Pali-Marwar (West Rajasthan IMD sub-division): The area has extensive agricultural expansion and decreased natural forest and shrubland, with urban growth further impacting these habitats. The conversion of natural lands to agricultural and urban uses has disrupted local ecosystems and reduced biodiversity. The Pali-Marwar project should be developed as a brownfield initiative, utilizing converted agricultural and urban lands to minimize additional habitat loss. Implementing sustainable agricultural practices and creating conservation areas within the development plan will help preserve remaining natural habitats and support local biodiversity.

The assessment of these DMIC projects highlights the significant negative impacts on natural land cover due to LULC changes. Adopting brownfield development strategies for these projects will help minimize further environmental degradation. By focusing on sustainable urban planning, green infrastructure, and conservation efforts tailored to each project's unique context, these developments can balance economic growth with ecological preservation, ensuring long-term environmental sustainability and resilience.

The LULC changes within the DMIC highlight a clear trend of the depletion of natural land cover due to anthropogenic activities. The observed trends call for urgent policy interventions to mitigate the adverse impacts on biodiversity and ecosystem services. Strategies such as implementing conservation corridors, enforcing land-use regulations, and promoting sustainable agriculture practices are essential to preserve the remaining natural habitats and ensure the long-term ecological health of the DMIC region.

4.2. Ecologically Sensitive Areas

The DMIC encompasses several ESAs in the 121 other districts, where the analysis of LULC changes reveals significant trends in the depletion of natural resources due to anthropogenic expansions. A complete list of the ESAs, their profiles, locations, and LULC changes is included in Supplementary Section C (Table S3 Summary of time series for Ecologically Sensitive Areas in the DMIC). This study utilizes Kendall's tau and Sen's slope statistics to quantify these changes, highlighting the transition from natural land covers to anthropogenic land uses. The key findings are noted below:

Forest to Agriculture Conversion: In Ranthambhore National Park (Sawai Madhopur district in East Rajasthan IMD sub-division) and the Gulf of Khambhat (Districts of Anand, Bharuch, and Surat in the Gujarat Region), forests exhibit a statistically significant decline. Simultaneously, agriculture in these regions has shown substantial growth, indicating that forested areas are being converted to agricultural land, driven by the demand for food production and economic development. The decline in forest cover in Ranthambhore National Park could lead to habitat loss for Bengal tigers and leopards, reducing their available hunting grounds and increasing human–wildlife conflict. The conversion of forests to agriculture in the Gulf of Khambhat may affect the habitat of diverse bird species and mangrove ecosystems, which are crucial for many fish species' breeding and nursery grounds.

Shrubland and Grassland Depletion: The depletion of shrubland and grassland is particularly notable in several regions. In the Nalsarovar Bird Sanctuary (Districts of Ahmedabad in Gujarat Region and Surendranagar in Saurashtra and Kacchh IMD sub-division), shrubland has declined significantly, paralleled by a substantial decrease in grassland. Concurrently, agricultural expansion suggests that these natural habitats are being transformed into cropland. This trend is consistent in the Kumbhalgarh Wildlife Sanctuary (Districts of Rajsamand, Udaipur, Pali, and Sirohi in West Rajasthan IMD sub-division), where shrubland and grassland are being supplanted by agriculture. The loss of shrubland and grasslands in Nalsarovar Bird Sanctuary could impact the habitat of migratory birds, such as flamingos, disrupting their breeding and feeding grounds. The reduction in shrubland and savanna in Kumbhalgarh Wildlife Sanctuary may lead to a decline in herbivorous species like the Indian gazelle and negatively affect predator species like the Indian wolf that rely on these prey species.

Impact on Waterbodies: While natural land cover classes such as forests, shrublands, and savannas are declining, there is a concurrent increase in waterbodies driven by the development of irrigation and other water management projects. For example, in the Rann of Kutch (Districts of Kachchh, Surendranagar, Banas Kantha, in the Saurashtra and Kacchh IMD sub-division), the area covered by waterbodies has increased significantly, reflecting efforts to enhance water resources for agricultural and urban use. Increased waterbodies can benefit aquatic plants and provide habitat for species like the Indian wild ass and various migratory birds. However, it may also lead to the displacement of terrestrial flora and fauna due to changing landscapes.

Bareland Reduction: In Velavadar Blackbuck National Park (Bhavnagar district in Saurashtra and Kutch IMD sub-division), bareland has significantly decreased while agriculture has increased. This suggests that previously uncultivated lands are being brought under agricultural production, driven by the expansion needs of the agrarian economy. The conversion of bareland to agricultural fields can disrupt the habitat of the blackbuck and other grassland species, leading to reduced grazing areas and increased human-wildlife conflict in the national park.

4.3. Comparison of LULC Changes with Other Large-Scale Infrastructure Projects

Our observation of a substantial increase in agricultural land, a trend also found in studies such as Meiyappan et al. (2016) [47], underscores the significance of the DMIC and BRI in the context of global infrastructure projects. Like the DMIC, the BRI has significantly altered land use patterns, particularly by converting natural habitats into agricultural and industrial zones. Both projects have grappled with the challenge of balancing economic development with environmental sustainability. However, the BRI's scale and geographic diversity present unique challenges, such as transnational coordination and varying environmental regulations across countries, that are not as prominently seen in the DMIC.

Our findings on the DMIC underscore the role of economic development in driving land use changes—a crucial point that needs to be addressed in sustainable planning. Increased trade activities and economic development are key drivers of these changes. This is further supported by Zhang et al. (2022) [21], who demonstrated that the BRI significantly influences agricultural land use, confirming earlier findings. Their study explored the spatio-temporal dynamics of LULC in BRI regions, utilizing the Landscape Ecological Risk Index to assess how the initiative impacts land cover and ecological risks over time, which can be the subject of further research in the DMIC.

The decreases in shrubland and grassland observed in our study are not isolated incidents; they are consistent with global patterns seen in other large-scale infrastructure projects. For instance, due to infrastructure development, the BRI regions have experienced similar reductions in natural landscapes. Our findings, which corroborate these trends and highlight the impacts of the DMIC's industrial activities on local vegetation, contribute to the global urbanization and environmental change discourse, adding a valuable regional perspective.

Our observations on the DMIC underscore the importance of strategic planning to minimize negative ecological impacts. Teo et al. (2019) [27] supports this by characterizing the impacts of several BRI infrastructure projects on local ecosystems and proposing a typology to understand better and mitigate environmental disruptions.

Additionally, our study on the DMIC highlights challenges in LULCC mapping, emphasizing the need for improved methodologies to address issues in accuracy and consistency. This is consistent with findings by Naboureh et al. (2020) [23], who reviewed similar challenges and underscored the importance of accurate mapping for effective land use planning and the development of robust environmental policies.

The modest increase in forest areas observed in our study indicates the success of afforestation and reforestation efforts by smart cities such as Pune and Surat. This is consistent with similar global initiatives, such as China's Green Great Wall project, which aims to combat desertification through large-scale tree planting. The difference is that the Green Great Wall is federally funded in China, while it is enabled as a local city-level reforestation initiative in India.

The European Union's reforestation efforts, part of their broader environmental sustainability goals, also provide a useful comparison. The EU has implemented extensive programs to restore natural habitats and enhance biodiversity. These efforts in the EU are driven by strong policy frameworks and significant investment in environmental restoration, demonstrating the importance of governance in successful reforestation initiatives.

Similar trends of urban expansion and increased built-up areas have been noted in the context of the European Union's Trans-European Transport Network (TEN-T). However, the EU's regulatory framework and environmental impact assessments provide a more structured approach to mitigating negative impacts than the DMIC. Our findings highlight the need for enhanced regulatory mechanisms governing the construction and operational phases of the DMIC to balance development with sustainability. This can be achieved through enhancing the SCM Livability Index and aligning DMIC with the Sustainable Development Goals (SDGs).

4.4. Implications for Policy and Planning

The observed trends in LULC changes, particularly the expansion of agricultural land and the reduction of natural landscapes, call for targeted policies that balance economic growth with environmental conservation. Promoting sustainable agricultural practices, enhancing green urban spaces, and implementing stringent land-use regulations are essential strategies to mitigate the adverse effects of urbanization and industrialization. Two policies can regulate the construction and operational phases of the DMIC.

Our findings emphasize the need for sustainable land management practices in the DMIC, highlighting significant fluctuations in land use over recent decades. Although our approach utilized MODIS data, this finding aligns with the work of Chen et al. (2018) [24], who examined spatio-temporal changes in agricultural land in China using alternative approaches such as the Food and Agriculture Organization (FAO) statistics and GlobeLand30 data.

4.4.1. The SCM City Livability Index

Integrating LULC studies into the SCM Livability Index Model significantly impacts urban sustainability and compliance monitoring. Specifically, incorporating LULC metrics into indices such as the Mixed Use and Compactness Index, Open Space Index, Housing and Inclusiveness Index, Pollution Index, and Mobility Index can support a better understanding of land transformation's effects on ecosystems. These indices track changes like green space conversion and habitat fragmentation, essential for developing balanced land use strategies, preserving biodiversity, and reducing urban sprawl.

4.4.2. Sustainable Development Goals

In India, Niti Aayog is the nodal governmental agency that sets the data collection and reporting framework, whereas the individual states have annual implementation responsibilities. While SDGs are not applied to large infrastructure projects, several elements of the program, particularly in the context of monitoring LULC changes, can be considered:

- Sustainable Urbanization (SDG 11.3): Monitoring the ratio of the land consumption rate to the population growth rate and ensuring participatory urban planning processes to create inclusive, sustainable communities.
- Climate Resilience (SDG 13.1 and 13.2): Integrating climate change adaptation measures and disaster risk reduction strategies into the project planning and implementation phases. This would include time-series analysis, as conducted in this study.
- Conservation of Terrestrial Ecosystems (SDG 15.1 and 15.3): Ensuring the conservation and sustainable use of forests and wetlands, combating land degradation, and striving for land degradation neutrality by monitoring changes in forest cover and protecting important biodiversity sites.
- Protection of Water-related Ecosystems (SDG 6.6): Protecting and restoring water-related ecosystems impacted by the DMIC projects, such as rivers and wetlands, to maintain biodiversity and water quality.
- Sustainable Infrastructure (SDG 9.1): Developing resilient infrastructure that supports economic development while minimizing environmental impact, ensuring equitable access to infrastructure for all communities.

By aligning the DMIC project with these SDGs, the initiative can foster sustainable development that balances economic growth with environmental conservation and social well-being.

4.5. Limitations and Future Research Directions

Future research should address several limitations of our study. The dominance of rural areas within the DMIC ZoI and the 500 m spatial resolution of MODIS data may necessitate further research with higher resolutions (30 m) to determine the extent of urban LULC changes. Utilizing higher-resolution data and more advanced machine-learning models would also enhance the accuracy and granularity of LULC analyses. Future research should explore the causal mechanisms behind the correlations and develop targeted interventions to address the district-specific identified challenges. Extending the temporal scope beyond 2021 would provide a more comprehensive view of the ongoing changes and their long-term impacts. Additionally, more comparative studies with other large-scale infrastructure projects globally could enrich our understanding of LULC dynamics and further validate our findings.

5. Conclusions

This study examines LULC changes over 2 decades in 121 districts and 24 smart cities under the DMIC's zone of influence. The methodology provides a robust framework for LULC analysis over a two-decade span, enabling the high-accuracy assessment of changes and trends. Integrating satellite imagery and GIS digitization, overlaid upon the IMD sub-divisional framework and advanced statistical tests, ensures that the results are scientifically reliable and can inform future land management and policy decisions.

While agricultural land has expanded year-on-year, shrubland and grassland areas have declined. Major infrastructure projects developing built-up and urban areas in the DMIC over the next few decades must consider this baseline of LULC changes observed over the last two decades.

Our study documents the effectiveness of reforestation efforts within the specific context of smart cities in the DMIC, providing a model for similar initiatives elsewhere. This regional success story can inform policy and practice in other regions pursuing similar environmental restoration goals.

This study's regional analysis goes beyond administrative boundaries by using the IMD sub-divisions as a framework. The IMD sub-divisions represent contiguous, cohesive, and homogenous tracts of land, transcending administrative boundaries. This approach enables our LULC analysis and is useful for the short-term and long-term monitoring of LULC changes in the DMIC. The IMD sub-divisional approach is unique compared to other research that uses political boundaries, such as states, provinces, or countries.

Further, the IMD framework as an analytical unit is foundational for other agroecological, climatic, meteorological, and environmental research. The IMD also publishes internationally accepted datasets that researchers can use to integrate with this DMIC study for additional outcomes. The dominance of rural areas within the DMIC ZoI and the 500 m spatial resolution of MODIS data may necessitate further research with higher resolutions to determine the extent of urban LULC changes accurately and accordingly, further research is needed to refine the LULC classification at higher spatial resolutions. Additionally, more in-depth studies should explore the relationship between LULC changes and land surface temperature, precipitation, and relative humidity to better understand the potential effects on population settlements in the DMIC study area.

This study offers significant insights and methodologies that can be applied to other infrastructure projects globally, such as the Chennai Bengaluru Industrial Corridor in India and the Eastern Economic Corridor in Thailand. By systematically analyzing the LULC dynamics, this study provides critical insights into the anthropogenic drivers of land cover changes. These findings serve as a valuable reference for policymakers, conservationists, and researchers engaged in sustainable development planning within the DMIC.

The balance between urban growth and conservation remains crucial for the sustainable development of these rapidly evolving smart cities.

Supplementary Materials: The following supporting information can be downloaded at: https://www.mdpi.com/article/10.3390/land13070957/s1, Supplementary Section A. Sen's slope graphs for the DMIC, smart cities, and other districts. Supplementary Section B. Summary of time series for smart cities and other districts. Table S1. Summary of Kendall's tau and Sen's slope for time series analysis of decadal LULC changes for smart cities in the DMIC. Table S2. Summary of Kendall's tau and Sen's slope for time series analysis of decadal LULC changes for districts in the DMIC. Supplementary Section C. Table S3. Summary of time series for Ecologically Sensitive Areas in the DMIC.

Author Contributions: A.K. conceptualized and wrote the manuscript. V.N. guided the methodology. N.K.T. recommended the manuscript's visual presentation and mapping content. E.W. recommended the statistical tests in the trend analysis. R.R.M. processed MODIS data via Google Earth Engine and ran the ground truthing exercise. A.K. processed the LULC data using IBM SPSS, developed all maps using ArcGIS, and ran the statistical tests using IBM SPSS, Excel, and XLSTAT. All authors have read and agreed to the published version of the manuscript.

Funding: This research received no external funding.

Data Availability Statement: The original contributions presented in the study are included in the article; further inquiries can be directed to the corresponding author.

Conflicts of Interest: There are no potential conflicts of interest.

References

1. National Industrial Corridor Development Corporation. Available online: https://www.nicdc.in/index.php/delhi-mumbai-industrial-corridor-dmic.html (accessed on 6 February 2021).
2. Jain, S.; Panda, J.; Kant, S. Possible socio-scientific issues of Land-use and Land-cover change impact and associated tools of study with a special reference to Delhi-Mumbai Industrial Corridor Region. *Int. J. Earth Atmos. Sci.* **2014**, *1*, 58–70.
3. Jain, S.; Roy, S.; Panda, J. Modeling land-use and land-cover change impact on summertime near-surface temperature variability over the Delhi-Mumbai Industrial Corridor. *Model. Earth Syst. Environ.* **2021**, *7*, 1309–1319. [CrossRef]
4. Smart City Mission. Available online: http://164.100.161.224/content/ (accessed on 1 March 2024).
5. Concept Paper of DMIC. Available online: https://dmaicindia.files.wordpress.com/2013/07/concept-paper-of-dmic-2007.pdf (accessed on 25 September 2023).
6. Mukhopadhyay, C. Is the Delhi Mumbai Industrial Corridor (DMIC) an Emerging 'Megaregion' in India. *Plan. Theory Pract.* **2018**, *19*, 305–309. [CrossRef]
7. Jat, M.K.; Garg, P.K.; Khare, D. Monitoring and modelling of urban sprawl using remote sensing and GIS techniques. *Int. J. Appl. Earth Obs. Geoinf.* **2008**, *10*, 26–43. Available online: https://www.sciencedirect.com/science/article/pii/S0303243407000335 (accessed on 25 November 2023). [CrossRef]
8. Sulla-Menashe, D.; Gray, J.M.; Abercrombie, S.P.; Friedl, M.A. Hierarchical mapping of annual global land cover 2001 to present: The MODIS Collection 6 Land Cover product. *Remote Sens. Environ.* **2019**, *222*, 183–194. [CrossRef]
9. Lal, P.; Dubey, A.K.; Kumar, A. Measuring the control of landscape modifications on surface temperature in India. *Geocarto Int.* **2022**, *37*, 15736–15753. [CrossRef]
10. Mathan, M.; Krishnaveni, M. Monitoring spatio-temporal dynamics of urban and peri-urban land transitions using ensemble of remote sensing spectral indices-a case study of Chennai Metropolitan Area. *Environ. Monit. Assess.* **2020**, *192*, 15.
11. Sahana, R.; Ahmed, R.; Sajjad, H. Analyzing land surface temperature distribution in response to land use/land cover change using split window algorithm and spectral radiance model in Sundarban Biosphere Reserve. *Model. Earth Syst. Environ.* **2016**, *2*, 1–11. [CrossRef]
12. Nayak, S. Land use and land cover change and their impact on temperature over central India. *Lett. Spat. Resour. Sci.* **2021**, *14*, 129–140. [CrossRef]
13. Duraisamy, V.; Bendapudi, R.; Jadhav, A. Identifying hotspots in land use land cover change and the drivers in a semi-arid region of India. *Environ. Monit. Assess.* **2018**, *190*, 535. [CrossRef]
14. Garai, D.; Narayana, A.C. Land use/land cover changes in the mining area of Godavari coal fields of southern India. *Egypt. J. Remote Sens. Space Sci.* **2018**, *21*, 375–381. [CrossRef]
15. Moulds, S.; Buytaert, W.; Mijic, A. A spatio-temporal land use and land cover reconstruction for India from 1960–2010. *Sci. Data* **2018**, *5*, 1–11. [CrossRef]
16. Jain, M.; Korzhenevych, A. Detection of urban system in India: Urban hierarchy revisited. *Landsc. Urban Plan.* **2019**, *190*, 103588. [CrossRef]

17. Ramachandran, H. Review of Industrial and Development Corridors in India. Available online: https://isid.org.in/publication/h-ramachandran-review-of-industrial-and-development-corridors-in-india-wp217-december-2019/ (accessed on 25 November 2023).
18. Williams, G.; Mahadevia, D.; Schindler, S.; Chattaraj, S. Megaprojects, mirages, and miracles: Territorializing the Delhi-Mumbai Industrial Corridor (DMIC) and state restructuring in contemporary India. *Territ. Politics Gov.* **2023**, *11*, 456–477. [CrossRef]
19. Kumar, J.; Sharma, R. Highway peripheral urbanization, industrialization, and land use change: A case study of NH-48 in National Capital Region. *GeoJournal* **2023**, *88*, 2969–2981. [CrossRef]
20. Biswas, P.; Dey, P. Spatio-temporal impact of industrial corridor development on built-up growth: A case of Delhi-Mumbai industrial corridor, India. *GeoJournal* **2023**, *88*, 3791–3815. [CrossRef]
21. Zhang, J.; Ren, M.; Lu, X.; Li, Y.; Cao, J. Effect of the Belt and Road Initiatives on Trade and Its Related LUCC and Ecosystem Services of Central Asian Nations. *Land* **2022**, *11*, 828. [CrossRef]
22. Zhang, X.; Yao, Y.; Luo, J.; Liang, W. Exploring Changes in Land Use and Landscape Ecological Risk in Key Regions of the Belt and Road Initiative Countries. *Land* **2022**, *11*, 940. [CrossRef]
23. Naboureh, A.; Bian, J.; Lei, G.; Li, A. A review of land use/land cover change mapping in the China- Central Asia-West Asia economic corridor countries. *Big Earth Data* **2020**, *5*, 237–257. [CrossRef]
24. Chen, D.; Yu, Q.; Hu, Q.; Xiang, M.; Zhou, Q.; Wu, W. Cultivated land change in the Belt and Road Initiative region. *J. Geogr. Sci.* **2018**, *28*, 1580–1594. [CrossRef]
25. Dong, S.; Li, Y.; Li, Y.; Li, S. Spatiotemporal patterns and drivers of Land Use and Land Cover Change in the China-Mongolia-Russia Economic Corridor. *Pol. J. Environ. Stud.* **2021**, *30*, 2527–2541. [CrossRef]
26. Zuo, Q.; Li, X.; Hao, L.; Hao, M. Spatiotemporal evolution of Land-Use and ecosystem services valuation in the Belt and Road Initiative. *Sustainability* **2020**, *12*, 6583. [CrossRef]
27. Teo, H.C.; Lechner, A.M.; Walton, G.W.; Chan, F.K.S.; Cheshmehzangi, A.; Tan-Mullins, M.; Chan, H.K.; Sternberg, T.; Campos-Arceiz, A. Environmental impacts of infrastructure development under the Belt and Road Initiative. *Environments* **2019**, *6*, 72. [CrossRef]
28. Hansen, M.C.; Potapov, P.V.; Moore, R. High-resolution global maps of 21st-century forest cover change. *Science* **2013**, *342*, 850–853. [CrossRef] [PubMed]
29. Cortes, C.; Vapnik, V. Support-Vector Networks. *Mach. Learn.* **1995**, *20*, 273–297. [CrossRef]
30. Breiman, L. Random Forests. *Mach. Learn.* **2001**, *45*, 5–32. [CrossRef]
31. Breiman, L.; Friedman, J.; Olshen, R.A.; Stone, C. *Classification and Regression Trees*, 1st ed.; Chapman and Hall/CRC: Boca Raton, FL, USA, 1984.
32. Ouma, Y.; Nkwae, B.; Moalafhi. Comparison of Machine Learning Classifiers for Multitemporal and Multisensor Mapping of Urban LULC Features, The International Archives of the Photogrammetry. *Remote Sens. Spat. Inf. Sci.* **2022**, *XLIII-B3-2022*, 681–689. [CrossRef]
33. Foody, G.M.; Mathur, A. Toward intelligent training data allocation for the classification of remotely sensed data. *Int. J. Remote Sens.* **2004**, *25*, 3031–3036. [CrossRef]
34. Congalton, R.G.; Green, K. *Assessing the Accuracy of Remotely Sensed Data: Principles and Practices*; CRC Press: Boca Raton, FL, USA, 2008.
35. Mountrakis, G.; Im, J.; Ogole, C. Support vector machines in remote sensing: A review. *ISPRS J. Photogramm. Remote Sens.* **2011**, *66*, 247–259. Available online: https://www.sciencedirect.com/science/article/pii/S0924271610001140 (accessed on 12 October 2023). [CrossRef]
36. Donchyts, G.; Baart, F.; Winsemius, H.; Gorelick, N.; Kwadijk, J.; Giesen, N.V.D. Earth's surface water change over the past 30 years. *Nat. Clim. Chang.* **2016**, *6*, 810–813. [CrossRef]
37. Schneider, D.P.; Deser, C.; Fasullo, J.; Trenberth, K.E. Climate Data Guide Spurs Discovery and Understanding. *Eos Trans. Am. Geophys. Union* **2013**, *94*, 121–122. [CrossRef]
38. The Climate Data Guide. Available online: https://climatedataguide.ucar.edu/climate-data/ceres-igbp-land-classification (accessed on 25 December 2023).
39. Friedl, M.; Sulla-Menashe, D.; Tan, B.; Schneider, A.; Ramankutty, N.; Sibley, A.; Huang, X. MODIS Collection 5 global land cover: Algorithm refinements and characterization of new datasets. *Remote Sens. Environ.* **2010**, *114*, 168–182. [CrossRef]
40. Jensen, J.R. *Introductory Digital Image Processing: A Remote Sensing Perspective*, 4th ed.; Prentice-Hall Inc.: Upper Saddle River, NJ, USA, 2005.
41. Landis, J.R.; Koch, G. The Measurement of Observer Agreement for Categorical Data. *Biometrics* **1977**, *33*, 159–174. [CrossRef]
42. XLSTAT Statistical and Data Analysis Solution. Available online: www.xlstat.com/en (accessed on 12 October 2023).
43. Roboaa, E.M.; Al-Barazanji, Z.; Mann. Kendall trend analysis of surface air temperatures and rainfall in Iraq. *Q. J. Hung. Meteorol. Serv.* **2015**, *119*, 493–514.
44. Alhaji, U.U.; Yusuf, A.S.; Edet, C.O.; Oche, C.O.; Agbo, E. Trend analysis of temperature in Gombe state using Mann Kendall trend test. *J. Sci. Res. Rep.* **2018**, *20*, 1–9. [CrossRef]
45. Kisi, O.; Ay, M. Comparison of Mann-Kendall and innovative trend method for water quality parameters of the Kizilirmak River. *Turk. J. Hydrol.* **2014**, *513*, 362–375. [CrossRef]

46. Sen, P.K. Estimates of the Regression Coefficient Based on Kendall's Tau. *J. Am. Stat. Assoc.* **1968**, *63*, 1379–1389. [CrossRef]
47. Meiyappan, P.; Roy, P.S.; Sharma, Y.; Ramachandran, R.M.; Joshi, P.K.; DeFries, R.S.; Jain, A.K. Dynamics and determinants of land change in India: Integrating satellite data with village socioeconomics. *Reg. Environ. Chang.* **2016**, *17*, 753–766. [CrossRef]

land

MDPI

Article

Harmonized Pan-European Time Series for Monitoring Soil Sealing

Christophe Sannier [1,*], Eva Ivits [2], Gergely Maucha [3], Joachim Maes [4] and Lewis Dijkstra [5,*]

1 GAF AG, Arnulfstr. 199, 80634 Munich, Germany
2 European Environment Agency, Kongens Nytorv 6, 1050 Copenhagen, Denmark; eva.ivits-wasser@eea.europa.eu
3 Lechner Non-Profit Ltd., 1114 Budapest, Hungary; gergely.maucha@lechnerkozpont.hu
4 European Commission, Directorate-General for Regional and Urban Policy, 1040 Brussels, Belgium; joachim.maes@ec.europa.eu
5 European Commission, Joint Research Centre, Unit of Territorial Development, Via Enrico Fermi, 2749, 21027 Ispra, VA, Italy
* Correspondence: christophe.sannier@gaf.de (C.S.); lewis.dijkstra@ec.europa.eu (L.D.)

Abstract: The European Copernicus Land Monitoring Service (CLMS) has been producing datasets on imperviousness every 3 years since 2006. However, for 2018, the input for the production of the imperviousness dataset was switched from mixed inputs to the Sentinel constellation. While this led to an improvement in the spatial detail from 20 m to 10 m, this also resulted in a break in the time series as the 2018 update was not comparable to the previous reference years. In addition, the European CLMS has been producing a new dataset from 2018 onward entitled CLC+ Backbone, which also includes a sealed area thematic class. When comparing both datasets with sampled reference data, it appears that the imperviousness dataset substantially underestimates sealed areas at the European level. However, the CLC+ dataset is only available from 2018 and currently does not include any change layer. To address these issues, a harmonized continental soil sealing combined dataset for Europe was produced for the entire observation period. This new dataset has been validated to be the best current dataset for monitoring soil sealing as a direct input for European policies with an estimated total sealed area of 175,664 km^2 over Europe and an increase in sealed areas of 1297 km^2 or 0.7% between 2015 and 2018, which is comparable to previous time periods. Finally, recommendations for future updates and the validation of imperviousness degree geospatial products are given.

Keywords: imperviousness; change detection; Copernicus Land Monitoring Service; pixel counting; area estimation

Citation: Sannier, C.; Ivits, E.; Maucha, G.; Maes, J.; Dijkstra, L. Harmonized Pan-European Time Series for Monitoring Soil Sealing. *Land* **2024**, *13*, 1087. https://doi.org/10.3390/land13071087

Academic Editor: Jūratė Sužiedelytė-Visockienė

Received: 29 May 2024
Revised: 4 July 2024
Accepted: 8 July 2024
Published: 19 July 2024

1. Introduction

Soil sealing is part of the land take process, i.e., when natural or agricultural land is converted to settlements, infrastructure and commercial or industrial use. The process covers the ground by an impermeable material such as asphalt or concrete. Imperviousness is the state of soil where it cannot be penetrated by air and water and in practice soil sealing and imperviousness are terms used synonymously. However, while soil sealing indicates an anthropogenic process, imperviousness also refers to natural impenetrable surfaces like rocks and glaciers. Hence, all sealed surfaces are impervious, but some impervious surfaces are not sealed.

Soil sealing often affects fertile agricultural land, and as sealed areas cannot be penetrated by air and water, the process puts biodiversity at risk. Soil sealing also increases the risk of flooding in urban areas in case of flash floods and it creates urban heat islands, which impacts human health. European policy making urges data providers to monitor soil sealing to be informed on development, agriculture, environment or sustainability at multiple levels of government. Given the importance of soils to the European Union's (EU) policy objectives on climate change mitigation and adaptation, food security and biodiversity, soils contribute to achieving the main objectives of the European Green Deal, as set out in the EU Soil Strategy for 2030. To support efforts to improve soil health, the European Commission has undertaken two key initiatives. Firstly, they established a mission on soil that utilizes a network of living laboratories to test solutions for improving soil. Secondly, in 2023, the commission proposed a Soil Monitoring Law. This law mandates monitoring of land take and soil sealing indicators across the EU [1]. While not directly regulating soil sealing itself, the data collected through this monitoring will be crucial for developing informed policies to address this issue.

Several regularly updated datasets based on land monitoring are available to measure imperviousness at the European scale. They are based on in situ monitoring, remote sensing or a combination of both. The Land Use/Cover Area frame Survey (LUCAS) is an in situ land monitoring program [2] that provides harmonized and comparable statistics on land use and land cover across the whole of the EU's territory. It provides every 3 years data on land cover and land use through some 300,000 field observations sampled from a 2 by 2 km grid covering the EU. However, LUCAS does not cover European Environment Agency (EEA) cooperating countries. In addition, the definition of the artificial thematic class in LUCAS is only partially related to sealed surfaces and there are too few LUCAS points over urban areas to make accurate estimates of changes in soil sealing.

Datasets based on remote sensing are available from the Copernicus Land Monitoring Service (CLMS) and include the CLMS CORINE Land Cover + Backbone (CLC+ BB) for 2018 [3] (2021; available at end of 2024) and the time series of the imperviousness degree layer [4], every 3 years from 2006 to 2018 (2021; available at end of 2024). Other global data sources exist, such as the Global Human Settlement Layer (GHSL) described in [5] and the World Settlement Footprint (WSF) [6]. However, both GHSL and WSF are global products and their respective definitions are more focused on assessing built-up areas, with a focus on buildings, whereas roads or parking spaces are not considered when disaggregating population census data into a gridded product.

There is a natural tendency from the remote sensing user community to extract area statistics from such spatial datasets (i.e., "pixel counting") from earth observation (EO)-based geospatial products to produce statistical indicators for various purposes as discussed in [7]. However, geospatial map products suffer from misclassification errors and "pixel counting" can be strongly biased unless the accuracy of such map products reaches a level when these misclassification errors can be considered negligible. There has been a considerable effort in the remote sensing community to assess the accuracy of map products against reference data to ensure that the maps could reach a sufficient level of accuracy. Despite substantial advances in this topic in the scientific literature in recent years notably with paper [8], this has yet to be fully implemented in operational projects. In addition, even if map accuracy assessment is performed correctly, high accuracy does not necessarily mean that area statistics can be directly extracted from a map as highlighted in [9]. This happens because the requirement level is not the same for mapping accuracy and for statistics. It may happen that an 80% mapping accuracy is considered good enough for a thematic map, but this may lead to up to an unacceptable 20% bias in the area estimates if commission and omission errors are strongly unbalanced with commission errors defined as areas classified as the theme of interest but should be classified as something else and omission errors defined as areas that should be classified as the theme of interest but were classified as something else. One way to address this is to apply a so-called Model Assisted Regression or Regression estimator as detailed in [10], which combines reference observations with EO-based products. However, applying such an approach is dependent on the availability of reference data at a disaggregated level.

In addition, even though CLMS products are probably the best available datasets to monitor soil sealing over Europe, the 2018 imperviousness degree layer was produced on a 10 m spatial resolution, taking advantage of the availability of Sentinel 2 data, whereas earlier layers are provided on 20 m resolution. Unfortunately, the upgrade to 10 m from 20 m spatial resolution has resulted in a break in the IMD-based areal statistics; the 20 m resolution IMD time series (2006–2009–2012–2015) could be successfully harmonized and have shown a credible evolution in sealed cover. The IMD2018 dataset exhibits significantly more sealed structures than previous layers; thus, the amount of sealed cover is showing an unrealistic growth compared to former status.

Therefore, the aim of this paper is to use the time series and available sample reference data [11] for the development of a reliable spatially explicit time series dataset of sealed area in Europe. In this context, Europe is the study area represented by all EEA member and cooperating countries and the UK and further referred to in this paper as EEA38 + UK. The total area covered represents 5.85 million km^2 with the exception of the French overseas regions located in south and central America and the Indian Ocean. The objectives of this study are three-fold: (i) assess the level of bias from existing sealing datasets as compared to available reference data, (ii) make use of the available reference data to develop a suitable approach for harmonization including the provision of a freely accessible and harmonized time-series dataset and (iii) make recommendations for future developments of earth observation-based products.

2. Data

2.1. Available Soil Sealing Geospatial Data Layers

The pan-European CLMS portfolio includes two kinds of image classifications containing detailed information about the extent and structure of artificially sealed areas: the High-Resolution Layer (HRL) Imperviousness [4] and the CLC+ Backbone raster [3] datasets.

2.1.1. CLMS Imperviousness Degree Layers

Imperviousness data are at the time of publication of this paper available for the reference years 2006, 2009, 2012, 2015 and 2018, and contain two types of products: status and change layers referred to as the HRL IMperviousness (IMD) and IMperviousness Classified Change (IMCC), respectively. The IMD layers capture the spatial distribution of artificially sealed areas, including the level of sealing of the soil per area unit. The level of sealed soil (imperviousness degree: 1–100%) is produced using an automatic algorithm based on the calibrated Normalized Difference Vegetation Index (NDVI) as described in [12,13]. The percentage of sealed area is mapped for each status layer for the five reference years. The IMD status layers are available in 10 m spatial resolution (2018) and 20 m spatial resolution (2006–2015) and as aggregated 100 m datasets. Two types of IMCC products are available for each of the 3-year periods between the five reference years (2006–2009, 2009–2012, 2012–2015, 2015–2018):

- A simple layer mapping the percentage of sealing increase or decrease for those pixels that show sealing change during the period covered. This product is available in 20 m and 100 m pixel sizes.
- A classified change product that maps the most relevant categories of sealing change (no sealing, new cover, loss of cover, unchanged sealed, increased sealing, decreased sealing). This product is available in the 20 m pixel size only.

2.1.2. CLMS CLC+ Backbone Layer

The CORINE Land Cover (CLC) product has been the flagship of the Copernicus Land Monitoring Service for almost thirty years (1990–2018), providing users with harmonized land cover/land use data at the continental scale. However, CLC has a relatively coarse spatial resolution, and a lack of detail in certain land cover categories. In response, the European Environment Agency has developed a suite of new products and applications, known collectively as CLC+. Among these new products, the CLC+ Backbone (BB) 2018 dataset provides a wall-to-wall land cover map of Europe on 10 m resolution, for each pixel showing the dominant land cover among the 11 basic land cover classes. The dataset is available as a 10 m raster. The semantic content of class 1 (sealed areas) is by definition very similar to the semantic content of IMD data. The planned update cycle of CLC+ BB data is a minimum of 2 years after the foreseen 2021 update—comparable to IMD.

Figure 1 shows the comparison of the IMD layer for 2015 and 2018 and the 2018 CLC+ BB artificial layer converted to 100 m against the Bing aerial dataset. Some of the differences between the IMD2015 and 2018 layers are not due to changes but more likely related to omission in the lower-resolution 2015 layer. For example, much of the road network was omitted in 2015 and some of it appears in 2018; these roads were not built in those 3 years but the low-resolution 2015 layer does not capture these landscape elements. In addition, the CLC+ BB layer appears to provide even more spatial details in 2018 when compared to the IMD 2018 layer.

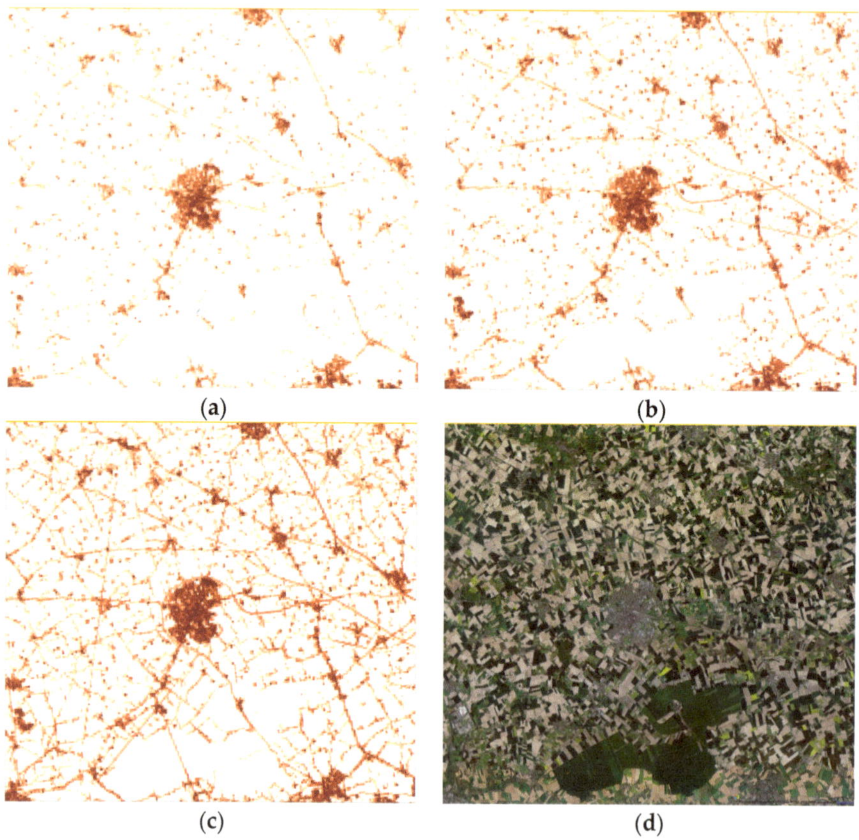

Figure 1. Extract over Hazebrouck in northern France of (**a**) CLMS IMD2015, (**b**) CLMS IMD2018, (**c**) CLC + BB 2018 artificial sealed and (**d**) Bing aerial. Grey = not impervious; red = impervious.

2.2. Reference Data

2.2.1. Sample Design and Stratification

An extensive validation exercise was performed as part of a dedicated CLMS project with more than 20,000 Primary Sample Units (PSUs, Figure 2) collected to assess the accuracy of imperviousness data layers. A two-stage sampling approach was implemented by selecting PSUs and Secondary Sampling Units (SSUs) within the PSUs. Two-stage sampling is considered suitable for accuracy assessment of land cover maps or area estimation [14] and can be adopted in certain cases because it represents a good compromise between the ease of data collection and a good geographic distribution. In our work, 1 ha square PSUs were selected and a grid of 5 × 5 SSUs was applied to facilitate the data collection as shown in Figure 2.

Figure 2. An example of a PSU (area of the image) and SSUs (red dots) within. SSUs are organized in a 20 m grid (25 sampling points) to assess impervious surfaces implemented by the CLMS service providers in the validation of the IMD layers.

The approach for drawing PSU locations was based on the LUCAS 2 × 2 km grid representing a total of 1,100,000 points across Europe [15]. The LUCAS point is then used to define the top left corner of the 1 ha PSU. However, considering the different characteristics and class definition of the imperviousness layers, the thematic information of LUCAS points was not used directly with the exception of the LUCAS landscape photos, which proved to be particularly useful for clarifying certain cases when they were available. The main advantage of using a LUCAS-based approach is that a systematic approach ensures full traceability and it is also possible that sampling units will be shared for several products, providing potential economies of scale.

Impervious areas remain relatively rare over Europe, covering less than 5% of the area (even though it has increased steadily over the years). Therefore, in case of the HRL images, a suitable stratification approach was required to ensure that sufficient reference observations are available. For both status and change layers, CORINE Land Cover artificial classes (CLC [16] and Open Street Map (OSM) road network) were used and converted to a binary layer to identify potential omission errors. Relevant OSM road types were selected and rasterized to 100 m (for example, abandoned, construction, cycleway, path, planned, trail, track. . .were removed) to obtain potential sealed area candidates. Using the relevant selection of OSM road types lead to a better spatialization of artificial and impervious areas.

CLC impervious classes were defined as follows based on CLC2018:

1.1.1 = continuous urban fabric;
1.1.2 = discontinuous urban fabric;
1.2.1 = industrial, commercial areas;
1.2.3 = ports;
1.2.4 = airports.

The following stratification approach was adopted based on omission/commission strata for the Imperviousness Status and change layers and is defined as follows:

- Stratum 1, commission status strata, where the imperviousness degree is between 1 and 100% in the 2015 layer.

- Stratum 2, High-Probability Omission strata, where the imperviousness degree is 0% in the 2015 reference year but where the Open Street Map and the traditional CLC 2015 layers indicate "impervious classes".
- Stratum 3, Low-Probability Omission strata covering the rest of the area.
- Stratum 4, commission Change Strata with all changes between 2006–2009, 2009–2012, 2012–2015 and 2015–2018 [gain, loss, increased and decreased were combined due to the very small area covered].

At first, the number of PSUs to allocate to each stratum was calculated as a function of the stratum area in a so-called proportional allocation. Then, for small strata, a minimum number of PSUs were allocated as suggested in [8]. This was achieved by densifying the LUCAS grid to achieve the required number. Similarly for very large strata, there may be too many PSUs and the required number was achieved by subsampling the LUCAS grid.

In the second phase, selected PSUs were overlaid with countries or groups of countries with an area greater than 90,000 km^2 (see Figure 3 below) to ensure that a required minimum of 50 PSUs were available for each stratum. If not, additional PSUs were selected for those countries or group of countries. The purpose was to be able to assess any heterogeneity in the quality of the data across different regions.

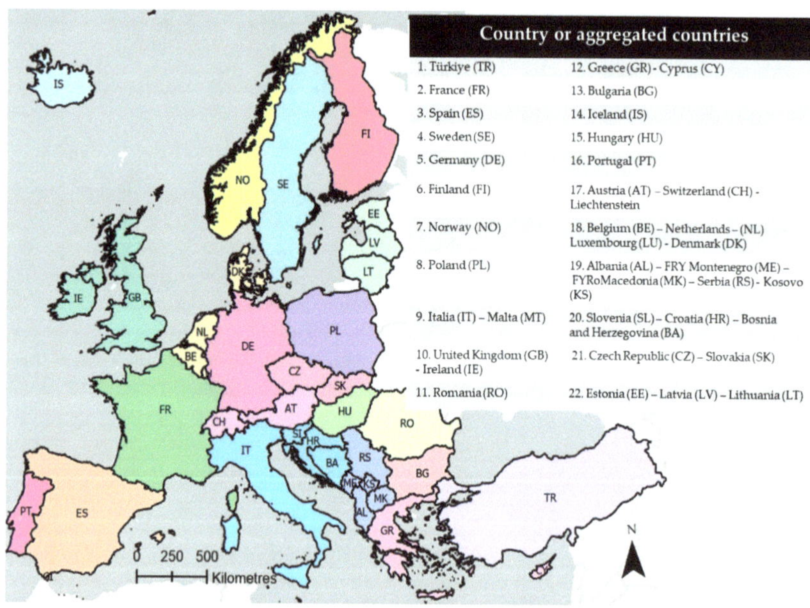

Figure 3. The level of reporting by country or aggregated countries (smaller countries were aggregated to ensure a minimum area of 90,000 km^2) as applied in the Copernicus Land Monitoring Service external validation contract.

As explained in [17] and applied in [18], this stratified systematic sampling approach leads to unequal sampling intensities between different strata, which need to be accounted for by applying a weight factor (w_i) to each sample unit based on the ratio between the number of sample units and the area of the stratum considered: $\hat{w}_i = \frac{n.N_i}{n_i.N}$, where i corresponds to a given stratum, N is the total number of possible units (population) and n is the number of PSUs. Not applying this correction would result in underestimating or overestimating map accuracies.

There were several iterations performed for the validation of IMD products, starting with the IMD 2012 layer for which the initial methodology was developed as described in [10]. When the 2015 IMD status layer was produced, the 2006, 2009 and 2012 status layers were reprocessed and the whole time series was validated with an initial total number of 18,005 PSUs across EEA38 + UK. In addition, change layers were also validated with an additional set of PSUs, resulting in a total of 22,777 PSUs as described in [19]. For the validation of the 2018 layers, the initial set of 18,005 PSUs were used but the interpretation was performed by a different team and for the validation of the 2015–2018 change layer, a subset of 10,005 PSUs was used with an overall total of 25,777 individual PSUs selected across all time periods as described in [20]. However, it should be noted that not all PSUs were used for any given time period due to missing data or cloud cover.

2.2.2. Response Design

Response design is the methodology used to obtain the reference data for the sample units [21]. Ref. [14] asserted that accuracy assessment is often made relative to some "higher quality determination of land cover". Ref. [22] indicated that visual interpretation is acceptable if the spatial resolution of EO data is sufficiently better compared to the thematic classification system. For this exercise, the visual interpretation of selected sample units will be used as a basis of the response design. The response design follows the definition of the products to be assessed both in terms of geometric and thematic characteristics. Therefore, the following characteristics are strictly followed and closely monitored as part of the visual interpretation:

- Minimum Mapping Unit (MMU);
- Minimum Mapping Width (MMW);
- Class definition (what is an impervious surface?);
- Ensure that the image data used are as close as possible temporally to that used for the map production.

Reference data can be collected with a double-blind protocol (without knowledge of the classes reported in the EO product (land cover map or classified image)) or following a plausibility approach. The plausibility approach considers that there can be errors in the reference data collected through visual interpretation as well as in the map itself. The plausibility approach reduces the amount of identified errors by comparing the sample with the map, thus adding another source of information for the photo-interpreter who then judges if it is reasonably coherent with the reality observed on finer spatial and/or temporal resolution images.

Statistical theory assumes that both EO and reference data are independent. Therefore a double-blind approach is preferable. However, double-blind data collection may produce major distortions in the presence of co-registration inaccuracy or conceptual uncertainty in the application of class labels. In the case of the imperviousness degree, conceptual uncertainty is low when a point has been perfectly located, but there may be a distance up to 10–20 m between the photo-interpreted reference point and the center of the corresponding pixel in a classified image. The apparent disagreement may be due to location inaccuracy rather than image classification errors. The use of 1 ha clusters as the PSUs (Figure 2) strongly reduces this anomaly.

Ref. [23] concluded that ignoring visual interpretation errors would result in underestimating the variance of the area estimate and developed an area estimator considering interpretation errors based on seven interpreters assessing the same sample. This would be difficult to implement over very large areas, but interpretation errors should be minimized as much as possible. Therefore, as a Quality Assurance and Control measure, at the start of the visual interpretation process, a certain number of samples (at least 10–20%) was interpreted twice and analyzed to identify the reasons for discrepancies. The percentage of samples interpreted twice decreases as the interpretation process progresses and as a result discrepancies decrease.

To ensure full traceability and transparency of the whole process, a double-blind approach was first applied, and a plausibility approach was applied at the end of the interpretation process on PSUs that disagreed with the map product.

In addition, some of the samples (around 10%) were interpreted by separate quality control interpretation teams as suggested in [23,24] and specific training sessions were organized to confront discrepancies occurring between different interpretation teams to ensure that a consensus is reached.

3. Methods

3.1. Accuracy Assessment

Thematic accuracy is usually presented in the form of an error matrix containing the results of the interpretation. To be valid, a confusion or error matrix should be derived from a probability sample and additionally an unequal sampling intensity resulting from the stratified systematic sampling approach should be accounted for. This requires applying a weight factor (w_i) to each sample unit based on the ratio between the number of samples and the size of the stratum considered as described at the end of Section 2.2.1.

A number of accuracy metrics can be computed from the error or confusion metrics such as overall, producer (related to omission errors) and user (related to commission errors) accuracies. In this case, we are dealing with a single theme.

Previously, the CLMS imperviousness degree layer was validated by applying a 30% threshold, thus transforming the impervious degree layer into a binary layer [11]. However, this does not consider the full range of values. Using continuous disagreement values requires an adaptation of the approach to compute confusion matrices [25,26].

Among the possible disagreement indicators, we propose to quantify the disagreement at the PSU level as the differences between the map value m_i and the reference r_i. The Mean Absolute Error (MAE), $MAE = \frac{\sum_i w_i |m_i - r_i|}{\sum_i w_i}$, where w_i is the extrapolation weight (inverse of the sampling intensity), and Root Mean Square Error (RMSE), $RMSE = \sqrt{\frac{\sum_i w_i (m_i - r_i)^2}{\sum_i w_i}}$, are common metrics used in this context and are currently used to assess the accuracy of the Global Human Settlement Layer (GHSL, [27] also now produced as part of the Copernicus Global Land initiative). In addition, if the map value is larger (lower) than the reference value for a PSU, it will contribute to the commission (omission) error as illustrated in Figure 4.

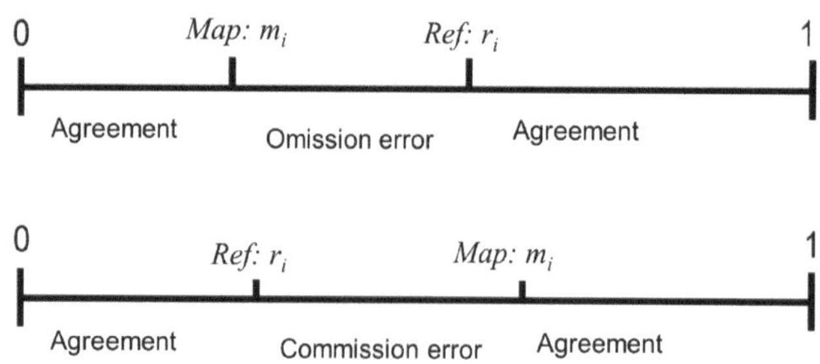

Figure 4. Commission and omission errors with continuous data between 0 and 1 (source: adapted from [28]).

For the MAE, the commission φ would be computed as $\varphi = \frac{\sum_i w_i \, pos(m_i - r_i)}{\sum_i w_i}$ and omission ψ as $\psi = \frac{\sum_i w_i \, pos(r_i - m_i)}{\sum_i w_i}$ with $pos(x)$ as the positive part function. The same approach can also be applied to RMSE to distinguish between omission and commission errors. This approach will allow us to assess commission and omission errors for the full range of the imperviousness degree layer values.

Using $pos(m_i - r_i)$ and $pos(r_i - m_i)$ as disagrerement indicators is similar to the concept of quantity disagreement proposed in [29] using the PSU as a measurement unit.

3.2. Harmonization of Sealing Layers

3.2.1. Biased and Unbiased Area Estimates

Biased area estimates are typically pixel counts where, in our case, area statistics are extracted from the HRL 100 m imperviousness degree layers and reported for the selected areas by taking the average imperviousness degree value of all pixels over a selected area. This value can then be multiplied by the total area in km^2 of the selected area to provide the biased area estimate. For the imperviousness change layers, the same procedure can be applied for pixels identified as gain or loss. It should also be noted that another important drawback of pixel counting is the absence of possibility of calculating the uncertainty of the resulting estimates.

Unbiased area estimates can be derived directly from the field data alone using the so-called direct expansion method [30], as long as the reference data have been collected based on a probabilistic sample and we assume that reference data are accurate. In the case of Simple Random Sampling (SRS) with equal probability, the estimate of proportion (y) of class (c) is given by $\overline{y}_c = \frac{1}{n} y_i$ and its variance is $var(\overline{y}_c) = \left(1 - \frac{n}{N}\right) \frac{1}{n(n-1)} \sum_{i=1}^{n} (y_i - \overline{y}_c)^2$, where y_i is the proportion of segment i (100 m pixels) covered by class c, N is the total number of segments in the region and n is the number of segments in the sample. The proportion of the study region sampled (*n/N*) is the sample fraction. The variance calculation above assumes a single-stage sampling; if a two-stage sampling is applied, this needs to be accounted for as described in [31]. The estimate of class area (Z) in the study area (D) is as follows: $\hat{Z}_c = D * \overline{y}_c$; and variance is $var(\hat{Z}_c) = D^2 * var(\overline{y}_c)$, where D is the area of the region. The direct expansion estimators are calculated for each stratum present in the AoI (area of interest) and the total estimate just corresponds to the weighted average of the proportions according to the area covered by each stratum. The standard error for the whole area is then as follows: $\sigma_{Total} = \sqrt{D_h^2.Var_h}$, where D_h is the stratum area. The 95% confidence interval (CI) is usually computed as $+/-$ 1.96* σ_{Total}., although the level of confidence can be smaller if sample size in each stratum is not large enough [32].

As a starting point for the harmonization procedure, the direct estimator applied on the 2018 sample validation data (described in Section 2.2.1) provides a total estimated sealed area of 160,434 ± 7495 km^2 (95% CI) at the EEA38 + UK level as shown in Table 1. The total sealed area mapped by CLMS IMD2018 is close to 109,000 km^2, while CLC+ BB reports 175,000 km^2. Therefore, CLMS IMD2018 is underestimating sealed areas by approximately 51,000 km^2 (or 44,00 km^2 if we consider the 95% CI). Even larger differences are observed for the previous years. However, the difference between CLC+ BB and the direct estimate is much smaller with CLC+ BB slightly overestimating sealed areas by ca. 15,000 km^2 (or 7000 km^2 with the 95%CI). It should be noted that two estimates are available for each year because years were assessed as pairs and not all sample units are always covered due to missing data or cloud cover from the IMD layer. However, both estimates are not statistically different with overlapping 95% confidence intervals (Table 1). For 2015, the differences are greater, because the IMD layer for 2015 was reprocessed following the 2018 production. In addition, the validation data for the 2015–2018 assessment were based on a smaller and different number of PSUs as compared to the previous years.

As explained in Section 2.2.1, there were different versions of the reference dataset for the status layer validation and the change layer validation. For consistency, the results for the status layer validation presented in Table 1 were based on the change validation dataset, which was deemed to be more accurate considering the greater number of sample units and the fact that it was produced after the reference dataset for the validation of the status layer. The results for gains presented in Table 2 are also based on the same dataset.

For 2006–2015, the change layers are either within the bounds of the 95% confidence interval of the CLMS validation data change area estimates or less than 20% above (Table 2). For 2015–2018, the differences are greater than 150%. This confirms the expectation that there is a discontinuity in the dataset with 2018 and even though 2006–2009 is also potentially problematic, this is far less critical by a factor close to 9 (268 km^2 vs. 2316 km^2). Therefore, our efforts concentrated on correcting the 2015–2018 IMCC data, and the change data from previous time periods especially for 2009–2012 and 2012–2015 was considered as acceptable. In addition, the focus was on correcting the gain data because losses can be considered negligible in comparison (there are very few areas where sealed surfaces a re-naturalized).

Table 1. Assessment of difference in area (km^2) between available status layers and CLMS validation data for the EEA countries + UK.

Area (km^2)	2006	2009	2012	2015	2018
HRL IMD pixel count estimate	82,289	83,872	85,410	86,453	108,996
CLC+ Backbone pixel count estimate	n/a	n/a	n/a	n/a	175,664
CLMS validation estimate and 95% confidence interval (CI)	n/a	158,370 ± 4247	159,212 ± 4262	161,382 ± 4275	160,434 ± 7495
Second CLMS validation estimate and 95% CI	157,177 ± 4242	157,659 ± 4242	160,544 ± 4271	159,258 ± 7492	not available
Difference with upper/lower limit of 95% CI for IMD	70,646	69,545	69,540	65,313	43,943
Difference with upper/lower limit of 95% CI for CLC + BB					7735

n/a: not available.

Table 2. Assessment of differences in area (km^2) between the IMCC dataset and CLMS Validation data for gains for the EEA38 countries and the UK (Red and amber, respectively, indicate large and smaller difference with upper limit of 95% CI).

Area (km^2)	2006–2009	2009–2012	2012–2015	2015–2018
HRL IMCC pixel count estimate	1715	1627	1126	3808
CLMS validation estimate and 95% confidence interval (CI)	1197 ± 251	1553 ± 509	1226 ± 369	1176 ± 314
Difference with upper limit of 95% CI	267	0	0	2318

3.2.2. Harmonization Approach

The following subsections describe the approach that was adopted to produce what is referred to as the harmonized imperviousness change/time series.

Based on the results above, the CLC+ BB sealed class appears to be more accurate than the IMD2018 layer. Therefore, a first step to correct overall impervious areas for 2018 relied on the extraction of CLC+ BB artificial class 1 into a binary 10 m layer.

The harmonization of the 2015–2018 change layer is based on calculating the areas of gain in dedicated calibration units and a subsequent filtering out of omission errors in gains through a reclassification procedure of the 2015–2018 IMCC as detailed as follows.

Step 1—Calculating the areas of gain for each of the 91 combinations of production/bioregions:

Based on experience, it is known that there is potentially substantial spatial variability in the quality of the IMCC layer. Therefore, an initial approach was to assess the level of change based on the CLMS validation data for each of the 91 combinations of production units (ca 300 × 300 km^2) and environmental zones [33], which we can refer to as calibration units (Figure 5). This represents a reasonable compromise between the level of spatial details and a sufficient number of PSUs available for each calibration unit (the total number is 95, but for the time being, the French overseas regions were excluded from the analysis).

This approach is similar to that of [34] for soybean mapping in the United States in which reference data were collected within 20 × 20 km blocks and the resulting soybean area estimates within each block were used to calibrate the soybean map. In our case, sealing areas and sealing change are much rarer phenomena to map. Therefore, our calibration units are larger with a smaller number of observations. Adjustments were estimated for all calibration units with two exceptions, (i) calibration units for which no gain was detected based on the validation data; and (ii) calibration units for which the level of gain from the map layer is less than for the validation data, and therefore would require us to add missing gain area, which is not possible with the adopted procedure.

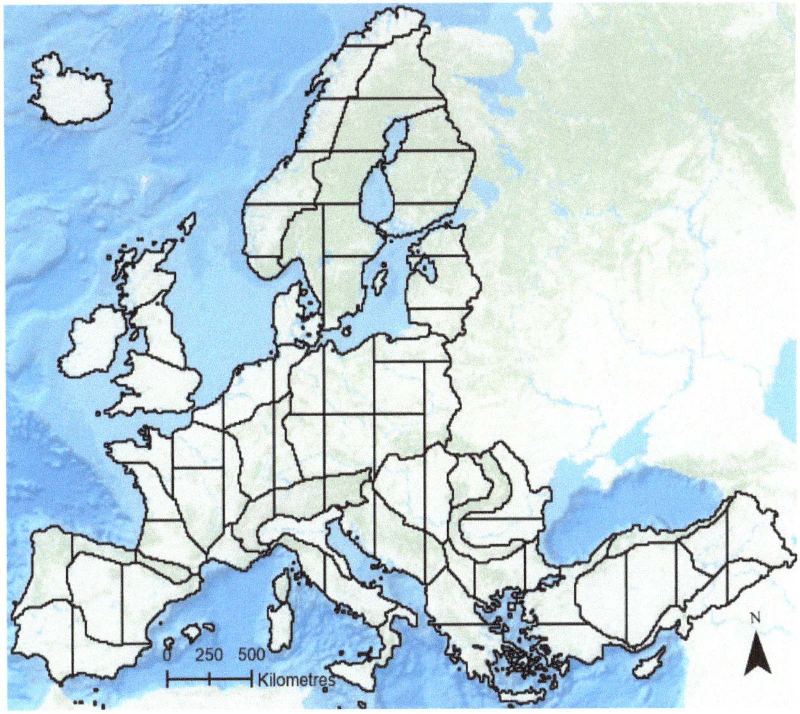

Figure 5. The description of calibration units based on the combination of 300 by 300 km production units and environmental zones [33].

Step 2—Discriminating function to separate actual gain from omission from previous periods.

In order to correct the large differences seen at the EEA38 + UK level and at the calibration units' level, some discriminating function needs to be applied to the gain data to separate real changes from other detected gain areas. There are several ways this could be achieved notably based on Machine/Deep Learning approaches, but this would require the input satellite imagery used in the production and considerable resources that were not available for this work.

Ref. [35] applied different thresholds locally to the density values to adjust the area covered by the Global Forest Change dataset from [36] and reduce the amount of bias, concluding that at the national level, a 70% threshold applied to the dataset was a closer depiction of the actual 30% threshold from the national forest definition, but for some areas, a more localized threshold adaptation was needed. Ref. [34] used the probability layer from an initial soybean classification in the USA to apply a separate threshold for each stratum to produce a binary layer that matched the soybean area estimates from collected field data.

As mentioned earlier, in our case, no reliable density or probability layers were available and a different approach was adopted based on the assumptions that most gains typically occur in the vicinity of existing sealed areas and that many of the detected omissions are often linked to the road network, which are further away from existing sealed objects. Therefore, an approach based on the distance of objects from the current vectorized change layer and the 100 m 2015 sealing status layer was applied as illustrated in Figure 6. The average Euclidian distance of each gain area polygon (red in Figure 6) from the IMCC layer for 2015–2018 to existing sealed area in 2015 was calculated and stored as an attribute.

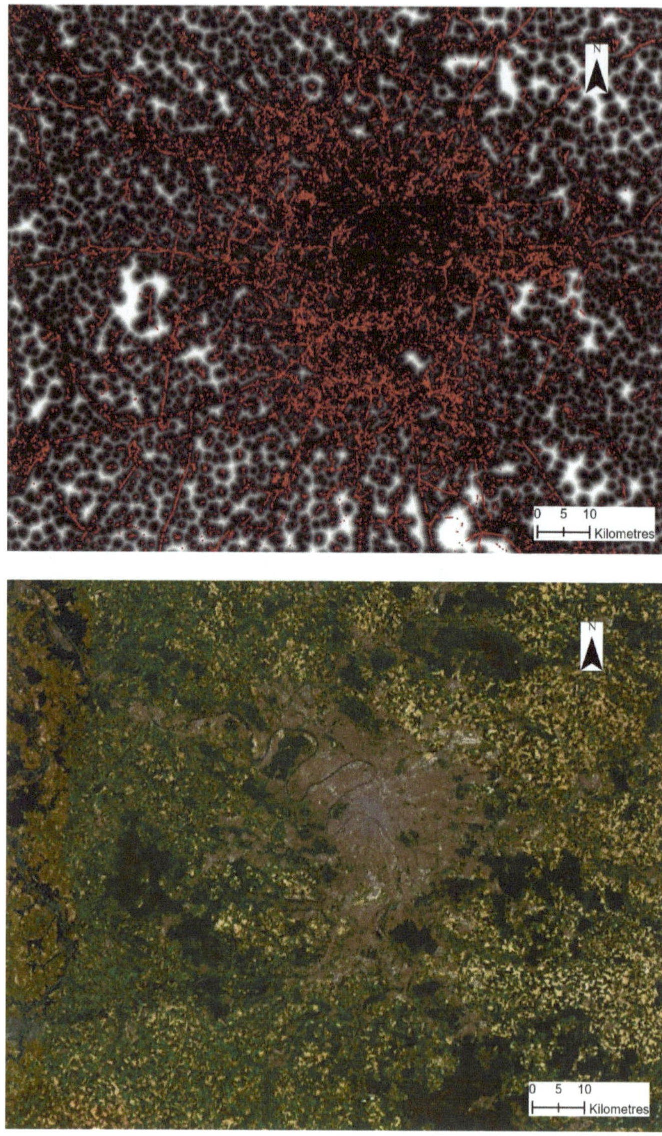

Figure 6. (**Top**): distance layer from 2015 imperviousness degree status layer (gray tones, with darker areas representing smallest distance) and candidate gain areas in red from 2015–2018 imperviousness classified change layer. (**Bottom**): a corresponding is Sentinel 2 image.

A distance threshold was then applied interactively to gain objects for each calibration unit to obtain, as close as possible, the required adjustment. Gain objects within the distance threshold to existing sealed areas (in the 2015 layer) were classified as actual gain and objects outside were classified as omissions as shown in Figure 7. Most omissions appear to represent either road segments now classified using the higher spatial resolution of the Sentinel 2 input data or some large objects. In the enlarged extract shown, green objects are classified as omissions because the mean Euclidian distance of these objects is above the threshold set for these calibration units, whereas red objects are classified as change.

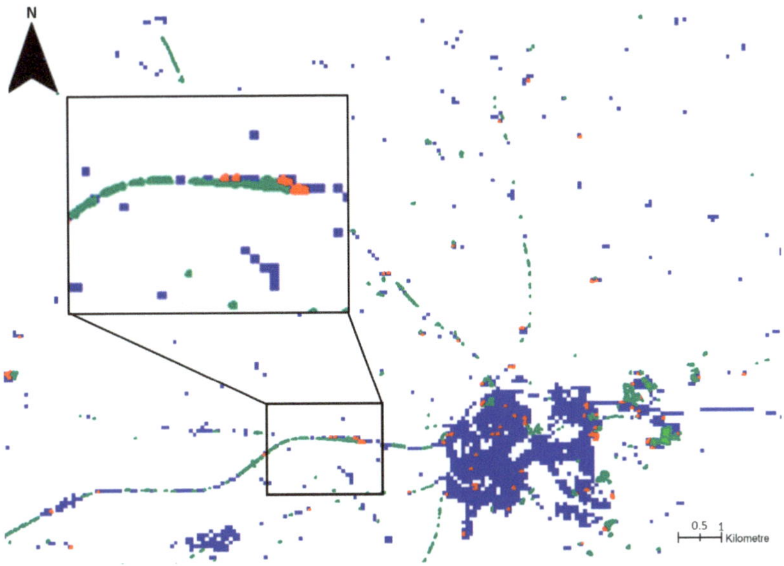

Figure 7. Existing sealed areas are shown in blue whilst potential gains are classified as actual gain in red and omission in green.

The main output from this stage is a reclassified vector IMCC1518 sealing gain dataset in which gain polygons were either classified as 'omissions' or 'actual gains'.

Step 3—Production of revised sealed 100 m spatial resolution time series 2006–2018:

The harmonized gain IMCC layers for 2015–2018 and the binary sealed layer derived from the 2018 CLC+ BB raster layers are used to produce a new 2015 status layer at 10 m resolution and layers for previous years by applying the following procedure:

1. The reclassified vector IMCC1518 layer is rasterized as a 10 m layer, only keeping values classified as 'actual sealing' gains.
2. This new sealing gain layer is combined with the losses and stable areas from the original IMCC1518 layer at 10 m to create a revised IMCC1518 layer.
3. Gain areas are removed from and losses added to the binary sealed layer derived from the CLC+ BB sealed 2018 10 m layer to create a new revised 2015 status layer.
4. Subsequent status layers for year n (e.g., year 2012) are produced from combining the status layer from the consecutive year (e.g., 2015) with the corresponding IMCC layer (e.g., IMCC 2012–2015) at 10 m resolution.
5. All 10 m layers are aggregated to 100 m spatial resolution, thus producing the harmonized dataset.

It should be stated that the 10 m layers produced from the procedure above should not be used directly because they result from a statistical calibration procedure, and the discrimination between omission and actual change even if correct at the production unit level may not be spatially correct. In addition, the reference dataset PSUs used for the calibration procedure are at 100 × 100 m. Therefore, it is thought that the harmonized imperviousness change 100 m aggregated layers are expected to be more accurate.

4. Results and Discussion

4.1. Validation of Status and Change Layers

Both datasets, the CLMS IMD and the harmonized imperiousness change at 100 m resolution were validated based on the available CLMS validation data using the blind interpretation because the plausibility analysis was performed on the IMD and not on the harmonized imperiousness change data (see Table 3).

Table 3. Comparison of validation results for blind interpretation CLMS IMD layers and harmonized imperiousness change datasets (Red and Green, respectively, indicate the worst and best results whilst amber indicates intermediate result or a result relatively close to the best one).

		CLMS IMD			Harmonized Imperiousness Change		
		RMSE	MAE	MAE > 0	RMSE	MAE	MAE > 0
2006	Overall	4.48	1.76	37.22	3.80	1.37	8.22
	Commission	0.81	0.11	2.30	2.57	0.66	3.96
	Omission	4.40	1.65	34.92	2.80	0.71	4.27
	Diff Com-Om	−3.59	−1.54	−32.63	−0.24	−0.05	−0.31
2009	Overall	4.46	1.75	36.57	3.82	1.37	8.17
	Commission	0.82	0.11	2.30	2.60	0.67	3.99
	Omission	4.38	1.64	34.27	2.80	0.70	4.18
	Diff Com-Om	−3.56	−1.53	−31.97	−0.20	−0.03	−0.19
2012	Overall	4.45	1.75	36.16	3.78	1.36	8.06
	Commission	0.85	0.11	2.37	2.57	0.66	3.94
	Omission	4.37	1.64	33.80	2.77	0.69	4.12
	Diff Com-Om	−3.52	−1.52	−31.43	−0.20	−0.03	−0.18
2015	Overall	4.60	1.77	36.29	3.75	1.35	8.01
	Commission	0.84	0.11	2.32	2.42	0.65	3.85
	Omission	4.52	1.66	33.97	2.86	0.70	4.16
	Diff Com-Om	−3.68	−1.55	−31.65	−0.44	−0.05	−0.32
2018	Overall	4.13	1.60	27.64	3.50	1.27	7.84
	Commission	1.07	0.15	2.65	2.09	0.56	3.47
	Omission	3.99	1.44	24.99	2.81	0.71	4.37
	Diff Com-Om	−2.92	−1.29	−22.35	−0.73	−0.15	−0.90

From the table above, the harmonized imperiousness change dataset provides the most balanced results overall as compared with CLMS IMD with much more omission for CLMS IMD. This is particularly highlighted when PSUs classified as non-sealed in both the validation and map layers are excluded from the analysis (MAE > 0). This is because non-sealed areas dominate the landscape with more than 97% of the area identified as non-sealed in the reference data. For the validation of thematic classes covering a relatively small proportion of a study area such as impervious areas, the MAE > 0 is preferable for the validation of continuous thematic layers and clearly identifies the most accurate dataset. However, if the dataset is to be used to extract area statistics, a balance between omission and commission errors is necessary as exhibited by the harmonized impervious change dataset.

The conclusion was that the harmonized imperiousness change dataset provided a more realistic picture of the spatial–temporal distribution of sealed areas across Europe and was therefore used in subsequent steps.

4.2. Change Area Estimation

Regarding total sealed areas, the production of the 100 m harmonized imperviousness change dataset by applying the harmonization process as described in Section 3.2 appears to produce pixel count estimates that are much closer to the change validation dataset by almost a factor 9 as shown in Table 4.

However, the picture is slightly more contrasted when assessing area statistics for member states or a group of member states for smaller countries as illustrated in Figure 8. The harmonized imperiousness change 2018 dataset is still much closer to the reference data overall except for three countries/a group of countries (Benelux + Denmark, Czech Republic + Slovakia and Germany) for which the CLMS IMD2018 dataset is closer to the reference data. Otherwise, the harmonized imperiousness change 2018 dataset is always closer to the reference data and 12 out of 22 countries/country groupings are in fact within the 95% confidence intervals of the reference data. This was true for only 4 countries in case of the CLMS IMD2018 dataset. These results would suggest that even though the CLC+ artificial class (used as a basis for constructing the harmonized imperiousness change 2018 dataset) is more accurate overall than the IMD2018, this is not always the case locally and CLC+ appears to be substantially overestimating sealed areas in the three countries/groups of countries for which the IMD2018 dataset is closer. In all other cases, the IMD2018 is underestimating sealed areas as compared with the reference dataset. This considers that the level of change is still relatively small.

Table 4. Assessment of differences in area (km^2) between CLMS IMD and harmonized imperiousness change status layers and CLMS Validation data for the EEA38 countries and the UK (Red and Green, respectively, indicate the worst and best results).

(km^2)	2006	2009	2012	2015	2018
CLMS IMD pixel count estimate	82,289	83,872	85,410	86,453	108,996
Harmonized sealing time series pixel count	169,969	171,684	173,311	174,437	175,664
CLMS validation estimate and 95% confidence interval (CI)	n/a	158,370 ± 4247	159,212 ± 4262	161,382 ± 4275	160,434 ± 7495
Second CLMS validation estimate and 95% CI	157,177 ± 4242	157,659 ± 4242	160,544 ± 4271	159,258 ± 7492	n/a
Status layer blind validation dataset	170,488	171,035	172,162	174,552	181,093
Difference with upper/lower limit of 95% CI for CLMS IMD	70,646	69,545	69,540	65,313	43,943
Difference with upper/lower limit of 95% CI for harmonized imperviousness change	8550	9067	9837	8780	7735

With respect to change areas between 2015 and 2018 for which there was a major discontinuity in the IMD dataset as illustrated in Table 2, the harmonized imperiousness change dataset also appears to provide a better characterization of gain as illustrated in Table 5. The total gain area resulting from the adjustment procedure resulted in 1297 km^2, close to the estimated 1273 km^2 estimated based on the validation data, as opposed to 3806 km^2 from the IMCC dataset.

Table 5. Comparison between imperviousness gain area estimates from CLMS validation data, initial CLC+ BB/IMCC layers and revised harmonized CLC+ BB/IMCC layer. (Red and amber, respectively, indicate large and smaller difference with upper limit of 95% CI).

km^2	2006–2009	2009–2012	2012–2015	2015–2018
CLMS validation	1197 ± 251	1553 ± 509	1226 ± 369	1176 ± 314
Initial CLC+ BB/IMCC	1715	1626	1126	3808
Difference with upper limit of 95% confidence interval	267	**0**	**0**	2318
Harmonized imperviousness change	1716	1626	1126	1297
Difference with upper limit of 95% confidence interval	267	**0**	**0**	**0**

The gain areas are now in line with the validation data for 2015–2018. It was already the case for 2009–2012 and 2012–2015, but the status layers for 2015, 2012, 2009 and 2006 should now exhibit less omission errors. The 2006–2009 layer is slightly overestimating gains and would benefit from a similar approach to that of the initial 2015–2018 IMCC, but the difference is not as large (268 km^2 versus 2316 km^2). These results are broadly confirmed when disaggregated at the level of countries/groups of countries as illustrated in Figure 9 with a few exceptions.

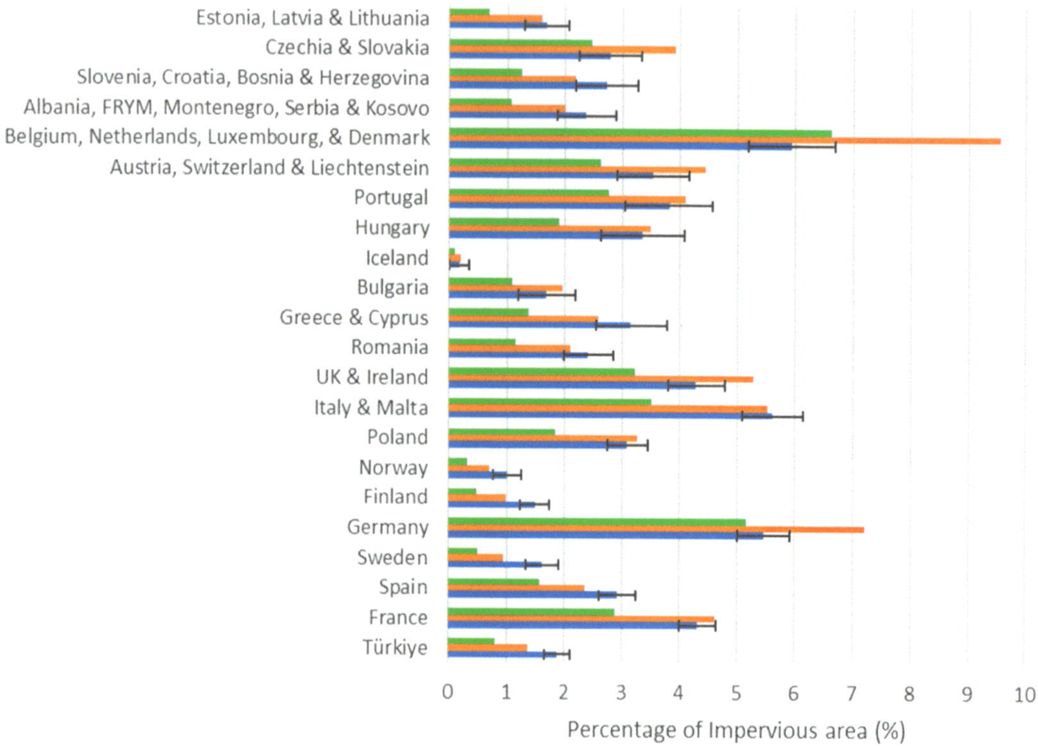

Figure 8. Percentages of impervious areas for large countries and groups of countries from EEA38 + UK for the validation reference, imperviousness degree and harmonized imperiousness change datasets for 2018 with the 95% confidence interval for the reference data.

Overall, the harmonized imperiousness change time series is much closer to the reference dataset with exceptions in Türkiye, Bulgaria, Austria, Switzerland and Liechtenstein and Iceland for which the CLMS IMD dataset is closer. In all other cases, the harmonized imperiousness change dataset is much closer to the reference data with 16 out of 22 countries/groups of countries within the 95% confidence interval range against only 8 out of 22 for the CLMS IMD dataset. This should also consider that the confidence intervals are quite large due to the relatively small number of samples classified as change for most countries/groups of countries. Nevertheless, this may suggest that the discriminating function applied to correct the 2015–2018 IMCC layer based on distance does not always appear to be sufficiently accurate and alternative approaches based on using confidence layers (which were not at our disposal) or a more sophisticated procedure based on Machine Learning/Deep Learning should further improve the quality of the results.

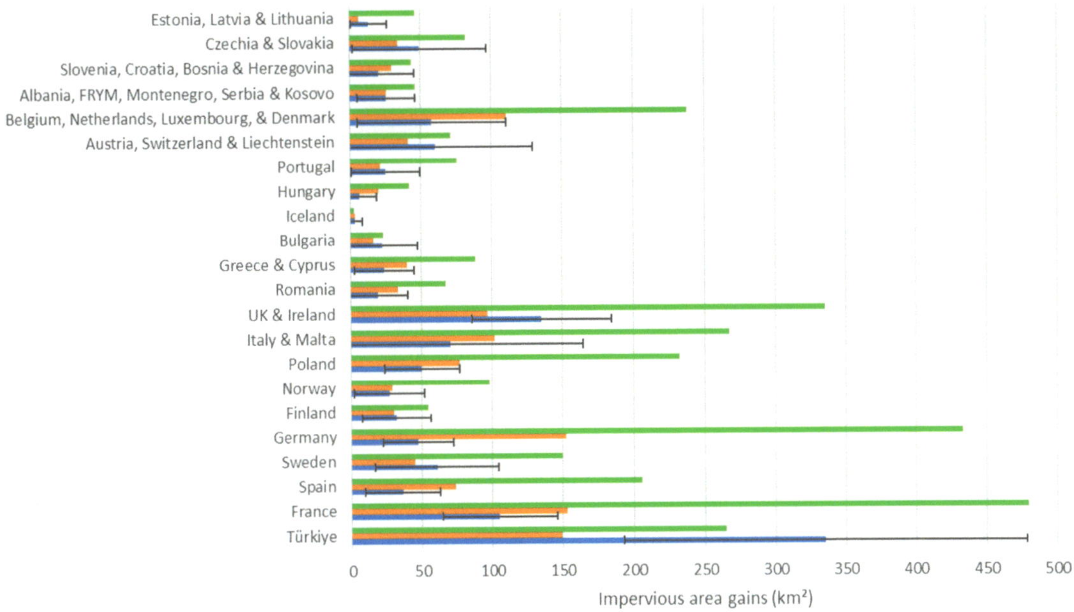

Figure 9. Total gain in impervious areas for countries/group of countries for 2015–2018 from EEA38 + UK for validation reference, CLMS IMD and harmonized imperiousness change datasets for 2018 with 95% confidence interval for reference data.

5. Conclusions and Recommendations

The development and application of a harmonization methodology based on the combination of CLMS datasets with available reference sample observations were applied to the IMD dataset for the purpose of developing robust environmental indicators. The resulting datasets provide a much more temporally and spatially consistent time series than the original CLMS dataset:

- The level of omission errors is reduced for all status layers.
- The level of commission errors is slightly increased but remains low and at a similar level to the level of omission errors, meaning that area statistics as extracted from the datasets should be close to reality, which was not the case previously.
- The discontinuity in the time series between 2015 and 2018 is now resolved with changes in line with the expected level of increase over EEA38 + UK.

The use of CLC+ BB to improve the CLMS IMD status layers was a major improvement, but there are still some areas/countries for which the CLMS IMD areas appear closer to the reference data. Therefore, some more sophisticated data fusion approaches should be applied combining the best of the two datasets. The approach developed in this study is applicable to other datasets whenever spatio-temporal consistency is required and can be applied as long as suitable reference data (i.e., observations obtained from a probabilistic sample) are available to calibrate the changes.

Even if the area statistics from the harmonized change layer are more reliable, there is a substantial amount of error for both omission and commissions and the future production of more accurate 10 m change layers should be prioritized as part of the CLMS production. More sophisticated approaches based on confidence layers, Machine Learning/Deep Learning methods, image-to-image change detection methods and/or manual enhancements if required should be applied. Status layers should be produced from the combination of the previous status layer with the new change detection and not the other way around, meaning that the change layer should be produced first and detailed product specifications should be defined that ensure that the product is accurate but also that there is a balance between omission and commission errors. In addition, the historical dataset should be reproduced every time that a new status layer becomes available as long as its quality is improved. The historical layers can easily be reprocessed based on the accounting approach starting from the most recent layer and integrating the change layers for each time period as done in this study.

In addition, a reliable reference dataset should be produced and made available to provide a basis for calibrating the production of the change and status layer. The LUCAS dataset has the potential to provide this, but the specifications of LUCAS should be more closely related to those of the CLMS IMD layer in terms of the thematic definition as well as spatial support. Finally, the issue of calibrating EO-based layers at a more disaggregated level such as NUTS (Nomenclature des Unités Territoriales Statistiques/Nomenclature of Territorial Units for Statistics) level 3 for Europe is necessary to produce relevant statistical indicators and the development of small-area estimators as described in [37] should be considered in the production of a suitable reference dataset.

Author Contributions: Conceptualization, C.S. and G.M.; methodology, C.S.; software, G.M.; validation, C.S.; formal analysis, C.S. and G.M.; investigation, C.S. and E.I.; resources, E.I.; data curation, E.I.; writing—original draft preparation, C.S. and E.I.; writing—review and editing, E.I., J.M., L.D. and G.M.; supervision, E.I.; project administration, E.I. and J.M.; funding acquisition, E.I. All authors have read and agreed to the published version of the manuscript.

Funding: This research was funded by the European Environment Agency, contract number 3506/R0-REGIND/EEA.59425; the validation reference dataset was produced from the Copernicus Land Monitoring Service EEA/MDI/14/010/Lot1 contract implemented by the European Environment Agency; and the APC was funded by the European Commission Joint Research Centre.

Data Availability Statement: The data produced as part of this work are available from the following link: https://sdi.eea.europa.eu/catalogue/srv/eng/catalog.search#/metadata/7db5272f-5825-442b-9fa0-8b1c424f4ff1 (accessed on 28 May 2024).

Conflicts of Interest: Author Christophe Sannier was employed by the company GAF AG, Author Gergely Maucha was employed by the company Lechner Non-Profit Ltd. The remaining authors declare that the research was conducted in the absence of any commercial or financial relationships that could be construed as a potential conflict of interest. The views expressed in this article are solely those of the authors and its content does not necessarily represent the views or position of the European Environment Agency.

References

1. European Commission. *Proposal for a Directive of the European Parliament and the Council on Soil Monitoring and Resilience (Soil Monitoring Law). COM(2023) 416 Final*; European Commission: Brussels, Belgium, 2023.
2. European Commission; Eurostat. LUCAS—Land Use and Land Cover Survey. 2023. Available online: https://ec.europa.eu/eurostat/statistics-explained/index.php?title=LUCAS_-_Land_use_and_land_cover_survey (accessed on 28 May 2024).
3. European Environment Agency. CLC+ Backbone 2018 (raster 10 m), Europe, 3-yearly. February 2023. Available online: https://doi.org/10.2909/cd534ebf-f553-42f0-9ac1-62c1dc36d32c (accessed on 28 May 2024).
4. European Environment Agency. Imperviousness Density 2018 (raster 100 m), Europe, 3-yearly. August 2020. Available online: https://doi.org/10.2909/524fa72f-61d7-4364-801e-3e271d7b10bc (accessed on 28 May 2024).
5. Pesaresi, M.; Guo, H.; Blaes, X.; Ehrlich, D.; Ferri, S.; Gueguen, L.; Halkia, M.; Kauffmann, M.; Kemper, T.; Lu, L.; et al. A Global Human Settlement Layer From Optical HR/VHR RS Data: Concept and First Results. *IEEE J. Sel. Top. Appl. Earth Obs. Remote Sens.* **2013**, *6*, 2102–2131. [CrossRef]
6. Marconcini, M.; Metz-Marconcini, A.; Üreyen, S.; Palacios-Lopez, D.; Hanke, W.; Bachofer, F.; Zeidler, J.; Esch, T.; Gorelick, N.; Kakarla, A.; et al. Outlining where humans live, the World Settlement Footprint 2015. *Sci. Data* **2020**, *7*, 242. [CrossRef] [PubMed]
7. Waldner, F.; Defourny, P. Where can pixel counting area estimates meet user-defined accuracy requirements? *Int. J. Appl. Earth Obs. Geoinf.* **2017**, *60*, 1–10. [CrossRef]
8. Olofsson, P.; Foody, G.M.; Herold, M.; Stehman, S.V.; Woodcock, C.E.; Wulder, M.A. Good practices for estimating area and assessing accuracy of land change. *Remote Sens. Environ.* **2014**, *148*, 42–57. [CrossRef]
9. McRoberts, R.E. Satellite image-based maps: Scientific inference or pretty pictures? *Remote Sens. Environ.* **2011**, *115*, 715–724. [CrossRef]
10. Gallego, J. Remote sensing and land cover area estimation. *Int. J. Remote Sens.* **2004**, *25*, 3019–3047. [CrossRef]

11. Sannier, C.; Gallego, J.; Dahmer, J.; Smith, G.; Dufourmont, H.; Pennec, A. Validation of Copernicus High Resolution Layer on Imperviousness Degree for 2006, 2009 and 2012. In Proceedings of the 12th International Symposium on Spatial Accuracy Assessment in Natural Resources and Environmental Sciences, Montpellier, France, 5–8 July 2016. Available online: http://www.spatial-accuracy.org/system/files/Accuracy%202016%20Proceedings.pdf (accessed on 28 May 2024).
12. Gangkofner, U.; Weichselbaum, J.; Kuntz, S.; Brodsky, L.; Larsson, K.; De Pasquale, V. Update of the European High-resolution Layer of Built-up Areas and Soil Sealing 2006 with Image2009 Data. In Proceedings of the 30th EARSeL Symposium 2010: Remote Sensing for Science, Education and Culture, Paris, France, 31 May–4 June 2010.
13. Lefebvre, A.; Sannier, C.; Corpetti, T. Monitoring urban areas with Sentinel-2A data: Application to the update of the Copernicus high resolution layer imperviousness degree. *Remote Sens.* **2016**, *8*, 606. [CrossRef]
14. Stehman, S.V. Sampling designs for accuracy assessment of land cover. *Int. J. Remote Sens.* **2009**, *30*, 5243–5272. [CrossRef]
15. Gallego, J.; Delincé, J. The European land use and cover area-frame statistical survey. In *Agricultural Survey Methods*; John Wiley & sons: Hoboken, NJ, USA, 2010; pp. 149–168.
16. Büttner, G.; Kosztra, B.; Maucha, G.; Pataki, R. *Implementation and Achievements of CLC2006*; ETC-LUSI, EEA; European Environment Agency: Copenhagen, Denmark, 2012; 65p. Available online: https://www.eea.europa.eu/data-and-maps/data/corine-land-cover-3/clc-final-report/clc-final-report (accessed on 28 May 2024).
17. Selkowitz, D.J.; Stehman, S.V. Thematic accuracy of the National Land Cover Database (NLCD) 2001 land cover for Alaska. *Remote Sens. Environ.* **2011**, *115*, 1401–1407. [CrossRef]
18. Olofsson, P.; Foody, G.M.; Stehman, S.V.; Woodcock, C.E. Making better use of accuracy data in land change studies: Estimating accuracy and area and quantifying uncertainty using stratified estimation. *Remote Sens. Environ.* **2013**, *129*, 122–131. [CrossRef]
19. European Environment Agency. HRL Imperviousness 2015 Validation Report. 2018. Available online: https://land.copernicus.eu/en/technical-library/hrl-imperviousness-2015-validation-report/@@download/file (accessed on 28 May 2024).
20. European Environment Agency. HRL Imperviousness 2018 Validation Report. 2020. Available online: https://land.copernicus.eu/en/technical-library/hrl-imperviousness-2018-validation-report/@@download/file (accessed on 28 May 2024).
21. Stehman, S.V.; Czaplewski, R.L. Design and analysis for thematic map accuracy assessment: Fundamental principles. *Remote Sens. Environ.* **1998**, *64*, 331–334. [CrossRef]
22. Czaplewski, R.L. Chapter 5: Accuracy assessment of maps of forest condition: Statistical design and methodological considerations. In *Remote Sensing of Forest Environments: Concepts and Case Studies*; Wulder, M.A., Franklin, S.E., Eds.; Kluwer Academic Publishers: Boston, MA, USA, 2003; pp. 115–140, 515p.
23. McRoberts, R.E.; Stehman, S.V.; Liknes, G.C.; Næsset, E.; Sannier, C.; Walters, B.F. The effects of imperfect reference data on remote sensing-assisted estimators of land cover class proportions. *ISPRS J. Photogramm. Remote Sens.* **2018**, *142*, 292–300. [CrossRef]
24. Stehman, S.V.; Mousoupetros, J.; McRoberts, R.E.; Næsset, E.; Pengra, B.W.; Xing, D.; Horton, J.A. Incorporating interpreter variability into estimation of the total variance of land cover area estimates under simple random sampling. *Remote Sens. Environ.* **2022**, *269*, 112806. [CrossRef]
25. Pontius, R.G., Jr.; Cheuk, M.L. A generalized cross-tabulation matrix to compare soft-classified maps at multiple resolutions. *Int. J. Geogr. Inf. Sci.* **2006**, *20*, 1–30. [CrossRef]
26. Silván-Cárdenas, J.L.; Wang, L. Sub-pixel confusion-uncertainty matrix for assessing soft classifications. *Remote Sens. Environ.* **2008**, *112*, 1081–1095.
27. European Commission. *Joint Research Centre. Call for Tenders JRC/IPR/2022/OP/1365; Implementation and Operation of the Copernicus Emergency Management Service Exposure Mapping Component Copernicus Global Human Settlement Layer (GHSL) Part 2: Technical Specifications*; European Commission: Brussels, Belgium, 2022.
28. Gallego, J.; Sannier, C.; Pennec, A. Validation of Copernicus Land Monitoring Services and Area Estimation. In Proceedings of the Seventh International Conference on Agricultural Statistics (ICAS VII), Rome, Italy, 26–28 October 2016. Available online: https://www.istat.it/storage/icas2016/g42-gallego.pdf (accessed on 28 May 2024).
29. Pontius, R.G., Jr.; Millones, M. Death to kappa: Birth of quantity disagreement and allocationdisagreement for accuracy assessment. *Int. J. Remote Sens.* **2011**, *32*, 4407–4429. [CrossRef]
30. Gallego, J. *Sampling Frames of Square Segments*; EUR 16317; Official Publication of the European Communities: Luxembourg, 1995; 72p.
31. Sannier, C.; McRoberts, R.E.; Fichet, L.V.; Makaga, E.M.K. Using the regression estimator with landsat data to estimate proportion forest cover and net proportion deforestation in gabon. *Remote Sens. Environ.* **2014**, *151*, 138–148. [CrossRef]
32. Stehman, S.V.; Xing, D. Confidence Intervals for Proportion of Area Estimated from a Stratified Random Sample. *Remote Sens. Environ.* **2022**, *280*, 113193. [CrossRef]
33. Metzger, M.J.; Bunce RG, H.; Jongman, R.H.; Mücher, C.A.; Watkins, J.W. A climatic stratification of the environment of Europe. *Glob. Ecol. Biogeogr.* **2005**, *14*, 549–563. [CrossRef]
34. Song, X.P.; Potapov, P.V.; Krylov, A.; King, L.; Di Bella, C.M.; Hudson, A.; Khan, A.; Adusei, B.; Stehman, S.V.; Hansen, M.C. National-scale soybean mapping and area estimation in the United States using medium resolution satellite imagery and field survey. *Remote Sens. Environ.* **2017**, *190*, 383–395. [CrossRef]
35. Sannier, C.; McRoberts, R.E.; Fichet, L.V. Suitability of Global Forest Change data to report forest cover estimates at national level in Gabon. *Remote Sens. Environ.* **2016**, *173*, 326–338. [CrossRef]

36. Hansen, M.C.; Potapov, P.V.; Moore, R.; Hancher, M.; Turubanova, S.A.; Tyukavina, A.; Thau, D.; Stehman, S.V.; Goetz, S.J.; Loveland, T.R.; et al. High-Resolution Global Maps of 21st-Century Forest Cover Change. *Science* **2013**, *342*, 850–853. Available online: http://earthenginepartners.appspot.com/science-2013-global-forest (accessed on 28 May 2024). [CrossRef] [PubMed]
37. Rao, J.N.K. *Small Area Estimation*; Wiley: New York, NY, USA, 2003.

land

MDPI

Article

Optimization of Land Use Structure Based on the Coupling of GMOP and PLUS Models: A Case Study of Lvliang City, China

Zhen Wang †**, Anya Zhong** † **and Quanzhi Li** *

School of Geographical Sciences and Surveying Engineering, China University of Mining and Technology, Beijing 100083, China
* Correspondence: bqt2300204053@student.cumtb.edu.cn
† These authors contributed equally to this work.

Abstract: Reasonable land use planning and management efficiently allocates land resources, promotes socio-economic development, protects the ecological environment, and fosters sustainable development. It is a crucial foundation for achieving harmonious coexistence between humans and nature. Optimizing land use is key to land use planning and management. Four scenarios are established: an economic development scenario (EDS), an ecological protection scenario (EPS), a natural development scenario (NDS), and a coordinated development scenario (CDS). This study simulates land use patterns under these scenarios through the coupling of the GMOP and PLUS models. It analyzes the land use efficiency transformation index, landscape ecological index, comprehensive land use benefits, and ecosystem service value (ESV) for each pattern. The optimal land use pattern is determined by balancing these factors. The results indicated that under the CDS, the areas of wasteland, grassland, forest land, water bodies, construction land, and unused land in Lvliang City were 6724.29 km^2, 6664.74 km^2, 6581.84 km^2, 126.94 km^2, 1017.33 km^2, and 0.42 km^2, respectively. This represented the optimal land use plan for Lvliang City. The plan minimized human interference with the landscape pattern, achieved the highest land use efficiency transformation index, and reached a reasonable balance between land use benefits and ESV. The research findings provide valuable insights and decision support for regional land use planning, territorial space planning, and related policy formulation.

Keywords: land use optimization; MOP model; PLUS models; ecosystem service values; Lvliang city

Citation: Wang, Z.; Zhong, A.; Li, Q. Optimization of Land Use Structure Based on the Coupling of GMOP and PLUS Models: A Case Study of Lvliang City, China. *Land* **2024**, *13*, 1335. https://doi.org/10.3390/land13081335

Academic Editors: Jūratė Sužiedelytė-Visockienė and Bindong Sun

Received: 15 July 2024
Revised: 16 August 2024
Accepted: 21 August 2024
Published: 22 August 2024

1. Introduction

China possesses extensive land resources; however, the significant population leads to a limited availability of land per capita. Additionally, inefficiencies in land use structures are evident, manifesting as low land use efficiency, loss of farmland, and an imbalance between urban and rural land distribution [1–3]. The land use structure illustrates the distribution and proportion of diverse land use types within a defined area, thereby reflecting the composition of land designated for various purposes in a specific spatial context. An inadequate land use structure directly hinders regional economic development and disrupts the harmonious functioning of ecosystems, thereby affecting the sustainability of the region [4]. Land use structure optimization involves the utilization of scientific methodologies to improve the allocation and management of finite land resources. This process requires the consideration of multiple development objectives and strives to attain an optimal balance among different regions and land uses. By doing so, the efficiency of land resource utilization is enhanced, the ecological environment is preserved, the quality of life for residents is improved, and sustainable development is promoted across economic, social, and environmental dimensions [5,6]. With the rapid progression of urbanization, land utilization and development are essential components of urban construction. Nevertheless, the intensification of land development inevitably exerts a negative impact on the environment. Activities such as deforestation, the encroachment on farmland, overgrazing, and coal mining significantly disrupt the balance between economic advancement and environmental conservation [7–10]. Severe impacts on the ecosystem are inevitably caused by inefficient land use, and significant depletion of ESV is likewise occasioned by excessive land development and inappropriate land use type conversions. The conflict between economic development and environmental preservation is mitigated through the optimization of land use structures and layouts. This approach simultaneously addresses multiple objectives, such as promoting economic growth, ensuring ecological protection, and fostering social harmony, thus contributing to sustainable regional development.

There has been extensive research on land use optimization both domestically and internationally. Land use structure optimization is primarily divided into quantitative structure optimization and spatial layout simulation optimization [11,12]. The main tools for quantitative structure optimization include GM (1, 1), system dynamics models, and Markov chains. However, these models have drawbacks such as high data requirements, simple model assumptions, sensitivity to external factors, and limited prediction ranges [13–15]. The GMOP model, on the other hand, can simultaneously consider multiple objectives and factors, has strong data processing capabilities and flexibility, and can simulate land use changes under different policy implementations, providing more scientific and reasonable predictions for land use planning [16,17]. Many scholars have conducted extensive research on using GMOP for land use structure optimization. For example, Zhu et al. [18] developed a multi-objective, constrained regional land use structure model known as PLUS-GMOP, in which three scenarios were hypothesized. This model facilitated the selection of the optimal land use strategy for the Wuhan metropolitan area, providing an innovative technical framework for the sustainable development of large urban regions; Li et al. [19] developed a grey linear programming optimization model designed to improve the land use structure in the Sichuan-Yunnan ecological barrier region. This model aimed to maximize ecological value while also enhancing economic benefits, utilizing the current land use conditions in the study area. This approach offered a clear methodology for optimizing ecological service value within large ecological functional zones; Mo et al. [20] employed the gray linear multi-objective programming approach, framed within the context of production–living–ecological space (PLES), to develop the GMOP-Markov-PLUS model. This model effectively predicted future land use patterns across various scenarios. Additionally, it proposed three distinct long-term land use strategies for the study area, which included considerations for ecological conservation, economic growth, and sustainable development prospects.

Multi-objective linear programming effectively addresses the quantitative aspects of land use optimization, generating appropriate schemes for land use quantities. The predominant models for simulating land structure optimization scenarios currently encompass the CA–Markov model [21], the CLUES model [22], and the FLUS model [23]. Nonetheless, these models are encumbered by challenges such as low spatial resolution, omission of local details, inadequate local adaptability of model parameters, and an absence of dynamic adaptability. While they attain high precision in simulations of small-scale land use, their performance is diminished in larger-scale simulations. The PLUS model primarily functions as a simulation tool for forecasting future alterations in land use. It integrates a module for analyzing land expansion strategies along with a cellular automata model that is influenced by various classes of stochastic patch seeds. It is distinguished by its user-friendly interface and simplicity of operation. The PLUS model is particularly effective for large-scale regional land use simulations. Its integration with GIS enables a detailed analysis of spatial heterogeneity, characteristics, and temporal dynamics within land use structures. As a result, this integration significantly improves the optimization of spatial configurations in land use patterns [24,25].

As urban development progresses and various social factors drive change, the pattern of land use is becoming increasingly complex and uncertain, making the study of land use structure optimization a hot topic. Zhong et al. [26] combined the GMOP and PLUS models, considering the existing land use layout and policy constraints, and concluded that the land pattern under the balanced development (BD) scenario ensured economic growth without compromising ecological benefits. Shu et al. [27] employed the ESV, GMOP, and PLUS models to forecast land use changes and ESV across various development scenarios. Their findings indicated that the sustainable development (SD) scenario was likely more appropriate for future regional advancements. While the ESV was marginally lower compared to the ecological land protection (ELP) scenario, there was a notable increase in economic benefits. Luan et al. [28] conducted simulations of land use patterns across different scenarios by integrating the NSGA-II and PLUS models. Their analysis revealed that the most significant variations among the scenarios were predominantly observed in the areas designated for forest land and cultivated land. Meng et al. [29] utilized a combination of the GMOP and PLUS models to simulate land use configurations across four distinct scenarios. Their analysis focused on changes in ecological benefits, economic advantages, and carbon emissions, enabling the identification of the most favorable land use arrangement.

Previous studies primarily used the PLUS model to simulate land use configurations under various development scenarios, and then conducted a basic analysis of ecosystem service value (ESV) and land use benefits to identify the optimal land use structure. However, these studies did not examine the conversion efficiency between ESV and land use benefits as land use patterns evolved, nor did they account for the impact of changing land use patterns on landscape configurations. This study is based on the coupling of the GMOP and PLUS models, establishing different scenarios and constraints to predict the land use structure of Lvliang City in 2035. A land use benefit conversion index is constructed to explore the conversion efficiency between land use benefits and ESV during the transition from the current land use pattern to future patterns. The sustainable development potential is evaluated under various development scenarios, and the optimal land use structure for Lvliang City is determined through a comprehensive analysis of the land use benefit conversion index, land use benefits, landscape pattern indices, and ESV across different scenarios.

2. Study Area and Data Sources

2.1. Study Area

Lvliang City, a prefecture-level city under the administration of Shanxi Province, boasts abundant mineral resources and favorable metallogenic conditions, with coal, iron, and dolomite being the primary minerals (Figure 1). It is distinguished by multiple metallogenic epochs, extensive distribution, varied mineral types, significant reserves, superior quality, and ease of extraction. Spanning latitudes from 36°43′ to 38°43′ North and longitudes from 110°22′ to 112°19′ East, Lvliang City is adjacent to Taiyuan and Jinzhong in the east, separated from Shaanxi Province by the Yellow River to the west, and borders Linfen and Xinzhou to the south and north, respectively, with a total area of 21,000 square kilometers. The city possesses a favorable climate, with local economic development primarily dependent on activities such as coal mining and agricultural cultivation. However, human activities like mineral resource extraction pose a threat to the ecological environment. To address the conflict between local economic growth and environmental protection, Lvliang City was chosen as the study area to optimize a land use pattern that considers both economic advancement and ecological preservation.

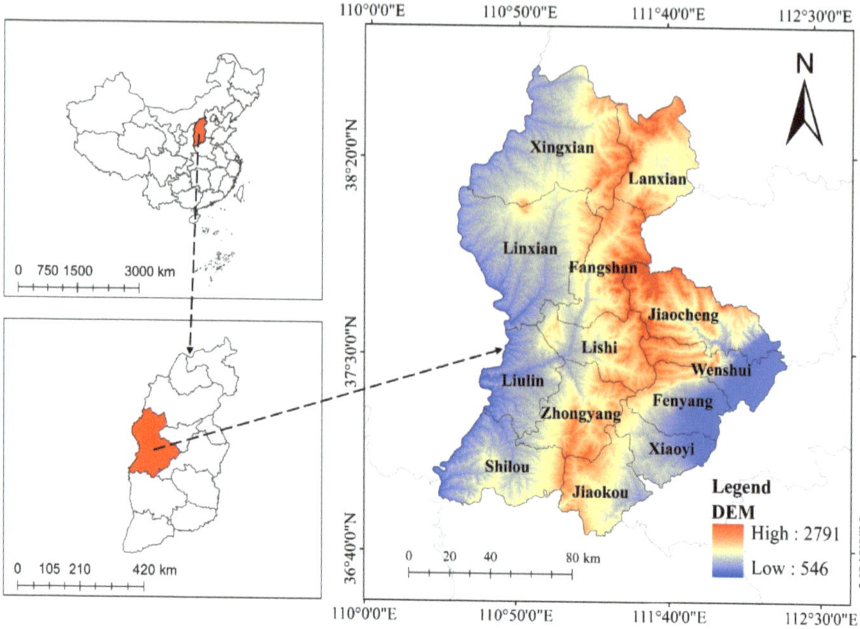

Figure 1. Map of the study area.

2.2. *Data Sources*

The data utilized in the study primarily consisted of spatial and textual information. Spatial data includes land use, topographical, and meteorological datasets, while textual data encompasses the Lvliang Statistical Yearbook (2017–2023) and the Land Spatial Planning (2021–2035) datasets. Based on the relevant classification criteria, the present land use data were classified into six main categories. Processing of the spatial data is conducted using the ArcGIS 10.8 platform, with further details provided in Table 1. The data employed in this study were sourced from reputable websites. Extensive research carried out by numerous scholars utilizing these datasets has confirmed their scientific rigor and objectivity, establishing their relevance for investigations into land use optimization.

Table 1. Data sources and description.

Data Type	Data Content	Data Description	Data Source
Land Use Type	Land use monitoring data for Lvliang City, 2005, 2010 and 2020	The spatial resolution is 30 m × 30 m, and it was divided into six categories according to the purpose of the study	https://www.resdc.cn/, accessed on 16 March 2024.
Topographic Data	Soil Type	Image element size is 1000 × 1000	https://www.resdc.cn/, accessed on 16 March 2024.
	DEM	Initial resolution of 250 m × 250 m	https://www.resdc.cn/, accessed on 16 March 2024.
	Slope	Initial resolution of 250 m × 250 m	Generated by DEM
	River Data	vector data	https://www.webmap.cn/, accessed on 16 March 2024.
Meteorological Data	Rainfall Data	Image element size is 1000 × 1000	https://www.resdc.cn/, accessed on 16 March 2024.
	Temperature Data	Image element size is 1000 × 1000	https://www.resdc.cn/, accessed on 16 March 2024.

Table 1. *Cont.*

Data Type	Data Content	Data Description	Data Source
Social Data	GDP	Image element size is 1000 × 1000	https://www.resdc.cn/, accessed on 16 March 2024.
	Population	Image element size is 1000 × 1000	https://www.resdc.cn/, accessed on 16 March 2024.
	County Government Locations	Image element size is 1000 × 1000	https://www.resdc.cn/, accessed on 16 March 2024.
	Primary roads Secondary roads Tertiary roads	Road vector data	https://www.webmap.cn/, accessed on 16 March 2024.

3. Design Framework and Methodology

3.1. Data Processing

3.1.1. Research Framework

The research methodology presented in this paper included three main components (Figure 2):

(1) Prediction of Land Use Quantity Structure: The prediction of land quantity structure involves forecasting the proportions of different land types within a specific area in the future. Four different development scenarios are initially set, as follows: a natural development scenario (NDS), an economic development scenario (EDS), an ecological protection scenario (EPS), and a coordinated development scenario (CDS). The prediction of land use structure under the NDS is based on the land use data of Lvliang City from 2005 and 2020. Using the Markov Chain model, the land area for each type of land use in Lvliang City in 2035 is predicted. This process is implemented in the PLUS model. The EDS and EPS fall under single-objective planning problems, as each of these scenarios requires maximizing either economic or ecological benefits as the sole objective. By calculating and predicting the economic and ecological benefit coefficients for each land type in 2035 and setting constraints for each type of land use, the solution is achieved using LINGO20.0 software. LINGO, developed by Lindo System, Inc. (Chicago, IL, USA) in the United States, is an interactive solver for both linear and general optimization. It effectively addresses nonlinear programming challenges in addition to solving various linear and nonlinear equations, making it a highly versatile tool and an optimal choice for tackling complex optimization models. The CDS falls under multi-objective planning problems, solved using the NSGA-II.

(2) Simulation of Land Use Structure Layout: Based on the predicted land use quantity structures under different scenarios, and using the 2020 land use data as the baseline, the spatial layout of land use in Lvliang City under various future development scenarios is simulated.

(3) Analysis of Land Use Layout: Based on the predicted areas of each land type under different scenarios, the benefits of land use are calculated using ecological and economic benefit coefficients. The ESV is calculated using the Xie Gaodi equivalent factor method. In this study, four landscape pattern indices are selected to analyze the land use layout under different scenarios. These indices include the aggregation index (AI), the largest patch index (LPI), the landscape division index (DIVISION), and the Shannon Diversity Index (SHDI), all calculated using Fragstats4.2 software. These landscape pattern indices reflect the aggregation and dispersion states of the landscape under different development scenarios. Additionally, a land use benefit conversion index is constructed to analyze the efficiency of conversion between ESV and land use benefits. Finally, a comprehensive analysis of land use benefits, ESV, landscape pattern indices, and land use benefit conversion indices under different scenarios is conducted to determine the optimal land use structure.

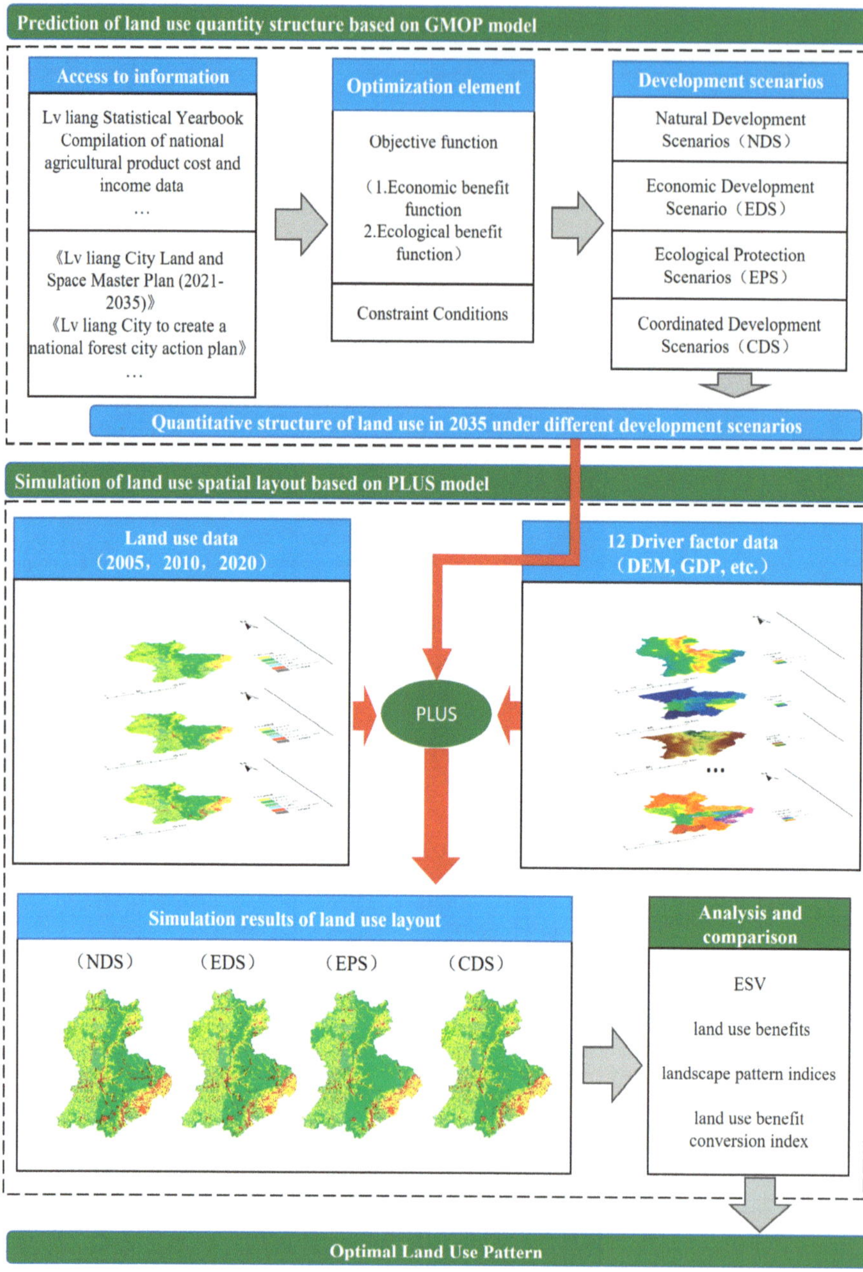

Figure 2. Diagram of the research idea.

3.1.2. Verification of Simulation Accuracy

The spatial distribution of land use in Lvliang City in 2020 was obtained by simulating land use data from 2010 (Figure 3). The analysis indicated a strong correspondence between the land use pattern simulated for 2020 by the PLUS model and the actual land use conditions. The arrangement of the six simulated land use categories demonstrated significant agreement with real-world observations. The model's accuracy was evaluated by comparing the simulated outcomes with the actual 2020 land use data, resulting in a Kappa coefficient of 0.88 and an overall accuracy of 91.60%. These findings underscore the high precision of the PLUS model, confirming its effectiveness in forecasting future land use changes within the study area.

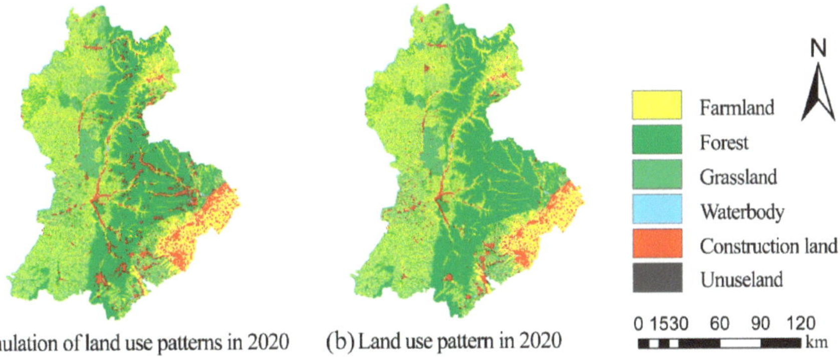

(a) Simulation of land use patterns in 2020 (b) Land use pattern in 2020

Figure 3. Comparison of land use modeling with actual data for 2020.

3.2. Methodology

3.2.1. The GMOP Model

The GMOP model is developed by integrating the GM (1, 1) model with the MOP model. Within this framework, the GM model predicts the eco-efficiency and economic coefficients of potential land use types, thereby establishing a foundation for the formulation of the objective function. Meanwhile, the MOP model addresses multi-objective optimization challenges across different scenarios [30], yielding the area allocated to each land use type by the year 2035 under various conditions. The MOP model, a pivotal model in the study of land use optimization, is predicated on constrained data and objective laws to facilitate scientifically grounded predictions. It provides a method for optimizing one or multiple objectives [31]. The detailed model is delineated below:

$$F_1(\mathrm{x}) = \max \sum_{j=1}^{n} a_j x_j \tag{1}$$

$$F_2(\mathrm{x}) = \max \sum_{j=1}^{n} b_j x_j \tag{2}$$

$$s.t = \begin{cases} \sum_{j=1}^{n} c_{ij} x_j = (\geq, \leq) d_i \ (i = 1, 2, 3 \ldots, m) \\ x_j \geq 0 \ (j = 1, 2, 3 \ldots, n) \end{cases} \tag{3}$$

where $F_1(x)$ and $F_2(x)$ denote the functions for land economic benefits and ecological benefits, respectively. Here, x_j represents the area allocated to each land use type, while n indicates the total number of variables. The coefficients a_j and b_j relate to the economic and ecological benefit coefficients for each respective land use type. The notation "s.t" indicates the constraints imposed on land use for each category, with c_{ij} representing the coefficient associated with the j-th variable in the i-th constraint, and m signifying the total number of constraint conditions. Finally, d_i pertains to the i-th constraint condition.

3.2.2. The PLUS Model

The PLUS model, operating as a grid-based cellular automaton (CA), is used to simulate land use and land cover (LULC) changes. This model integrates a land expansion analysis strategy (LEAS) within the CA framework, enabling the exploration of different land use transitions. It supports the creation of various scenarios to forecast and analyze future land use dynamics [32].

(1) Adaptation Probability. The LEAS module incorporates a stochastic sampling mechanism designed to reduce computational costs while simultaneously utilizing the random forest algorithm to assess the development probabilities associated with various land use types. The formula is presented as follows:

$$P_{i,k(X)}^{d} = \frac{\sum_{n=1}^{M} I[hn(X) = d]}{M} \tag{4}$$

where M denotes the total count of decision trees, X represents the vector that comprises the driving factors, $hn(X)$ indicates the predicted land use type generated by the n-th decision tree, and d takes on a value of either 0 or 1.

(2) Adaptive Inertia Coefficient. This coefficient is adaptively adjusted during repeated runs, based on the discrepancy between the expected land type data and the actual land type data. This mechanism effectively mitigates the uncertainties and complexities associated with natural processes and human activities involved in land use conversion. Consequently, it improves the accuracy of the simulation model and attains the intended outcomes regarding land use types [33]. The formula is as follows:

$$D_k^t = \begin{cases} D_k^{t-1}\left(\left|G_k^{t-1}\right|\right) \leq \left|G_k^{t-2}\right|\right) \\ D_k^{t-1} \times \frac{G_k^{t-2}}{G_k^{t-1}}\left(0 > G_k^{t-2} > G_k^{t-1}\right) \\ G_k^{t-1} \times \frac{G_k^{t-1}}{G_k^{t-2}}\left(G_k^{t-1} > G_k^{t-2} > 0\right) \end{cases} \tag{5}$$

where D_k^t signifies the inertia coefficient for the k-th land use type at time t. Additionally, G_k^{t-1} and G_k^{t-2} represent the discrepancies between the actual land amount and the demand at times $t-1$ and $t-2$, respectively.

(3) Optimization of Land Use Layout. Twelve factors, encompassing elevation, slope, population, soil type, GDP, road networks, rivers, and distance to the county government seat, are identified as driving forces for land use change. Concurrently, water bodies are designated as restricted areas during the optimization process. The precision of the model is assessed through two primary parameters: overall accuracy and the Kappa coefficient.

3.2.3. Constructing the Objective Function

(1) Economic Benefit Function. The economic advantages of land are primarily defined by the economic output per unit area for each category of land. This study utilizes statistical yearbooks from Luliang City spanning the years 2017 to 2023. The output for cultivated and forest land is indicated by the values of agricultural and forestry production, respectively. The output for grassland is denoted by the value derived from animal husbandry, while the output for aquatic areas is represented by the fishery output value. Furthermore, the economic impact of construction land is demonstrated through the output values generated by the secondary and tertiary industries.

(2) Ecological Benefit Function. The ecological benefit coefficients of land are primarily assessed using the equivalent factor method, as suggested by researchers such as Xie Gaodi. This methodology primarily captures the ESV provided by land. Since the supply services of land ecosystems are already incorporated into the economic benefits, the ecological advantages encompass the regulatory, supporting, and cultural services of ecosystems. This study employs the terrestrial ESV equivalent factor method, as recommended by scholars, including Xie Gaodi [34], for the purpose of evaluation. Additionally, the equivalent factor table is adjusted based on the ratio of

the NPP level in Lvliang City compared to the national average [35]. Data regarding the prices of local food crops and their yield per unit area are obtained by consulting the Lvliang Statistical Yearbook from 2016 to 2022. The ESV is quantified at one-seventh of the economic value linked to grain production per unit area of farmland. This value encompasses a range of ecosystem services, including supply, regulatory, supporting, and cultural services. Acknowledging that supply services are included within economic benefits, ecological benefits are delineated by regulatory, supporting, and cultural services. Annual coefficients for land ecological benefits are calculated, and these coefficients are projected for Lvliang City in 2035 using the GM (1: 1) model. To ensure the reliability of the revised eco-efficiency coefficients for land, a sensitivity index is employed to evaluate how variations in these coefficients affect the total ESV for each land type. The sensitivity of ESV to these coefficients is evaluated by modifying the eco-efficiency coefficients for each land type by ±50 percent. The formula is as follows:

$$C_s = \frac{(E_2 - E_1)/E_1}{(V_{2i} - V_{1i})/V_{1i}} \tag{6}$$

where C_S denotes the sensitivity index that measures the response of a specific land type to the value of land ecosystem services. E_1 and E_2 represent the ESV in Lvliang City prior to and following the adjustment, respectively. V_{1i} and V_{2i} indicate the ecological benefit coefficients for the i-th land type before and after the adjustment, respectively. A C_S value of less than 1 indicates that the ESV is inelastic regarding the ecological benefit coefficients of that land type. Lower C_S values suggest a diminished responsiveness of the assessment of land ESV to the precision of the ecological benefit coefficients, thereby indicating a higher degree of rationality in the coefficients [36]. Relevant parameters are shown in Tables 2 and 3.

Table 2. Parameters of economic and ecological benefits per unit area of land use type (unit: RMB/km^2).

Efficiency	Farmland	Forest	Grassland	Waterbody	Construction Land	Unused Land
economic efficiency	808.48	167.74	247.83	44.76	164,272.42	0.01
eco-efficiency	33.88	215.17	92.42	1316.57	0	2.47

Table 3. Sensitivity coefficients for each land use type.

Year	Farmland	Forest	Grassland	Waterbody	Construction Land	Unused Land
2005	0.0043	0.0268	0.0096	0.0036	0	0.0001
2020	0.0015	0.0093	0.0037	0.0011	0	0.0001

3.2.4. Restrictive Condition

The constraints are primarily established in accordance with a series of land use policies and regulations promulgated by Lvliang City and Shanxi Province, including but not limited to "The Overall Planning of Lvliang City's Territorial Space (2021–2035)", the State Council's approval of "The Territorial Spatial Planning of Shanxi Province (2021–2035)," and "The Action Plan for Lvliang City to Create a National Forest City". These constraints delineate a series of conditions for socio-economic development and ecological protection, as specified in Table 4.

Table 4. Constraint information.

Constraint	Prerequisite	Foundation
Total land area	$X_1 + X_2 + X_3 + X_4 + X_5 + X_6 = 21{,}115.57\ km^2$	Total land area constraint.
Farmland	$4165.34\ km^2 \leq X_1 \leq X_1^+$, X_1^+ is the area of farmland in Lvliang City in 2020	The Lvliang City Territorial Spatial Master Plan (2021–2035) calls for farmland holdings of 4165.342 km^2.
Forest	$6334.67\ km^2 \leq X_2 \leq 7528.98\ km^2$	Action Programme for the Creation of a National Forest City in Lvliang City: Lvliang citywide forest cover of more than 30 per cent, which is less than 1.1 times the current value in 2020.
Grassland	$6453.09\ km^2 \leq X_3 \leq 6906.55\ km^2$	Greater than projected under natural development conditions and less than 1.1 times the current value in 2020.
Waterbody	$121.17\ km^2 \leq X_4 \leq 129.76\ km^2$	Greater than the current value in 2020 and less than the projected value of natural development.
Construction land	$898.41\ km^2 \leq X_5 \leq 1167.94\ km^2$	The State Council's approval of the "Shanxi Province Land Space Planning (2021–2035)" requires that the expansion of the urban development boundary be controlled within 1.3 times the size of the urban construction land based on 2020, with 1.3 times the existing construction land in Lvliang City as the upper boundary, and the lower boundary as the status quo value in 2020.
Unused Land	$0.37\ km^2 \leq X_5 \leq 0.52\ km^2$	Greater than the current value in 2020 and less than the projected value of natural development.

4. Results

4.1. Analysis of the Quantitative Structure of Land Use

According to the statistical data presented in Table 5, the NDS indicated a substantial increase in built-up land, which rose by 55.05% compared to 2020. Furthermore, grassland, water bodies, and unused land experienced increases of 27.78%, 7.09%, and 39.13%, respectively. In contrast, farmland and forest land decreased by 4.39% and 5.43%, respectively, compared to 2020. The NDS encompassed land use changes from 2005 to 2020, which were used to predict the areas of various land types in 2035. During this period, the areas of farmland and forest land decreased, while the areas of built-up land, grassland, water bodies, and unused land increased. Consequently, the anticipated changes in land areas by 2035 reflected the trends observed between 2005 and 2020. In this context, the unrestricted expansion of built-up land encroached upon both farmland and forest land, highlighting a distinctly unreasonable land use strategy.

Table 5. Changes in land quantity structure under different development scenarios (unit: km^2).

Land Type	2020	Development Scenarios			
		Natural Development	Economic Development	Ecological Protection	Coordinated Development
Farmland	6972.40	6666.23	6972.40	5651.49	6724.29
Forest	6844.53	6473.01	6334.67	7528.98	6664.74
Grassland	6278.68	6453.08	6519.02	6906.55	6581.84
Waterbody	121.17	129.76	121.17	129.76	126.94
Construction land	898.41	1392.96	1167.94	898.41	1017.33
Unused Land	0.37	0.52	0.37	0.37	0.42

In the EDS, forest land experienced a notable reduction, declining by 7.45% relative to 2020. Areas set aside for cropland, water bodies and unused land remain unchanged. In contrast, grassland saw an increase of 3.83%, while construction land expanded significantly by 30%. Construction land is a vital component supporting economic development, hence the substantial increase. Lvliang City's agricultural economy is well-developed, so under the EDS, efforts were made to maintain farmland. Additionally, a slight increase in grassland area was observed, attributed to the high output value of animal husbandry in Lvliang City. In this development scenario, although the growth of built-up land slowed, it continued to increase at an unsustainable rate. The area of farmland did not exhibit a significant decrease; however, the area of forest land declined markedly, which posed serious challenges to ecological protection efforts.

In the EPS, the farmland area underwent a significant decrease of 18.95% compared to 2020. In contrast, both forest land and grassland increased by 10% relative to the same year. The areas designated for construction land and unused land remained stable, while water bodies experienced an increase of 7.09%. As demonstrated in Table 2, forest land and water bodies were associated with greater ecological benefits. Consequently, under the EDS, these areas were expanded, while construction land and unused land, which have low ecological benefits, were maintained at their existing levels. In comparison to economic benefits, the ecological benefits of farmland are negligible, leading to its substantial reduction. In this scenario, although the expansion of built-up land area was minimal, substantial increases were noted in both forest and grassland areas. Conversely, the cultivated land area faced a notable decline. Considering the importance of construction and agricultural lands for the economic development of Lvliang City, this development scenario, although advantageous for ecological preservation, was evidently insufficient to support the local economy.

In the CDS, both farmland and forest land areas underwent slight reductions, decreasing by 3.69% and 2.70%, respectively, relative to 2020. Conversely, grassland and water body areas demonstrated increases of 4.61% and 4.55%, respectively. The area allocated for construction experienced a significant growth of 11.69%, whereas the extent of unused land expanded by 12.54%. When compared to the EDS, the increase in construction land was more modest, and the areas of grassland and water bodies saw modest increases. In comparison to the EPS, the decrease in farmland was less substantial, and the change in forest land area was minimal. Consequently, the land use pattern established in this development scenario effectively addressed the requirements for economic growth while also reducing the conflict between economic development and ecological preservation.

4.2. Simulation Analysis of Spatial Layout of Land Use Structure

The land-use pattern simulation of the four scenarios is shown in Figure 4. In the NDS, a significant increase in construction land was observed, primarily reflecting an outward expansion from the city center in all directions. The most pronounced growth occurred in the southern region, where forested areas and farmland were notably reduced to accommodate this expansion. Additionally, the southeastern sector experienced considerable growth in construction land, mainly resulting from the encroachment upon farmland. Significant reductions were observed in both forest and farmland areas, whereas the grassland area experienced an increase, predominantly concentrated in the northern region, characterized by the encroachment on forest land.

In the EDS, a rapid expansion of construction land was observed, predominantly characterized by an outward extension from the city center in all directions. The southern region of Lvliang City experienced particularly significant growth in construction land, largely at the expense of forested areas. Conversely, the southeastern region did not exhibit notable changes in construction land, as it was primarily composed of farmland, which remained stable under this scenario. Consequently, no substantial expansion of construction land occurred in that area. Additionally, an increase in grassland was documented, especially in the central-western and western regions, with the latter showing a more pronounced growth, largely due to encroachment on forested land.

In the EPS, notable increases were recorded in both forest and grassland areas. The northwestern region experienced substantial expansion of forest land, which was characterized by diffusion in various directions. Similarly, the central-eastern area also saw a rise in forest land, primarily due to encroachment upon farmland. Grassland expansion was predominantly concentrated in the northwest, largely at the expense of agricultural fields. Overall, the area of farmland diminished significantly, with major reductions documented in the northwest and central-eastern regions, attributed to the encroachment of both forest and grassland. In contrast, the southeastern region did not display significant contraction in farmland. Additionally, the area designated for construction remained stable, while the water body area increased.

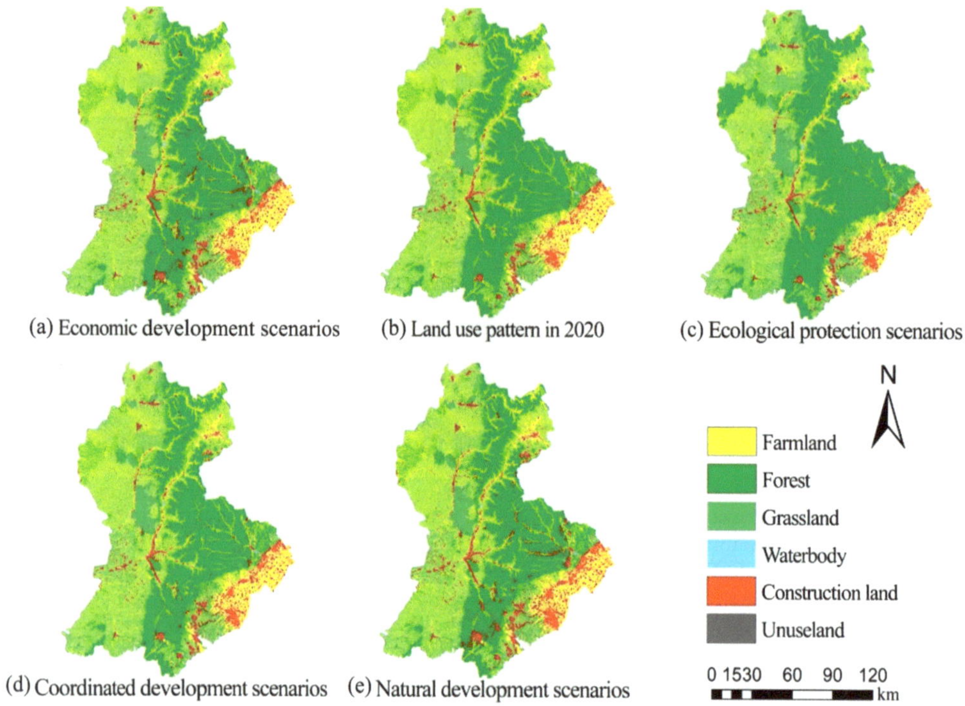

Figure 4. Land use simulation of Lvliang City under different development scenarios in 2035.

In the CDS, an expansion of construction land was observed, predominantly marked by the urban area's extension in multiple directions. In the southern region, although construction land expanded, the pace of growth was relatively gradual. Importantly, there were no substantial decreases in the areas designated for arable or forest land. Conversely, an increase in grassland was noted in both the central and northern regions.

Figure 5 indicates that the EPS demonstrated the highest LPI, whereas the EDS recorded the lowest LPI. Generally, a higher LPI indicates the presence of a greater number of large patches in the landscape, with less human disturbance to natural landscapes. The LPI for the CDS was more like that of 2020 and exceeded the values observed in both the NDS and EDS. This finding suggests that the landscape in the CDS experienced less impact from human activities than that in the EDS. This result indicated a greater degree of landscape patch aggregation in the CDS relative to the EPS. With regard to the SHDI, the NDS demonstrated the highest value. In contrast, the CDS was found to be comparable to both the EDS and the land use pattern of 2020, while surpassing the EPS. This finding indicated that the CDS supported greater species diversity. Regarding the DIVISION, no notable differences were observed across the various development scenarios. Overall, in the CDS, the land use pattern was characterized by lower levels of human disturbance. The landscape patches were more cohesively clustered, biodiversity was richer, and the quality of the ecological environment was higher.

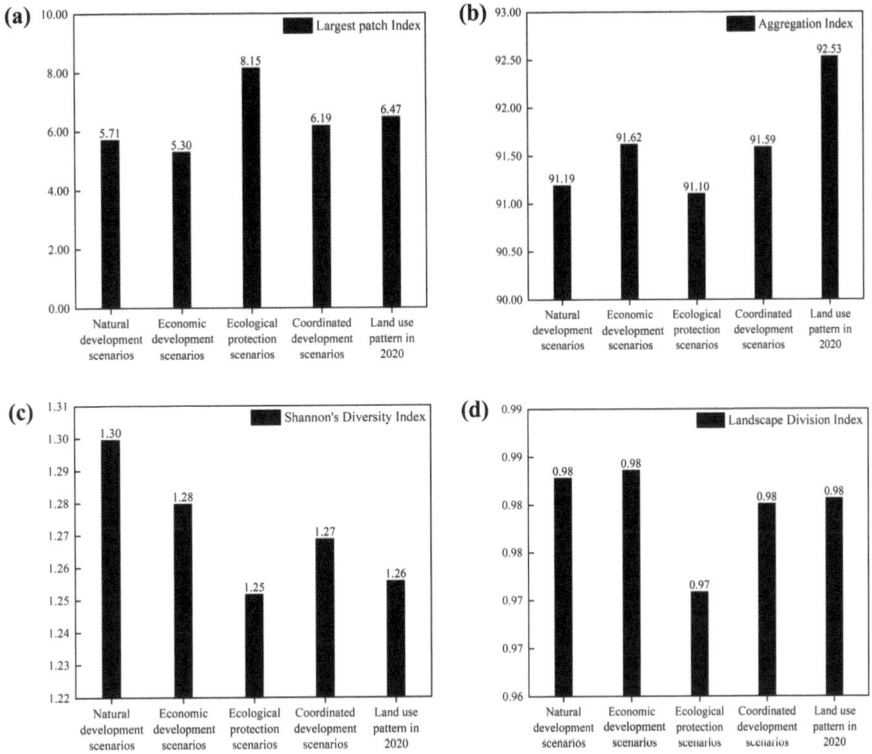

Figure 5. Landscape index for different development scenarios in Lvliang, 2035. (**a**) Largest patch index; (**b**) Aggregation index; (**c**) Shannon Diversity Index; (**d**) Landscape division index.

4.3. Land Use Benefits and ESV

Using the projected economic and ecological benefit coefficients for land use in Lvliang City in 2035, as detailed in prior sections, the benefits of land use across different development scenarios for that year are calculated. The ESV equivalents from the years 2016 to 2022 were applied, and the GM (1: 1) model was used to predict the ESV equivalents for Lvliang City in 2035. These predictions, along with the updated equivalent factor table, facilitated the calculation of ESV for the simulated land use patterns across different scenarios in 2035.

As indicated in Table 6, the NDS yielded the highest land use benefits; however, the corresponding ESV were comparatively low. In contrast, the EDS demonstrated increased land use benefits but recorded the lowest ESV. Conversely, the EPS achieved the highest ESV, albeit with the lowest land use benefits. This outcome, while satisfying ecological protection standards, proved detrimental to economic development. The CDS demonstrated both high land use benefits and high ESV, capable of balancing economic growth with environmental quality.

Table 6. Land use benefits and ESV under different scenarios in 2035 (Unit: million yuan).

Development Scenario	Land Use Benefits	Ecosystem Services Values
Natural Development Scenario	23,929.02	263.08
Economic Development Scenario	20,254.13	260.87
Ecological Protection Scenarios	15,775.43	286.65
Coordinated Development Scenario	17,553.64	268.61
2020 Land Use Pattern	15,837.95	270.14

Numerous studies indicate that changes in land use patterns within a region significantly affect ESV and land use benefits [37,38]. Improvements and optimizations in land use enhance the capacity for ecosystem services. However, during the transition of land use patterns, economic development and ecological protection within a region often appear mutually exclusive, with economic development frequently occurring at the expense of the environment. To examine the economic benefits that can be obtained at the cost of ecological benefits, this study uses land use benefits to represent economic benefits and ESV to represent ecological benefits. An index for land use benefit conversion (I) is established to explore the conversion efficiency between land use benefits and ESV when land use patterns transition to different scenarios in 2020. A larger index (I) indicates that for each unit of ecological benefit sacrificed, greater economic benefits are achieved, suggesting that the transformation of land use patterns in these scenarios has greater developmental potential and is more conducive to sustainable development. The formula is as follows:

$$I = \left| \frac{LUE_2 - LUE_1}{E_4 - E_3} \right| \tag{7}$$

where LUE_2 denotes the land use benefits associated with various development scenarios, while LUE_1 reflects the land use benefits derived from the 2020 land use pattern as projected for 2035. Similarly, E_4 indicates the ESV corresponding to different development scenarios, and E_3 represents the ESV of the 2020 land use pattern projected for 2035.

Figure 6 illustrates that the conversion rate associated with the NDS was the highest. However, this scenario's land use pattern underwent significant expansion of construction land, surpassing the established constraints and failing to align with the future development objectives of Lvliang City. The conversion rate observed in the CDS approached that of the NDS, while it was markedly higher than those recorded for both the EDS and EPS. This indicated that transitioning from the 2020 land use pattern to the pattern under the CDS would yield the highest land use benefits per unit sacrifice of ESV, suggesting a high potential for land development and greater sustainability under this pattern.

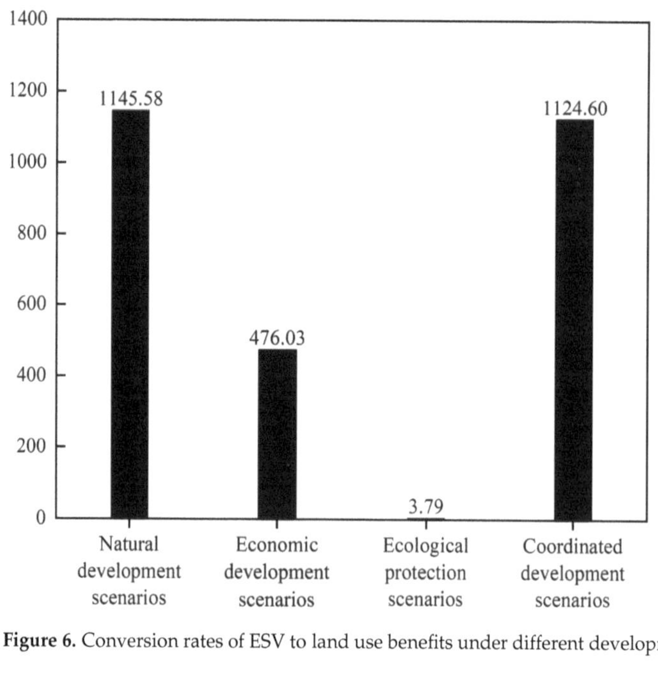

Figure 6. Conversion rates of ESV to land use benefits under different development scenarios.

5. Discussion

This study used land use data from 2005 and 2020 along with GMOP and PLUS models to simulate both the quantity and spatial distribution of land use, predicting land use patterns for 2035 under various development scenarios. Previous research primarily identified optimal land use patterns by analyzing ecosystem service value (ESV) and land use benefits across different development scenarios [39,40]. However, these studies did not account for the conversion efficiency between ESV and land use benefits during the transition from the current land use pattern to future scenarios, nor did they consider the impact of land use pattern changes on landscape configurations [41]. Consequently, this study introduced a land use benefit conversion index to represent the conversion efficiency between ESV and land use benefits for the first time. A comprehensive analysis of ESV, land use benefits, landscape pattern indices, and the land use benefit conversion index was performed to evaluate these indicators and identify the optimal land use pattern. The research findings provide scientific theoretical support for land use optimization and ecological protection in Lvliang City and offer a methodological example for land use optimization in other regions.

However, the study has some limitations. The land use categories were divided into six primary categories, which, while reflecting changes in various land types under different scenarios in Lvliang City, do not fully explain the relationship between economic development, ecological protection, and mining development in a city rich in mineral resources. The construction land category should be further detailed into types such as mining land to explore its relationship with the economy and ecology. The ESV in this study was calculated based on the equivalent factor table by Xie Gaodi. Although the equivalent factors are simple, widely applicable, and have low data requirements [42], they have certain limitations in accuracy, dynamism, complexity, and regional applicability. A more in-depth analysis combined with other methods is needed to improve the comprehensiveness and accuracy of the assessment.

Given the limitations of this study, future research should develop a more targeted land use classification system and conduct a thorough analysis of the relationships among economic development, ecological protection, and mining activities in mineral resource-based cities. Additionally, the objective and comprehensive assessment of ESV within a region, along with the reduction of errors introduced by subjective factors, necessitates further exploration and analysis in future studies to improve the accuracy and objectivity of the evaluation results.

6. Conclusions

In this study, a multi-objective structural optimization model that integrates GMOP and PLUS was developed to achieve a balance among the economic, social, and ecological benefits of land use. Various scenarios regarding land use structures were derived for different development contexts in Lvliang City, based on specific constraints. The key findings of this research are summarized as follows:

(1) The GMOP and PLUS coupling models are used to determine land use structure and spatial distribution under various development objectives. By employing this coupled approach, land use patterns for Lvliang City are derived across four distinct development scenarios.

(2) From the perspective of landscape patterns, the land use pattern under the CDS is characterized by minimal human disturbance, enhanced patch aggregation, greater species diversity, and improved ecological quality.

(3) The land use benefits under the NDS, EDS, EPS, and CDS are 2392.902 billion yuan, 2025.413 billion yuan, 1577.543 billion yuan, and 1755.364 billion yuan, respectively. The ESV are 26.308 billion yuan, 26.087 billion yuan, 28.665 billion yuan, and 26.861 billion yuan, respectively. This shows that the CDS can meet economic development needs while also considering ecological environment protection.

(4) Under the CDS, when the areas of cultivated land, grassland, forest land, water bodies, construction land, and unused land are 6724.29 km^2, 6664.74 km^2, 6581.84 km^2, 126.94 km^2, 1017.33 km^2, and 0.42 km^2, respectively, the land use benefit conversion index is at its highest.

The research method outlined in this paper provides a valuable reference for optimizing urban land use structures. The findings establish a foundation for developing land use planning and management policies in resource-based cities.

Author Contributions: Conceptualization, methodology, writing—review and editing, and supervision, Z.W. and A.Z.; methodology, software, visualization, and writing—original draft preparation, A.Z.; investigation and validation, Z.W.; supervision, Q.L. All authors have read and agreed to the published version of the manuscript.

Funding: This research was funded by the Key R&D project for school local cooperation in Lvliang City, grant number 2022XDHZ12.

Institutional Review Board Statement: Not applicable.

Informed Consent Statement: Not applicable.

Data Availability Statement: The data presented in this study are available on request from the corresponding author. The data are not publicly available due to privacy restrictions.

Conflicts of Interest: The authors declare no conflicts of interest.

References

1. Zhao, X.; Pu, J.; Wang, X.; Chen, J.; Yang, L.E.; Gu, Z. Land-Use Spatio-Temporal Change and Its Driving Factors in an Artificial Forest Area in Southwest China. *Sustainability* **2018**, *10*, 4066. [CrossRef]
2. Rimal, B.; Zhang, L.; Keshtkar, H.; Wang, N.; Lin, Y. Monitoring and Modeling of Spatiotemporal Urban Expansion and Land-Use/Land-Cover Change Using Integrated Markov Chain Cellular Automata Model. *ISPRS Int. J. Geo-Inf.* **2017**, *6*, 288. [CrossRef]
3. Zscheischler, J.; Rogga, S.; Lange, A. The success of transdisciplinary research for sustainable land use: Individual perceptions and assessments. *Sustain. Sci.* **2018**, *13*, 1061–1074. [CrossRef]
4. Zhang, H.B.; Zhang, X.H. Land use structural optimization of Lilin based on GMOP-ESV. *Trans. Nonferrous Met. Soc. China* **2011**, *21*, S738–S742. [CrossRef]
5. Chuai, X.W.; Huang, X.J.; Wu, C.Y.; Li, J.B.; Lu, Q.L.; Qi, X.X.; Zhang, M.; Zuo, T.H.; Lu, J.Y. Land use and ecosystems services value changes and ecological land management in coastal Jiangsu, China. *Habitat Int.* **2016**, *57*, 164–174. [CrossRef]
6. MohanRajan, S.N.; Loganathan, A.; Manoharan, P. Survey on Land Use/Land Cover (LULC) change analysis in remote sensing and GIS environment: Techniques and Challenges. *Environ. Sci. Pollut. Res.* **2020**, *27*, 29900–29926. [CrossRef]
7. Torres, P.; Rodes-Blanco, M.; Viana-Soto, A.; Nieto, H.; García, M. The Role of Remote Sensing for the Assessment and Monitoring of Forest Health: A Systematic Evidence Synthesis. *Forests* **2021**, *12*, 1134. [CrossRef]
8. Liu, H.; Wang, Y.; Sang, L.L.; Zhao, C.S.; Hu, T.Y.; Liu, H.T.; Zhang, Z.; Wang, S.Y.; Miao, S.X.; Ju, Z.S. Evaluation of Spatiotemporal Changes in Cropland Quantity and Quality with Multi-Source Remote Sensing. *Land* **2023**, *12*, 1764. [CrossRef]
9. Ma, Q.Q.; Chai, L.R.; Hou, F.J.; Chang, S.H.; Ma, Y.S.; Tsunekawa, A.; Cheng, Y.X. Quantifying Grazing Intensity Using Remote Sensing in Alpine Meadows on Qinghai-Tibetan Plateau. *Sustainability* **2019**, *11*, 417. [CrossRef]
10. Zhong, A.Y.; Hu, C.M.; You, L. Evaluation and Prediction of Ecological Restoration Effect of Beijing Wangping Coal Mine Based on Modified Remote Sensing Ecological Index. *Land* **2023**, *12*, 2059. [CrossRef]
11. Ma, S.H.; Wen, Z.Z. Optimization of land use structure to balance economic benefits and ecosystem services under uncertainties: A case study in Wuhan, China. *J. Clean Prod.* **2021**, *311*, 127537. [CrossRef]
12. Wu, R.; Lan, H.F.; Cao, Y.X.; Li, P.Y. Optimization of low-carbon land use in Chengdu based on multi-objective linear programming and the future land use simulation model. *Front. Environ. Sci.* **2022**, *10*, 989747. [CrossRef]
13. Zhang, P.; Liu, L.; Yang, L.; Zhao, J.; Li, Y.; Qi, Y.; Ma, X.; Cao, L. Exploring the response of ecosystem service value to land use changes under multiple scenarios coupling a mixed-cell cellular automata model and system dynamics model in Xi'an, China. *Ecol. Indic.* **2023**, *147*, 110009. [CrossRef]
14. Shi, Q.; Gu, C.-J.; Xiao, C. Multiple scenarios analysis on land use simulation by coupling socioeconomic and ecological sustainability in Shanghai, China. *Sustain. Cities Soc.* **2023**, *95*, 104578. [CrossRef]
15. Huang, Y.; Nian, P.; Zhang, W. The prediction of interregional land use differences in Beijing: A Markov model. *Environ. Earth Sci.* **2014**, *73*, 4077–4090. [CrossRef]

16. Du, Y.; Li, X.; He, X.; Li, X.; Yang, G.; Li, D.; Xu, W.; Qiao, X.; Li, C.; Sui, L. Multi-Scenario Simulation and Trade-Off Analysis of Ecological Service Value in the Manas River Basin Based on Land Use Optimization in China. *Int. J. Environ. Res. Public Health* **2022**, *19*, 6216. [CrossRef]
17. Zhang, W.; Wang, H.; Han, F.; Gao, J.; Nguyen, T.; Chen, Y.; Huang, B.; Zhan, F.B.; Zhou, L.; Hong, S. Modeling urban growth by the use of a multiobjective optimization approach: Environmental and economic issues for the Yangtze watershed, China. *Environ. Sci. Pollut. Res. Int.* **2014**, *21*, 13027–13042. [CrossRef]
18. Zhu, L.Z.; Huang, Y.P. Multi-Scenario Simulation of Ecosystem Service Value in Wuhan Metropolitan Area Based on PLUS-GMOP Model. *Sustainability* **2022**, *14*, 13604. [CrossRef]
19. Li, C.; Wu, Y.M.; Gao, B.P.; Zheng, K.J.; Wu, Y.; Li, C. Multi-scenario simulation of ecosystem service value for optimization of land use in the Sichuan-Yunnan ecological barrier, China. *Ecol. Indic.* **2021**, *132*, 108328. [CrossRef]
20. Mo, J.X.; Sun, P.L.; Shen, D.D.; Li, N.; Zhang, J.Y.; Wang, K. Simulation Analysis of Land-Use Spatial Conflict in a Geopark Based on the GMOP-Markov-PLUS Model: A Case Study of Yimengshan Geopark, China. *Land* **2023**, *12*, 1291. [CrossRef]
21. Firozjaei, M.K.; Sedighi, A.; Argany, M.; Jelokhani-Niaraki, M.; Arsanjani, J.J. A geographical direction-based approach for capturing the local variation of urban expansion in the application of CA-Markov model. *Cities* **2019**, *93*, 120–135. [CrossRef]
22. Verburg, P.H.; Soepboer, W.; Veldkamp, A.; Limpiada, R.; Espaldon, V.; Mastura, S.S.A. Modeling the spatial dynamics of regional land use: The CLUE-S model. *Environ. Manag.* **2002**, *30*, 391–405. [CrossRef] [PubMed]
23. Zhang, Y.; Yu, P.H.; Tian, Y.S.; Chen, H.T.; Chen, Y.Y. Exploring the impact of integrated spatial function zones on land use dynamics and ecosystem services tradeoffs based on a future land use simulation (FLUS) model. *Ecol. Indic.* **2023**, *150*, 110246. [CrossRef]
24. Liang, X.; Guan, Q.F.; Clarke, K.C.; Liu, S.S.; Wang, B.Y.; Yao, Y. Understanding the drivers of sustainable land expansion using a patch-generating land use simulation (PLUS) model: A case study in Wuhan, China. *Comput. Environ. Urban Syst.* **2021**, *85*, 101569. [CrossRef]
25. Zhong, Y.; Zhang, X.; Yang, Y.; Xue, M. Optimization and Simulation of Mountain City Land Use Based on MOP-PLUS Model: A Case Study of Caijia Cluster, Chongqing. *ISPRS Int. J. Geo-Inf.* **2023**, *12*, 451. [CrossRef]
26. Shu, R.; Wang, Z.; Guo, N.; Wei, M.; Zou, Y.; Hou, K. Multi-Scenario Land Use Optimization Simulation and Ecosystem Service Value Estimation Based on Fine-Scale Land Survey Data. *Land* **2024**, *13*, 557. [CrossRef]
27. Luan, C.; Liu, R.; Zhang, Q.; Sun, J.; Liu, J. Multi-objective land use optimization based on integrated NSGA-II-PLUS model: Comprehensive consideration of economic development and ecosystem services value enhancement. *J. Clean Prod.* **2024**, *434*, 140306. [CrossRef]
28. Meng, F.; Zhou, Z.; Zhang, P. Multi-Objective Optimization of Land Use in the Beijing-Tianjin-Hebei Region of China Based on the GMOP-PLUS Coupling Model. *Sustainability* **2023**, *15*, 3977. [CrossRef]
29. Chen, N.; Xin, C.L.; Tang, D.B.; Zhang, L.; Xin, S.J. Multi-scenario land use optimization and carbon stock assessment in Northwest China. *Chin. J. Environ. Sci.* **2023**, *44*, 4655–4665.
30. Guo, P.F.; Wang, H.Y.; Qin, F.; Miao, C.H.; Zhang, F.F. Coupled MOP and PLUS-SA Model Research on Land Use Scenario Simulations in Zhengzhou Metropolitan Area, Central China. *Remote Sens.* **2023**, *15*, 3762. [CrossRef]
31. Wang, Z.M.; Guo, M.D.; Zhang, D.; Chen, R.Q.; Xi, C.F.; Yang, H.B. Coupling the Calibrated GlobalLand30 Data and Modified PLUS Model for Multi-Scenario Land Use Simulation and Landscape Ecological Risk Assessment. *Remote Sens.* **2023**, *15*, 5186. [CrossRef]
32. Wu, X.X.; Liu, X.P.; Zhang, D.C.; Zhang, J.B.; He, J.Y.; Xu, X.C. Simulating mixed land-use change under multi-label concept by integrating a convolutional neural network and cellular automata: A case study of Huizhou, China. *GISci. Remote Sens.* **2022**, *59*, 609–632. [CrossRef]
33. Zhu, C.M.; Yuan, S.F.; Yang, L.X. Optimization and trade-off analysis of land use pattern in Hangzhou coupled with MOP and FLUS models. *Trans. Chin. Soc. Agric. Eng.* **2023**, *39*, 235–244.
34. Xie, G.; Zhen, L.; Lu, C.; Xiao, Y.; Chen, C. An expert knowledge-based approach to ecosystem service valorisation. *J. Nat. Resour.* **2008**, *5*, 911–919.
35. Li, X.; Fu, J.Y.; Jiang, D.; Lin, G.; Cao, C.L. Land use optimization in Ningbo City with a coupled GA and PLUS model. *J. Clean Prod.* **2022**, *375*, 134004. [CrossRef]
36. Jiang, Y.; Du, G.M.; Teng, H.; Wang, J.; Li, H.L. Multi-Scenario Land Use Change Simulation and Spatial Response of Ecosystem Service Value in Black Soil Region of Northeast China. *Land* **2023**, *12*, 962. [CrossRef]
37. Hasan, S.S.; Zhen, L.; Miah, M.G.; Ahamed, T.; Samie, A. Impact of land use change on ecosystem services: A review. *Environ. Dev.* **2020**, *34*, 100527. [CrossRef]
38. Tan, Z.; Guan, Q.; Lin, J.; Yang, L.; Luo, H.; Ma, Y.; Tian, J.; Wang, Q.; Wang, N. The response and simulation of ecosystem services value to land use/land cover in an oasis, Northwest China. *Ecol. Indic.* **2020**, *118*, 106711. [CrossRef]
39. Dong, K.; Wang, H.; Luo, K.; Yan, X.; Yi, S.; Huang, X. The Use of an Optimized Grey Multi-Objective Programming-PLUS Model for Multi-Scenario Simulation of Land Use in the Weigan-Kuche River Oasis, China. *Land* **2024**, *13*, 802. [CrossRef]
40. Tang, H.; Halike, A.; Yao, K.; Wei, Q.; Yao, L.; Tuheti, B.; Luo, J.; Duan, Y. Ecosystem service valuation and multi-scenario simulation in the Ebinur Lake Basin using a coupled GMOP-PLUS model. *Sci. Rep.* **2024**, *14*, 5071. [CrossRef]

41. Zhang, Y.; Naerkezi, N.; Zhang, Y.; Wang, B. Multi-Scenario Land Use/Cover Change and Its Impact on Carbon Storage Based on the Coupled GMOP-PLUS-InVEST Model in the Hexi Corridor, China. *Sustainability* **2024**, *16*, 1402. [CrossRef]
42. He, S.; Wang, J.; Li, J.; Sha, J.; Zhou, J.; Jiao, Y. Quantification and Simulation of the Ecosystem Service Value of Karst Region in Southwest China. *Land* **2024**, *13*, 812. [CrossRef]

land

MDPI

Article

Analysis of the Spatiotemporal Differentiation and Influencing Factors of Land Use Efficiency in the Beijing–Tianjin–Hebei Urban Agglomeration

Haixin Huang and Jiageng Yang *

School of Economics and Management, Northwest University, Xi'an 710127, China; huanghaixin@stumail.nwu.edu.cn
* Correspondence: 202220154@stumail.nwu.edu.cn

Abstract: Optimizing urban land use is of significant practical importance for promoting economic development, enhancing the standard of living for individuals residing in metropolitan areas, enhancing urban infrastructure and public services, driving urban transformation and upgrading, and attaining synchronized progress of the economy, society, and environment. This paper uses the super-efficiency SBM model to measure the urban land use efficiency (ULUE) of 13 cities in the Beijing–Tianjin–Hebei (BTH) urban agglomeration from 2005 to 2020 and explores the spatiotemporal evolution characteristics and influencing factors of ULUE in this urban agglomeration using analysis of spatial data and application of geographic detector methods. The results show that (1) from 2005 to 2020, the ULUE of the BTH urban agglomeration had an initial rise followed by a decline; however, the overall efficiency score is above 1, suggesting an overall effective state; (2) a distribution pattern with Beijing as its core was established, exhibiting greater ULUE in the northern region and poorer efficiency in the southern region, with significant correlation characteristics in efficiency values between adjacent cities; and (3) capital input, labor input, social welfare, and ecological environment are all influencing factors that promote the improvement in ULUE in the BTH region, and the interaction of any two factors explains the ULUE in this region better than a single factor. The empirical research results can provide useful references for improving the input–output ratio of land units and further spatial planning and policy formulation in the BTH region.

Keywords: sustainability; urban land use efficiency; SBM-DEA

Citation: Huang, H.; Yang, J. Analysis of the Spatiotemporal Differentiation and Influencing Factors of Land Use Efficiency in the Beijing–Tianjin–Hebei Urban Agglomeration. *Land* **2024**, *13*, 1508. https://doi.org/10.3390/land13091508

Academic Editor: Jūratė Sužiedelytė-Visockienė

Received: 15 August 2024
Revised: 11 September 2024
Accepted: 16 September 2024
Published: 17 September 2024

1. Introduction

With the advancement of modernization and industrialization, a large amount of farmland, forests, and other lands have been used for development and construction, exacerbating the continuous reduction of land resources and making the scarcity of land resources increasingly apparent. Especially in urban areas, the conflict between land resources and human social and economic development is more pronounced [1,2], specifically manifesting in issues such as insufficient urban land carrying capacity, soil erosion, ecological environment deterioration, and inefficient land use [3]. However, to attract capital investment and talent influx to promote economic growth, land expansion has become common in both developed and developing countries [4]. Therefore, under the constraint of scarce land resources, increasing the input–output ratio per unit of land is the key to achieving sustainable economic development [5]. By improving urban ULUE, more economic value and social benefits can be created in a limited space [6,7]. One of the main reasons China has been able to achieve the miracle of rapid economic growth is the significant increase in land resource investment [8]. As the capital of China, Beijing and its surrounding areas have historically held an important political and strategic position, with a deep cultural heritage and a long history of development, gathering a large amount of capital, technology, and talents, while also bearing the important responsibility of spreading traditional Chinese culture and modern civilization to the world.

With 2.4% of the country's land carrying 8.1% of the country's resident population and producing about 8.5% of the country's GDP, the Beijing–Tianjin–Hebei region is one of China's most vibrant, most open, most innovative, and most populous regions and is an important engine for pulling China's economy forward. In 2014, the strategy of Beijing–Tianjin–Hebei's coordinated development had risen to the height of national strategy. The realistic needs and strategic tasks of urbanization development have placed unprecedented pressure on land use in the Beijing–Tianjin–Hebei region. Therefore, strengthening the research on land use efficiency in the Beijing–Tianjin–Hebei region is of great value in addressing the problems and challenges faced in the new era and in realizing the sustainable and coordinated development of urban economic, social, and ecological benefits.

To rationally allocate land resources and improve the efficiency of land use, many theories have been gradually formed in the academic world. The basic theory on urban land use efficiency can be traced back to the early 20th century, when the ecological location school of thought established concentric circles, fan-shaped, and multi-core models based on human ecology [9]. In the 1960s and 1970s, the economic location school of thought built a monocentric, exogenous, dynamic model based on neoclassical economics [10–12]. The social location school of thought has built a theory of decision analysis and interaction based on behavioral science [13]. In the 1970s and 1980s, the political location school of thought emphasized social relations of production and government intervention [14,15], arguing that social relations of production, political power, and government intervention are key factors in determining changes in the spatial structure of urban land use [16]. From the 20th century to the present, modern concepts such as compact cities and smart growth have culminated in an effort to reduce transportation demand, increase resource efficiency, and improve environmental quality [17,18].

Based on the above theories, scholars have mainly focused on land green use efficiency [19], factors affecting land green use efficiency [20], the impact of land use on ecosystem services [21,22], and measurement methods of land use efficiency [23]. At present, a unified evaluation index system for urban land use efficiency has not yet been formed, but in general, the evaluation indexes are gradually developing from single indicators to multi-dimensional indicators, and non-desired output factors such as environmental constraints are included in the examination. In terms of research methods, the comprehensive evaluation method has developed into parametric and non-parametric methods, and DEA models [24,25] have been fully applied to the measurement of land use efficiency, using kernel density estimation, Terrell's index, exploratory spatial data analysis, and center of gravity models to analyze the spatial and temporal variations in urban land use efficiency [26–29]. The factors affecting land use efficiency were also studied through regression models such as the Tobit model, spatial panel econometric model, and geodetector model. The research scale mainly focuses on provincial units or single cities; for example, Kuang [30] measured the provincial arable land use efficiency of 31 provinces in China from 2000 to 2017, and Fu [31] analyzed and studied the urban land use efficiency of Jiangsu Province, China, from 2006 to 2017, by using the data envelopment analysis method and the information entropy method. The regional synergistic development of urban agglomerations is less involved. From the results of studies, it is clear that economic [2,32] and social factors [33,34] are the main factors affecting the efficiency of urban land use. For example, Masini studied 417 metropolitan areas in Europe and found a positive correlation between the level of economic development and land use efficiency [35]. Using a spatial panel model, Gao found that regional economic integration in metropolitan areas contributes to the optimal allocation of resources in socio-economic transformation, which improves urban land use efficiency [36]. Cao found a positive correlation between land use efficiency and factors such as the level of comprehensive economic development, fixed asset investment and environmental protection [37].

Therefore, this paper makes the following marginal contributions: Firstly, the land-averaged carbon emissions of cities are included in the rating index system of land use efficiency, and the SBM-DEA model is used to measure the land use efficiency of the Beijing–

Tianjin–Hebei city cluster. This method overcomes the subjectivity in the process of determining the weights of the comprehensive evaluation indexes, makes the determination of the efficiency boundary clearer, and can better identify the less efficient parts. Considering that the correlation between the independent and dependent variables may come from spatial similarity, this paper explores the spatial and temporal evolution characteristics of land use efficiency and its influencing factors in this urban agglomeration using exploratory spatial data analysis as well as geodetectors.

2. Study Area and Datasets

2.1. Overview of the Study Region

The Beijing–Tianjin–Hebei metropolitan agglomeration is situated in the North China Plain (longitude 113°04′–119°53′ E, latitude 36°01′–42°37′ N), covering a total area of 218,000 km^2. The cultivated land, forest land, grassland, and wetlands in BTH are 64,400 km^2, 75,500 km^2, 19,700 km^2, and 1800 km^2, respectively, encompassing 13 cities including Beijing, Tianjin, Baoding, Tangshan, and Shijiazhuang (as shown in Figure 1). In terms of economic development, the GDP of the BTH region in 2020 was 8639.3 billion yuan, accounting for 8.5% of the national GDP. In terms of technological research and development, the region had 37 invention patents per 10,000 permanent residents in 2020. In terms of transportation, the total length of expressways in the three provinces and cities of BTH reached 10,307 km in 2020. The coordinated division of labor and cooperation among Tianjin and Hebei ports has continued to deepen, with Tianjin Port focusing on container trunk transportation and optimizing the structure of bulk cargo transportation, Hebei ports consolidating functions for the transportation of energy and raw materials, and all nine planned airports in BTH, including Beijing Daxing International Airport, being put into operation. This region is one of the most dynamic, open, innovative, and populous areas in China. From 2005 to 2020, the area of construction land in the BTH region increased by approximately 6275.89 km^2. As the coordinated cooperation of the BTH urban agglomeration deepens, problems such as environmental degradation and soil damage will become increasingly prominent. Therefore, it is imperative to measure the ULUE of the BTH urban agglomeration, which can provide more targeted reference and guidance for improving the input–output ratio per unit of land and further spatial planning and policy formulation.

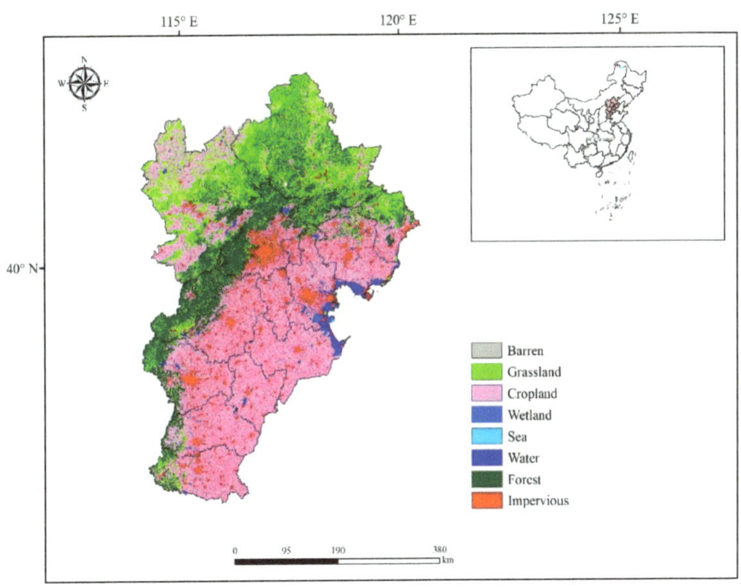

Figure 1. Location of BTH urban agglomeration.

2.2. Data Sources

This paper collected and preprocessed the input and output indicators such as capital, land, and labor of 13 cities in the BTH urban agglomeration from 2005 to 2020. The data are sourced from the China Statistical Yearbook, China City Statistical Yearbook, China Population and Employment Statistical Yearbook, China Urban Construction Statistical Yearbook, provincial and municipal statistical yearbooks, and the CNKI statistical database. The carbon dioxide emission data were obtained from the China Cities Greenhouse Gas Working Group (CCG).

3. Analysis Method

3.1. ULUE Evaluation Index System

Production efficiency refers to the amount of output achieved per unit of input [38]. Therefore, ULUE is the input–output ratio that includes production factors such as labor, capital, and land. This paper, referencing Liao [39], constructs an evaluation index system for ULUE in the BTH urban agglomeration based on inputs, desired outputs, and undesired outputs. While focusing on the economic and social benefits produced by land development in the BTH urban agglomeration, it also comprehensively considers the negative environmental impacts of land development and utilization. In terms of index selection, the per unit land fixed capital investment is used as the capital input index [24], the proportion of built-up land area as the land input index [39], and the number of urban employees per unit land area as the labor input index [40] (Table 1). Among them, the per unit land fixed asset investment can measure the economic development level and capital input situation of a region, reflecting the relationship between the accumulation of fixed assets and the economic development level of a region. The built-up area proportion is the ratio of the built-up area, which encompasses buildings, transportation facilities, and other urban planning elements, to the overall area of a city or region. This ratio serves as a measure of urbanization and urban land utilization. The number of urban employees per unit land area refers to the average number of urban employees within a unit area, which can measure the degree of urbanization and the employment situation of a region. For output index selection, economic benefits, social welfare, and ecological environment [41] are chosen as desired output indicators, while per unit land sewage discharge and per unit land carbon emissions are chosen as undesired output indicators. In terms of desired outputs, per unit land GDP and the average wages of on-the-job employees reflect the economic development level of a region from macro and micro perspectives, respectively, and are thus used to reflect the economic benefits of the BTH urban agglomeration. Healthcare and education are key to people's welfare, so the number of beds in public health institutions per unit land and the number of primary and secondary schools per unit land are selected to reflect the social welfare of the BTH urban agglomeration. The green coverage rate of built-up areas and per capita green area reflect the ecological livability from macro and micro dimensions [42], respectively, and hence, they are utilized to mirror the ecological conditions of the BTH urban agglomeration. In terms of undesired outputs, sewage discharge and greenhouse gas emissions are two major factors that negatively affect the environment during production activities. Therefore, per unit land sewage discharge and per unit land carbon emissions are chosen as indicators to measure undesired outputs.

Table 1. Construction of the ULUE evaluation index system for the BTH urban agglomeration.

Primary Indicators	Secondary Indicators	Tertiary Indicators	Calculation Method
Input Indicators	Capital	Fixed Capital Investment Per Unit Land Area (x1)	Total Social Fixed Asset Investment in the Urban Area/Area of the Urban Area

Table 1. *Cont.*

Primary Indicators	Secondary Indicators	Tertiary Indicators	Calculation Method
Desired Outputs	Land	Proportion of Built-up Land Area (x2)	Built-up Area of the Urban Area/Area of the Urban Area
	Labor	Number of Urban Employees Per Unit Land Area (x3)	Number of Urban Employees/Area of the Urban Area
	Economic Benefits	GDP Per Unit Land Area (x4)	GDP of the Urban Area/Area of the Urban Area
		Average Wages of On-the-job Employees (x5)	Total Wages of On-the-job Employees/Number of On-the-job Employees
	Social Welfare	Number of Beds in Public Health Institutions Per Unit Land Area (x6)	Number of Beds in Public Health Institutions in the Urban Area/Area of the Urban Area
		Education Level Per Unit Land Area (x7)	Number of Primary and Secondary Schools in the Urban Area/Area of the Urban Area
	Ecological Environment	Per Capita Public Green Area (x8)	Public Green Area in the Urban Area/Total Population of the Urban Area
		Green Coverage Rate of Built-up Areas (x9)	Total Green Area in the Urban Area/Area of the Urban Area
Undesired Outputs	Sewage Discharge	Sewage Discharge Per Unit Land Area (x10)	Sewage Discharge/Area of the Urban Area
	Carbon Emissions	Carbon Emissions Per Unit Land Area (x11)	Total Carbon Emissions/Area of the Urban Area

3.2. SBM-DEA Model

The Slacks-Based Measure (SBM) model, introduced by Tone, represents a sophisticated extension of the conventional Data Envelopment Analysis (DEA) approach. It addresses the inherent limitations of radial and angular measures employed in traditional DEA models by focusing on slack variables—indicators of inefficiency derived from the excess inputs required or the deficient outputs produced by a decision-making unit (DMU) to achieve optimal efficiency. This methodology offers a more nuanced and comprehensive assessment of efficiency, particularly in scenarios involving multiple inputs and outputs. Specifically, the super-efficiency SBM model extends the basic SBM framework to evaluate DMUs that already lie on the efficiency frontier, as determined by traditional DEA models. By excluding each DMU from its own reference set during evaluation, the model enables a more discriminating analysis of efficiency among these ostensibly efficient units. The following is the formula used for calculation:

$$\rho = \frac{1+\frac{1}{n}\sum_{i=1}^{n}\frac{r_i^-}{x_{ik}}}{1-\frac{1}{r_1+r_2}\left(\sum_{m=1}^{r_1}\frac{r_m^g}{y_{mk}^g}+\sum_{m=1}^{r_2}\frac{r_m^b}{y_{mk}^b}\right)} \text{s.t.} \quad (1)$$

$$\begin{cases} x_k \geq XA - r^- \\ y_k^g \leq Y^g A + r^g \\ z_k^g \geq Z^g A - r^b \\ r^- \geq 0, r^g \geq 0, r^b \geq 0, A \geq 0 \end{cases} \quad (2)$$

Among them, ρ is the measured ULUE value; n is the number of input indicators; r_1 is the number of desired output indicators; r_2 is the number of undesired output indicators; r^- is the slack variable for inputs; r^g is the slack variable for desired outputs; r^b is the slack variable for undesired outputs; x is the input value; y^g is the desired output value; z^b is the undesired output value; A is the weight vector; X is the input matrix; Y^g is the desired output matrix; and Z^b is the undesired output matrix.

3.3. Exploratory Spatial Data Analysis

The ESDA method involves describing and visualizing the spatial distribution patterns of phenomena or objects to discover spatial clustering in the data and reveal the spatial interaction mechanisms between the subjects of study, thereby providing reference and guidance for more effectively solving some current practical problems [43]. This paper uses the Moran's I index to explore whether there is spatial clustering or heterogeneity in the ULUE of the BTH urban agglomeration, analyzing and comparing the interaction relationships and degree of differentiation between various cities. The following is the formula used for calculation:

$$I = \frac{n\sum_{i=1}^{n}\sum_{j=1}^{n} W_{ij}\left(X_i - \overline{X}\right)\left(X_j - \overline{X}\right)}{\sum_{i=1}^{n}\sum_{j=1}^{n} W_{ij}\left(X_i - \overline{X}\right)} \quad (3)$$

In the formula, I is the global spatial autocorrelation coefficient; n represents the number of municipal administrative units; X_i and X_j are the observed values of city i and city j; $\overline{X}$ is the average value of the observed values X_i; and W_{ij} is the spatial weight matrix, using the Queen contiguity spatial weight matrix. When $I > 0$, there is a positive spatial correlation; when $I = 0$, there is no spatial autocorrelation, when $I < 0$ there is a negative spatial correlation.

Further, the local Moran's I index is used to analyze the spatial clustering characteristics of local areas within the BTH urban agglomeration, reflecting the interaction relationships and degree of differentiation between various cities. The following is the formula used for calculation:

$$I_i = \frac{X_i - \overline{X}}{\sum_{i=1, j\neq 1}^{n}\left(X_j - \overline{X}\right)^2 / n - 1}\sum_{j=1, i\neq 1}^{n} W_{ij}\left(X_j - \overline{X}\right) \quad (4)$$

In the formula, I_i is the local spatial autocorrelation coefficient, and the other variables have the same meaning as in Formula (3). When $I_i > 0$, it indicates a positive spatial correlation among local areas, and the larger the value is, the more evident the spatial correlation; when $I_i < 0$, it indicates a negative spatial correlation among local areas, and the larger the value is, the more evident the spatial differentiation; when $I_i = 0$, it indicates a random spatial distribution.

3.4. Efficiency Evolution Analysis

Early DEA models were used for evaluating enterprise efficiency, comparing the efficiency of similar enterprises in the same period. Therefore, they were mainly used for cross-sectional data processing. The emergence of panel data has expanded the application scenarios of DEA models, making them suitable not only for macro data analysis but also for the construction and decomposition of productivity indices. Therefore, this paper selects the GML model to measure the ULUE of the BTH urban agglomeration, addressing the shortcomings of the super-efficiency SBM model in dynamic efficiency analysis of time series data, thereby enabling a comprehensive analysis of the dynamic evolution of ULUE in the BTH urban agglomeration from 2005 to 2020. The specific formula is as follows:

$$\begin{aligned} M\left(x^t, y^t, x^{t+1}, y^{t+1}\right) &= \sqrt{(M_t * M_{t+1})} = \sqrt{\left[\frac{D^{t+1}\left(x^{t+1}, y^{t+1}\right)}{D^t\left(x^t, y^t\right)} \times \frac{D^{t+1}\left(x^{t+1}, y^{t+1}\right)}{D^t\left(x^t, y^t\right)}\right]} \\ &= \frac{D^{t+1}\left(x^{t+1}, y^{t+1}\right)}{D^t\left(x^t, y^t\right)} \times \sqrt{\left[\frac{D^t\left(x^t, y^t\right)}{D^{t+1}\left(x^t, y^t\right)} \times \frac{D^t\left(x^{t+1}, y^{t+1}\right)}{D^{t+1}\left(x^{t+1}, y^{t+1}\right)}\right]} \end{aligned} \quad (5)$$

In the formula, $D^t\left(x^t, y^t\right)$ and $D^t\left(x^{t+1}, y^{t+1}\right)$ respectively represent the distance functions of the decision-making unit between periods t and $t+1$ when the base period is taken as a reference; $D^{t+1}\left(x^t, y^t\right)$ and $D^{t+1}\left(x^{t+1}, y^{t+1}\right)$ respectively represent the distance functions of the decision-making unit between periods t and $t+1$ when period $t+1$ is taken as the base period; and $M(x^t, y^t, x^{t+1}, y^{t+1})$ represents the change in ULUE of the decision-making unit between periods t and $t+1$. When $M > 1$, it indicates an improvement in urban ULUE; otherwise, it indicates a decline. The GML index can be further decomposed into changes in catch-up efficiency and technological progress. EC reflects the efficiency changes caused by factors such as factor allocation and management under the existing technological level, while TC refers to the efficiency changes brought about by technological progress. When EC > 1, it signifies that the decision-making unit's technical efficiency has experienced a relative improvement; otherwise, it indicates that the technology has not been fully utilized. When TC > 1, It signifies that the decision-making entity has achieved advancements in technology; otherwise, it indicates technological regression.

3.5. Geographical Detector

The geographical detector can effectively explore spatial differentiation characteristics. Therefore, this paper uses the factor detector and interaction detector in the geographical detector to investigate the relationship between ULUE (Y) and influencing factors (X) in the BTH urban agglomeration. The factor detector is capable of identifying the impact of various factors on the spatial arrangement of specific items, thereby resolving the issue of causation in the heterogeneous spatial distribution among diverse objects. The results of the factor detector are measured by the q value, which is calculated using the following formula:

$$q = 1 - \frac{1}{N\sigma^2} \sum_{h=1}^{L} N_h \sigma_h^2 \tag{6}$$

In the formula, $h = 1, 2, \cdots, L$ represents the stratification of the driving factors; N_h is the number of layers on h; and $N\sigma^2$ and $N\sigma_h^2$ respectively represent the total variance of the entire observation area and the sum of the intra-layer variances. The value of q ranges from [0, 1], with a larger q value indicating a greater driving effect of the factor on the improvement of ULUE, and vice versa. Additionally, this paper introduces interaction detection to explore the interactive explanatory effect of factor combinations on the dependent variable, i.e., whether the explanatory power of two factors on the dependent variable is enhanced or weakened when they interact. The results of the interaction between two factors can be divided into five categories (as shown in Table 2).

Table 2. Types of interactions between two independent variables and dependent variables.

Judgement Criteria	Interaction Effect
$q(X_1 \cap X_2) < min(q(X_1), q(X_2))$	Nonlinear weakening
$min(q(X_1), q(X_2)) < q(X_1 \cap X_2) < max(q(X_1), q(X_2))$	Univariate nonlinear weakening
$q(X_1 \cap X_2) > max(q(X_1), q(X_2))$	Bivariate enhancement
$q(X_1 \cap X_2) = q(X_1) + q(X_2)$	Independent
$q(X_1 \cap X_2) > q(X_1) + q(X_2)$	Nonlinear enhancement

4. Results and Analysis

4.1. Spatial Temporal Analysis of Land Use Changes in BTH Urban Agglomeration

4.1.1. Temporal Evolution Characteristics

This paper utilizes the MAXDEA Ultra 8.0 software to measure the land use efficiency of the BTH urban agglomeration in 2005, 2010, 2015, and 2020 (as shown in Figure 2). Overall, the ULUE of the three regions of Beijing, Tianjin, and Hebei varies, with Beijing consistently leading and being significantly higher than the average of the BTH urban agglomeration. Specifically, Beijing's ULUE generally shows an inverted "N" pattern, where it is first decreasing, then increasing, and decreasing again, reaching a trough of 1.30 in 2010 and a peak of 1.32 in 2015. As Beijing's urbanization has accelerated, the expansion of urban construction land has encroached on ecological land, leading to environmental degradation, which in turn caused a decline in Beijing's ULUE from 2005 to 2010. With the transformation of the economic development model and the introduction of various environmental regulations, Beijing has gradually moved towards green development, resulting in an increase in ULUE from 2010 to 2015. However, due to the impact of the COVID-19 pandemic, economic activities were restricted, leading to a decline in Beijing's ULUE in 2020. The ULUE of Tianjin generally shows an inverted "U" pattern, first increasing and then decreasing, with a peak of 1.07 in 2010. From 2005 to 2010, with economic development, Tianjin's construction land was fully utilized, and the ULUE increased accordingly. However, from 2010 to 2015, land use efficiency declined, possibly due to the devastating explosion at Tianjin Port in 2015, which claimed over a hundred lives, damaged 304 buildings, 12,428 vehicles, and 7533 containers, significantly impacting Tianjin's social governance. In 2020, Tianjin's land use efficiency also declined due to the COVID-19 pandemic. The ULUE of Hebei Province generally shows an inverted "U" pattern, first increasing and then decreasing, reaching a peak of 1.09 in 2015, which is consistent with the overall trend of the BTH urban agglomeration. From 2005 to 2015, Hebei's ULUE significantly improved, surpassing Tianjin in 2015. However, it is worth noting that there was a sharp decline in the ULUE in 2020, indicating that Hebei's development resilience is still lacking. Overall, the BTH urban agglomeration was affected by the COVID-19 pandemic in 2020, resulting in an overall decline in ULUE.

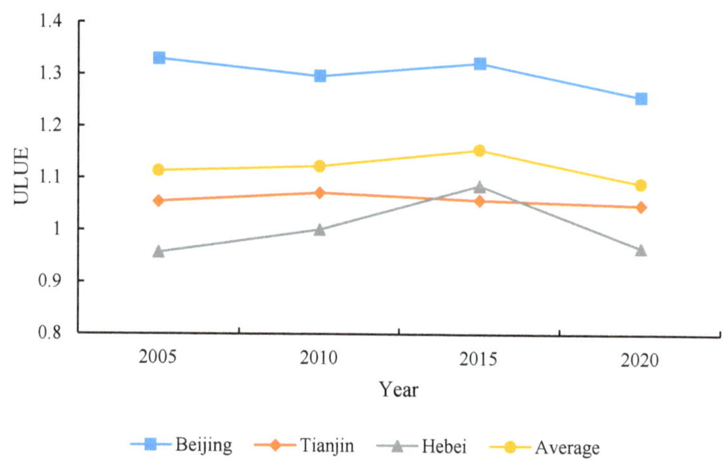

Figure 2. ULUE values of the BTH urban agglomeration from 2005 to 2020.

4.1.2. Spatial Evolution Characteristics

This paper uses ArcGIS10.8 and the natural breaks method to classify the ULUE of 13 cities in the BTH urban agglomeration in 2005, 2010, 2015, and 2020 into four categories: low-level areas, lower-level areas, higher-level areas, and high-level areas (as shown in Figure 3). The ULUE of the BTH urban agglomeration shows significant spatial distribution differences, presenting a spatial distribution pattern of high in the north and low in the south, forming a polarization of the development characteristic centered on Beijing. Among them, Beijing and Chengde in Hebei Province are the regions with high and relatively stable ULUE in the entire BTH urban agglomeration. Undoubtedly, as the capital of China, Beijing is far ahead of other cities in terms of development model, environmental regulations, and resource utilization efficiency. Chengde's "advantage lies in ecology, potential lies in ecology, hope lies in ecology". Currently, the city has a forest area of 35.56 million mu, accounting for more than one-third of the BTH region. The forest coverage rate stands at 60.03%, surpassing the national average by 36 percentage points and the province average by 25 percentage points, giving it a great advantage in green development.

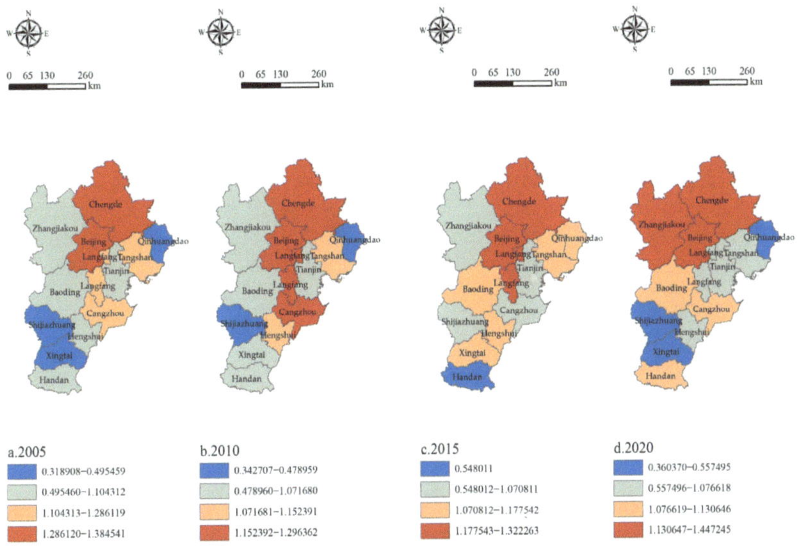

Figure 3. Spatial pattern of ULUE in the BTH urban agglomeration during 2005–2020.

From the temporal cross-section, the cities with high and higher ULUE levels in the BTH urban agglomeration in 2005 were Beijing, Chengde, Langfang, Tangshan, and Cangzhou. Due to its status as the political, cultural, and international exchange and the technological innovation center, Beijing leads in ULUE. Chengde has always adhered to the concept of green development and is a green barrier for the BTH urban agglomeration, hence its high ULUE. Cities with a certain industrial base require substantial capital, labor, and land investments in the early stages of development, and the extensive economic model leads to lower ULUE. In 2010, the cities with high and higher ULUE levels in the BTH urban agglomeration were Beijing, Chengde, Langfang, and Cangzhou, indicating that the urban scale of Cangzhou and Langfang had initially formed, no longer requiring substantial factor inputs, and effectively undertaking the industrial transfer from Beijing and Tianjin, which alleviated Beijing's non-capital functions. In 2015, The quantity of cities exhibiting elevated and even more elevated ULUE levels inside the BTH urban agglomeration has risen, including Beijing, Chengde, Langfang, Tangshan, Qinhuangdao, Baoding, Hengshui, and Xingtai. This indicates that the BTH urban agglomeration has begun to transform into an intensive development model, significantly improving overall ULUE. The cities in the BTH urban agglomeration that had elevated levels of ULUE in 2020 were Beijing, Chengde, Zhangjiakou, Baoding, Cangzhou, and Handan. The COVID-19 pandemic had a substantial impact on Qinhuangdao, Tangshan, Cangzhou, and Xingtai, leading to a decrease in ULUE levels. It is worth noting that in 2020, Zhangjiakou and Handan showed a leap in ULUE levels. Further exploration of the reasons is provided in the following text.

4.2. Spatial Agglomeration Characteristics of Urban ULUE

To further analyze the spatial correlation of the BTH urban agglomeration, Geoda 1.20 software was used to calculate the global Moran's I index values for the ULUE of the BTH urban agglomeration in 2005, 2010, 2015, and 2020, which were −0.134, −0.157, 0.016, and −0.182, respectively (Table 3). Except for 2005, all passed the 10% significance level test in statistical terms. The results show that urban ULUE in 2010 and 2020 exhibited a negative spatial correlation, while in 2015, it showed a positive spatial correlation. This indicates that the spatial distribution of ULUE in the BTH urban agglomeration exhibits certain spatial homogeneity or heterogeneity characteristics, but the stability is weak. The negative Moran's I index in 2010 indicates that the initial development level of the region was low, with significant differentiation. The development of high-level cities would create an echo effect on surrounding cities, attracting talent and capital inflow from neighboring areas, thus leading to the aggregation of high- and low-level regions. With the overall improvement in economic levels and infrastructure enhancement, the expansion diffusion effect would increase, leading to aggregation among regions of the same level. This is evidenced by the transition of the Moran's I index from negative to positive in the BTH urban agglomeration in 2015. In 2020, the Moran's I index once again shifted from positive to negative, indicating that the region was affected by external factors or that a particular city implemented institutional innovations, causing production factors to be attracted to high levels again, resulting in the echo effect surpassing the diffusion effect once more. While the global Moran's I index for 2005 did not meet the criteria for statistical significance, it does not necessarily imply that the ULUE in the BTH urban agglomeration has no spatial correlation at all. Spatial correlation may be a complex phenomenon that could exist in specific areas but not in others, or positive and negative correlations may offset each other, leading to insignificance overall. Therefore, it is not possible to simply judge the presence of spatial correlation based on the outcomes of the global Moran's I index. To more comprehensively understand the spatial relationships between different regions within the BTH urban agglomeration, it is necessary to calculate the local Moran's I index. By using the local Moran's I index, significant spatial correlation hotspots and cold spots can be identified, revealing the spatial distribution patterns of ULUE in local areas. This analytical method can help us deeply explore the spatial characteristics of ULUE within the BTH urban agglomeration, providing more targeted reference and guidance for further spatial planning and policy formulation.

Table 3. Moran' I index values.

Year	Global Moran' I	z-Value	*p*-Value
2005	−0.134	−1.0041	0.116
2010	−0.157	−1.6525	0.048
2015	0.016	2.4379	0.034
2020	−0.182	−1.9430	0.012

The local spatial association characteristics of ULUE in the BTH urban agglomeration are shown in Figure 4. In 2005, Tianjin was a high-high cluster, while Xingtai was a low-low cluster. Tianjin, with its significant seaport and airport, serves as an important transportation hub in the BTH urban agglomeration, thus forming a high-level cluster. Xingtai and neighboring cities are far from the central cities, with underdeveloped urban scale and insufficient resource allocation capabilities, resulting in a low-level cluster. In 2010, Tianjin was a high-high cluster, while Zhangjiakou was a low-high cluster. Located in the northern part of the BTH urban agglomeration, Zhangjiakou is an important ecological conservation area and environmental protection barrier for the urban agglomeration. Compared to cities in the central and southern parts with an industrial base, the economic benefits of green investments in Zhangjiakou may not be immediately apparent and require more input of production factors, thus forming a low-high cluster spatially. In 2015, Xingtai was a high-low cluster, while Zhangjiakou was a low-high cluster. In 2015, Xingtai ranked first in the province for reducing the number of days with severe or worse pollution and achieved an online monitoring data transmission efficiency of 99.45% for key national pollution sources, exceeding the national target of 75% and ranking first among prefecture-level cities in the province. Specific environmental protection measures, including strengthening air pollution control, promoting energy conservation and emission reduction in industrial enterprises, enhancing water resource protection and management, and improving the rural ecological environment, led to improvements in Xingtai's non-expected outputs and increased ULUE, thus forming a high-low cluster spatially. In 2020, Beijing exhibited a high-high cluster spatial distribution pattern, gradually forming the prototype of a world-class urban agglomeration centered on the capital, with green, smart, and livable urban development as the goal of the BTH urban agglomeration. However, overall, more than 80% of the cities did not show significant spatial clustering characteristics, indicating that there is still considerable room for improvement in the ULUE of the BTH urban agglomeration.

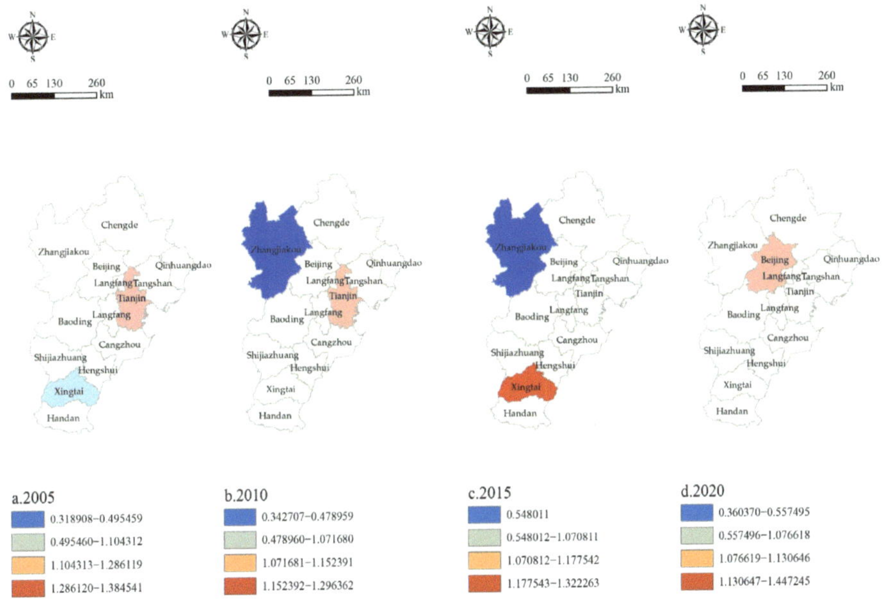

Figure 4. Local spatial association characteristics of ULUE in the BTH urban agglomeration during 2005–2020.

4.3. Spatial Evolution Characteristics of Urban Land Use Decomposition Efficiency

This paper uses the cumulative change values and geometric averages of the Global Malmquist–Luenberger (GML), efficiency change (EC), and technological progress (TC) indices from the Malmquist model for the years 2005–2020 to reflect the cumulative changes and annual average changes of each index (Table 4).

Table 4. Malmquist exponent and its decomposition value.

Area	Cumulative Change Value			Geometric Mean		
	GML	EC	TC	GML	EC	TC
Beijing	0.023243	0.05835	0.050747	1.041754	1.026142	1.061249
Tianjin	−0.03632	−0.02563	−0.01072	1.04487	1.006557	1.046793
Shijiazhuang	−0.43522	−0.60078	0.797856	1.327282	1.154614	1.274292
Tangshan	0.239769	0.077758	0.19746	1.174296	0.933307	1.254153
Qinhuangdao	−0.17902	−0.13814	1.595069	1.453172	1.008462	1.466284
Handan	2.012282	0.065909	0.497505	1.518747	1.032557	1.48238
Xingtai	−0.77352	−0.14715	0.672346	1.247653	1.009416	1.199544
Baoding	0.057511	0.045378	0.040746	1.122156	1.030264	1.08913
Zhangjiakou	0.576365	1.346611	0.094204	1.213028	1.287217	1.108462
Chengde	0.035983	0.199405	−0.09262	1.094533	0.966801	1.104921
Cangzhou	0.04915	0.002258	0.050815	0.992872	0.980843	1.016284
Langfang	−0.30318	−0.13193	−0.07892	1.119244	0.943493	1.183914
Hengshui	−0.02581	−0.14726	0.115111	1.187391	1.015924	1.172685
BTH region	0.095481	0.046522	0.302277	1.195154	1.03043	1.189238

From 2005 to 2020, the Global Malmquist–Luenberger in the BTH region increased cumulatively by 9.55%, and ULUE improved (as shown in Figure 5). Specifically, the ULUE of seven cities increased, while six cities saw a decrease in ULUE. Among them, Handan had the largest increase at 2.01, while Xingtai had the largest decrease at −0.77. Using ArcGIS 10.8 software, the spatial evolution maps of the Global Malmquist–Luenberger, catch-up efficiency, and technological progress in the BTH region from 2005 to 2020 were drawn. From the perspective of the Global Malmquist–Luenberger, during the period 2005–2010, there were more areas with efficiency decline and fewer areas with efficiency improvement; during the period 2010–2015, the areas with efficiency decline decreased, while the areas with efficiency improvement increased; during the period 2015–2020, the areas with efficiency decline further decreased, while the areas with efficiency improvement further increased. This indicates that the overall efficiency level of the BTH urban agglomeration is showing an upward trend. Specifically, from 2005 to 2010, 9 out of the 13 cities in the BTH urban agglomeration showed efficiency improvement, while Chengde, Zhangjiakou, Baoding, and Cangzhou experienced efficiency decline. From 2010 to 2015, 12 out of the 13 cities in the BTH urban agglomeration showed efficiency improvement, with only Cangzhou showing efficiency decline. From 2015 to 2020, 9 out of the 13 cities in the BTH urban agglomeration showed efficiency improvement, while Qinhuangdao, Langfang, Shijiazhuang, and Xingtai experienced efficiency decline.

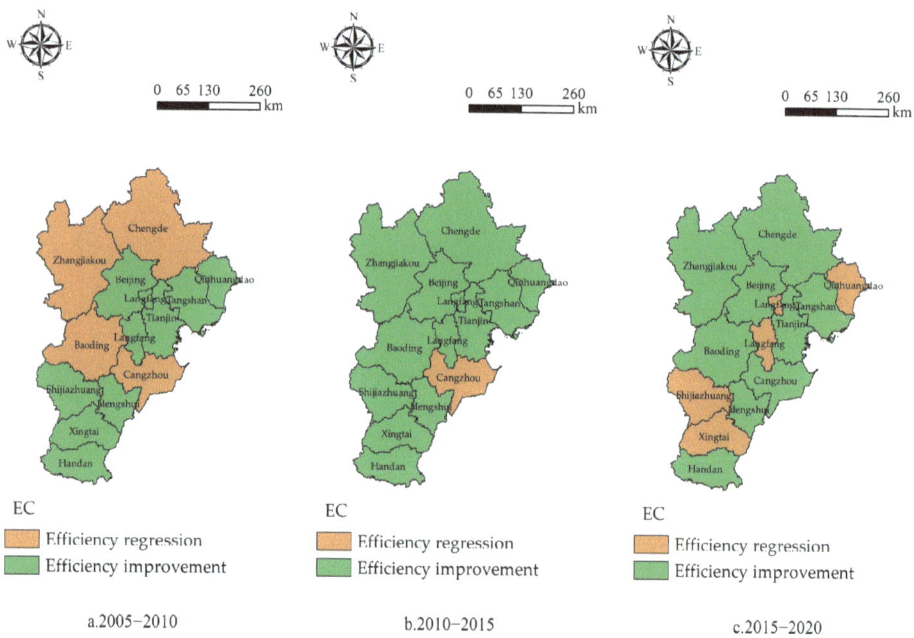

Figure 5. Spatiotemporal evolution of GML in BTH region from 2005–2020.

From the perspective of EC, during the period from 2005 to 2010, most regions in the BTH urban agglomeration showed a downward trend in efficiency, with only a few areas experiencing an increase, indicating an overall decline in the efficiency level of the BTH urban agglomeration (as shown in Figure 6). During the period from 2010 to 2015, the areas with declining efficiency decreased, while the areas with increasing efficiency increased, indicating that the overall efficiency level of the BTH urban agglomeration began to rise. During the period from 2015 to 2020, the areas with declining efficiency further decreased, while the areas with increasing efficiency further increased, indicating a significant upward trend in the overall efficiency level of the BTH urban agglomeration. This indicates that the ULUE of the urban agglomeration is highly dependent on scale effects, leading to issues of land resource waste. For cities with declining catch-up efficiency, it is necessary to strengthen the input of production factors and pay attention to the benefits brought by scale effects. It is worth noting that the trend of efficiency changes in the BTH urban agglomeration is not uniform. Some regions showed an upward trend in efficiency at an earlier stage, while others began to show an upward trend at a later stage. This may be related to the development strategy and industrial structure adjustment of the BTH urban agglomeration.

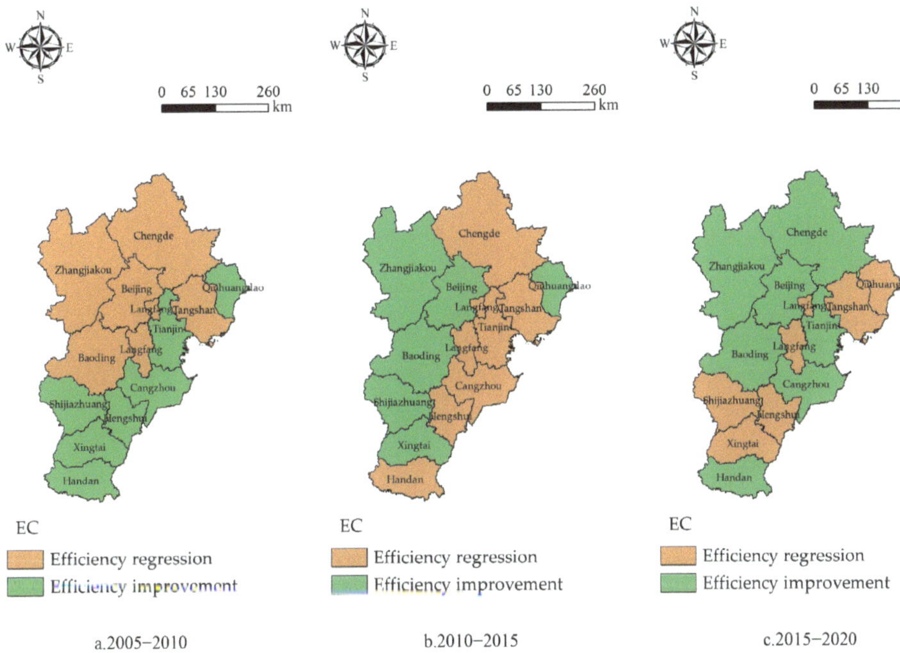

Figure 6. Spatiotemporal evolution of EC in BTH region from 2005–2020.

From the perspective of TC, from 2005 to 2010, most regions in the BTH urban agglomeration experienced a decline in efficiency, with only a few areas showing an improvement (as shown in Figure 7). From 2010 to 2015, the areas with declining efficiency decreased, while the areas with improving efficiency increased. From 2015 to 2020, the areas with declining efficiency further decreased, while the areas with improving efficiency further increased. This indicates that the overall efficiency level of the BTH urban agglomeration has shown an upward trend in recent years. Specifically, from 2005 to 2010, most cities in the BTH urban agglomeration experienced a decline in efficiency, mainly concentrated in the northern regions of Hebei Province and parts of Tianjin. This may be due to the over-reliance on traditional industries in these areas, leading to slow industrial structure adjustments and efficiency declines. From 2010 to 2015, the BTH urban agglomeration began implementing a series of industrial transformation and upgrading policies, promoting industrial structure adjustment and optimization, and strengthening regional coordinated development, which improved the overall efficiency of the urban agglomeration. By the period from 2015 to 2020, the trend of efficiency improvement in the BTH urban agglomeration became more evident, with further reductions in areas with declining efficiency and further increases in areas with improving efficiency. This indicates that the industrial structure adjustment and regional coordinated development in the BTH urban agglomeration have achieved significant results.

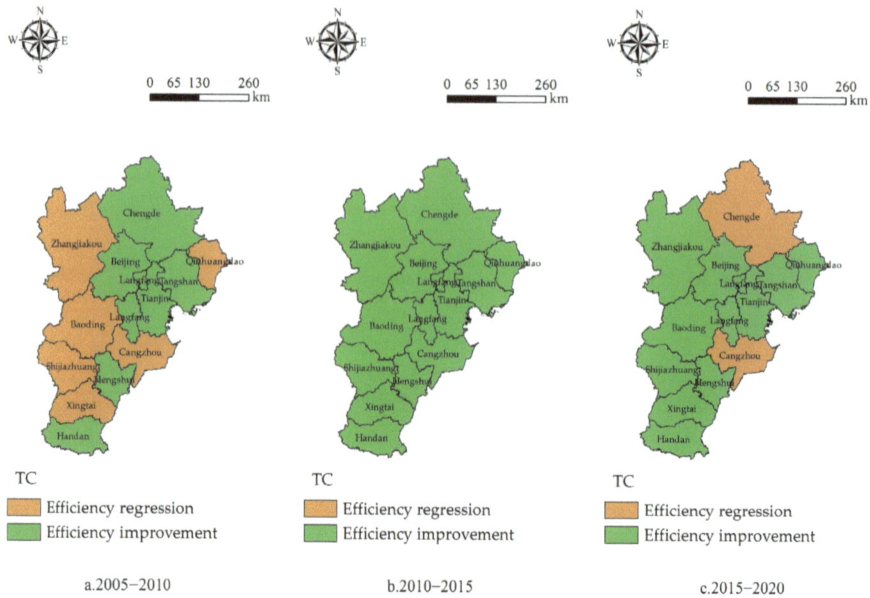

Figure 7. Spatiotemporal evolution of TC in BTH region during 2005–2020.

4.4. Geographical Detector

4.4.1. Factor Detection and Result Analysis

Using the geographical detector, ULUE is taken as the explained variable, and input indicators and expected outputs are taken as explanatory variables to study the impact of factors influencing the ULUE of the BTH region when acting alone. The results are shown in Table 5.

Table 5. Factor recognition results of geographical detector.

	x1	x2	x3	x4	x5	x6	x7	x8	x9
q statistic	0.7904	0.1165	0.7761	0.6771	0.9486	0.8764	0.3580	0.0477	0.1246
p value	0.000	0.1151	0.000	0.000	0.000	0.000	0.2571	0.5259	0.1360

The per capita fixed capital investment (x1), the per capita urban employment number (x3), the per capita GDP (x4), the average wage of on-the-job employees (x5), and the per capita number of beds in public health institutions (x6) all passed the 1% significance test in statistical terms. The ranking of the influence intensity of the spatiotemporal differentiation factors of ULUE in the BTH urban agglomeration is as follows: average wage of on-the-job employees (0.9486) > per capita number of beds in public health institutions (0.8764) > per capita fixed capital investment (0.7904) > per capita urban employment number (0.7761) > per capita GDP (0.6771). Specifically, the level of average wages for on-the-job employees affects ULUE. Higher wages mean higher productivity and innovation capability, which can promote more effective land use methods. The number of beds in public health institutions per capita also impacts ULUE; sufficient health resources can improve residents' quality of life, thereby influencing land use demand and methods. The amount of per capita fixed capital investment reflects the level of regional economic development. High levels of fixed capital investment drive industrial upgrading and technological innovation, thus affecting ULUE. The increase in urban employment numbers drives the urbanization process, impacting land use patterns and efficiency. Cities with higher per capita GDP have more funds and resource allocation capabilities, enabling more effective planning and utilization of land resources. Moreover, with the increase in per capita GDP, environmental awareness and investment in the region are enhanced. Regions that pay more attention to environmental protection and sustainable development tend to adopt more scientific and reasonable land use methods to ensure the sustainable use and protection of land.

4.4.2. Interaction Detection Results and Analysis

The results of the factor interaction detector indicate that the explanation of ULUE in the BTH region by single factors does not act independently. The interaction values of two factors are greater than the q values of single factors, indicating that the interaction of two factors enhances the explanatory power of ULUE in the BTH region, either through bivariate enhancement or nonlinear enhancement. Based on the thermal map resulting from the interactive influence of the driving factors in the geographic detector, the darker the color, the more intense the interaction effect (as shown in Figure 8). It can be observed that the interaction value between the wage level of on-the-job employees and the per capita public green space area is the highest, reaching 0.9674. This interaction result indicates that in the process of land use and development, it is essential to ensure that each unit of land brings economic benefits to every city resident while also considering the protection of the ecological environment, especially the green space resources each resident can enjoy. The interaction value between the proportion of built-up area and the per capita public green space area is the lowest, indicating that in the land development process in the BTH region, there may be development at the expense of environmental destruction. The expansion of construction land does not adequately meet the ecological needs of residents.

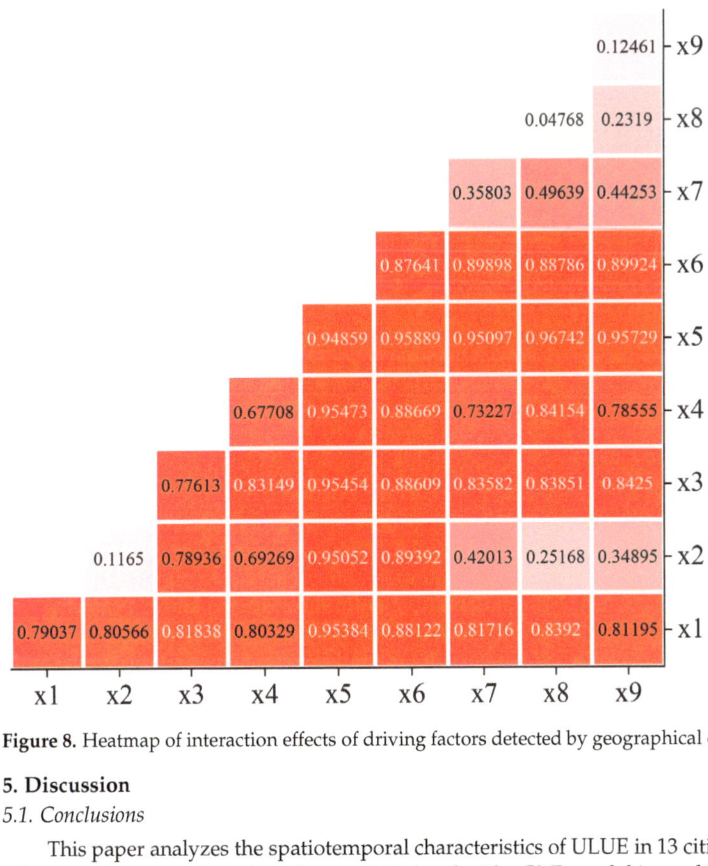

Figure 8. Heatmap of interaction effects of driving factors detected by geographical detector.

5. Discussion

5.1. Conclusions

This paper analyzes the spatiotemporal characteristics of ULUE in 13 cities of the BTH urban agglomeration based on the municipal scale. The GML model is used to decompose the evolution of efficiency, and the geographical detector is employed to analyze the factors driving the improvement in ULUE. The following conclusions are drawn:

First, from 2005 to 2020, the ULUE of the BTH urban agglomeration exhibited an inverted "U"-shaped evolution pattern, initially increasing and then decreasing. The overall level was relatively low, with significant regional differences, indicating substantial room for improvement. Second, the ULUE of the BTH urban agglomeration showed significant spatial distribution differences, presenting a pattern of higher efficiency in the north and lower in the south, forming a polarized development characteristic centered on Beijing. The northern cities of Chengde and Zhangjiakou, serving as green barriers for the region, had higher ULUE than the industrial cities in the central and southern parts of the BTH region. Local areas showed a high-high clustering evolution characteristic converging towards Beijing. Third, the ULUE of the BTH urban agglomeration cumulatively increased by 9.55%. The ULUE in the Beijing–Tianjin region was effectively improved through the combined effects of catch-up efficiency and technological progress. Fourth, the geographical detector indicated that the main factors driving the improvement in ULUE in the BTH region were the per unit fixed capital investment, per unit urban employment number, per unit GDP, average wage of on-the-job employees, and per unit number of beds in public health institutions, with the average wage of on-the-job employees having the greatest impact. Improving residents' income while ensuring that green space resources are available to each resident can better enhance ULUE.

5.2. Theorical Implications

The theoretical implications of this paper mainly have two aspects. Firstly, the measurement methods of ULUE mainly use the DEA model, SFA model, and SBM model [44–46]. The efficiency measured by the traditional DEA model is often affected by unexpected factors. The SFA model cannot match the real production process well and can only calculate the unique output efficiency. The SBM model can consider the expected output and non-expected output in the production process, and the introduction of relaxation variables makes the measurement of land use efficiency more accurate. In this paper, considering China's carbon peaking and carbon neutrality goals, the average carbon emission in the ground is introduced into the index of efficiency measurement, and the interaction between influencing factors is revealed through the geographical detector model, to improve the scientific evaluation results. Secondly, the development of urban agglomerations has a strong correlation in space, similar endowments in the natural environment, similar cultures in historical development, similar systems in transportation, and coordinated development strategies in policymaking. Research based on the scale of urban agglomerations is of great value to enrich, develop, and improve the evaluation of urban land use efficiency.

5.3. Practical Implications

Considering the conclusions above, the potential policy recommendations are proposed. Firstly, it can be deduced from the research conclusion that the Beijing–Tianjin–Hebei region is mainly developed centering around Beijing, and it is highly necessary to remove the non-capital functions of Beijing. The central urban area adheres to the characteristic development orientation of service economy, knowledge economy, and green economy. The suitability of natural conditions and the carrying capacity of resources and the environment should be fully considered. Attention should be paid to the orderly growth of the intensity of land development and the spatial and temporal scale. Secondly, it is essential to integrate the industrial layout and enhance the efficiency of land utilization. Considering the conclusions above, the main factors driving the improvement in ULUE in the BTH region were the per unit fixed capital investment, per unit urban employment number, per unit GDP, and so on. It is of great significance to evaluate the required land, capital, and manpower, as well as the jobs and GDP contribution that can be obtained by the determined industrial land scale to achieve sustainable development.

5.4. Limitations and Future Research

Firstly, this paper examines data from the Beijing–Tianjin–Hebei urban cluster covering the period from 2005 to 2020. Future studies may extend the analysis to additional cities, focusing on micro and meso levels including counties, cities, industrial parks, economic development zones, and various city types. The observation period could also be lengthened to identify more universally applicable patterns. Secondly, limited data availability means this paper has shortcomings in selecting evaluation indicators and influencing variables for land use efficiency. Finally, the use of geographic detectors to discuss the factors influencing land use efficiency in this paper may have resulted in omissions in the selection of explanatory variables. Therefore, further research is needed to enhance the selection of indicators and variables.

Author Contributions: Methodology, J.Y.; Writing—original draft, H.H. and J.Y.; Writing—review and editing, H.H. All authors have read and agreed to the published version of the manuscript.

Funding: This research received no external funding.

Data Availability Statement: The experimental data are mainly downloaded from the EPS data platform. The website of this data platform is: https://www.epsnet.com.cn/index.html#/Index, accessed on 13 August 2024. It is used to support the findings of this study, which are available from the corresponding author upon request.

Conflicts of Interest: The author declares that he has no conflicts of interest.

References

1. Ramachandra, T.V.; Aithal, B.H.; Sanna, D.D. Insights to Urban Dynamics through Landscape Spatial Pattern Analysis. *Int. J. Appl. Earth Obs. Geoinf.* **2012**, *18*, 329–343. [CrossRef]
2. Zou, L.; Liu, Y.; Wang, J.; Yang, Y.; Wang, Y. Land Use Conflict Identification and Sustainable Development Scenario Simulation on China's Southeast Coast. *J. Clean. Prod.* **2019**, *238*, 117899. [CrossRef]
3. Dong, G.; Liu, Z.; Niu, Y.; Jiang, W. Identification of Land Use Conflicts in Shandong Province from an Ecological Security Perspective. *Land* **2022**, *11*, 2196. [CrossRef]
4. Chen, D.; Hu, W.; Li, Y.; Zhang, C.; Lu, X.; Cheng, H. Exploring the Temporal and Spatial Effects of City Size on Regional Economic Integration: Evidence from the Yangtze River Economic Belt in China. *Land Use Policy* **2023**, *132*, 106770. [CrossRef]
5. Zhao, J.; Zhu, D.; Cheng, J.; Jiang, X.; Lun, F.; Zhang, Q. Does Regional Economic Integration Promote Urban Land Use Efficiency? Evidence from the Yangtze River Delta, China. *Habitat Int.* **2021**, *116*, 102404. [CrossRef]
6. Jingxin, G.; Jinbo, S.; Lufang, W. A New Methodology to Measure the Urban Construction Land-Use Efficiency Based on the Two-Stage DEA Model. *Land Use Policy* **2022**, *112*, 105799. [CrossRef]
7. Jiang, H. Spatial–Temporal Differences of Industrial Land Use Efficiency and Its Influencing Factors for China's Central Region: Analyzed by SBM Model. *Environ. Technol. Innov.* **2021**, *22*, 101489. [CrossRef]
8. Zhang, X.; Jie, X.; Ning, S.; Wang, K.; Li, X. Coupling and Coordinated Development of Urban Land Use Economic Efficiency and Green Manufacturing Systems in the Chengdu-Chongqing Economic Circle. *Sustain. Cities Soc.* **2022**, *85*, 104012. [CrossRef]
9. Bardo, J.W.; Hartman, J.J. *Urban Sociology: A Systematic Introduction*; F.E. Peacock: Itasca, IL, USA, 1982.
10. Koopmans, T.C.; Beckmann, M.J. Assignment Problems and the Location of Economic Activities. *Econometrica* **1957**, *25*, 53.
11. Wheaton, W.C. Urban Residential Growth under Perfect Foresight. *J. Urban Econ.* **1982**, *12*, 1–21. [CrossRef]
12. Capozza, D.R.; Helsley, R.W. The Fundamentals of Land Prices and Urban Growth. *J. Urban Econ.* **1989**, *26*, 295–306. [CrossRef]
13. Fava, S.F.; Chapin, F.S. Urban Land Use Planning. *Am. Sociol. Rev.* **1957**, *22*, 765.
14. Scott, A.J. *The Urban Land Nexus and the State*; Routledge: London, UK, 1980.
15. Russett, B.M. Locational Approaches to Power and Conflict. *Am. Political Sci. Rev.* **1977**, *71*, 1617–1618.
16. Williams, O.P. *Urban Politics as Political Ecology*; Palgrave Macmillan: London, UK, 1975.
17. Jenks, M.; Burton, E.J.; Williams, K. *The Compact City: A Sustainable Urban Form?* Routledge: London, UK, 1996.
18. Gabriel, S.A.; Faria, J.A.; Moglen, G.E. A Multiobjective Optimization Approach to Smart Growth in Land Development. *Socio-Econ. Plan. Sci.* **2006**, *40*, 212–248.
19. Irwin, E.G. New Directions for Urban Economic Models of Land Use Change: Incorporating Spatial Heterogeneity and Transitional Dynamics. *J. Reg. Sci.* **2009**, *50*, 65–91.
20. Verburg, P.H.; van Berkel, D.B.; van Doorn, A.M.; van Eupen, M.; van den Heiligenberg, H.A.R.M. Trajectories of Land Use Change in Europe: A Model-Based Exploration of Rural Futures. *Landsc. Ecol.* **2010**, *25*, 217–232. [CrossRef]
21. Thornton, P.E.; Calvin, K.; Jones, A.D.; Di Vittorio, A.V.; Bond-Lamberty, B.; Chini, L.; Shi, X.; Mao, J.; Collins, W.D.; Edmonds, J.; et al. Biospheric Feedback Effects in a Synchronously Coupled Model of Human and Earth Systems. *Nat. Clim. Change* **2017**, *7*, 496–500. [CrossRef]

22. Edwards, D.P.; Socolar, J.B.; Mills, S.C.; Burivalova, Z.; Koh, L.P.; Wilcove, D.S. Conservation of Tropical Forests in the Anthropocene. *Curr. Biol.* **2019**, *29*, R1008–R1020.
23. Kaur, H.; Garg, P. Urban Sustainability Assessment Tools: A Review. *J. Clean. Prod.* **2019**, *210*, 146–158. [CrossRef]
24. Zhu, X.; Zhang, P.; Wei, Y.; Li, Y.; Zhao, H. Measuring the Efficiency and Driving Factors of Urban Land Use Based on the DEA Method and the PLS-SEM Model—A Case Study of 35 Large and Medium-Sized Cities in China. *Sustain. Cities Soc.* **2019**, *50*, 101646. [CrossRef]
25. Tone, K.; Tsutsui, M. Dynamic DEA with Network Structure: A Slacks-Based Measure Approach. *Omega* **2014**, *42*, 124–131. [CrossRef]
26. Xie, H.; Wang, W. Exploring the Spatial-Temporal Disparities of Urban Land Use Economic Efficiency in China and Its Influencing Factors under Environmental Constraints Based on a Sequential Slacks-Based Model. *Sustainability* **2015**, *7*, 10171–10190. [CrossRef]
27. Wu, C.; Wei, Y.D.; Huang, X.; Chen, B. Economic Transition, Spatial Development and Urban Land Use Efficiency in the Yangtze River Delta, China. *Habitat Int.* **2017**, *63*, 67–78. [CrossRef]
28. Xie, H.; Chen, Q.; Lu, F.; Wang, W.; Yao, G.; Yu, J. Spatial-Temporal Disparities and Influencing Factors of Total-Factor Green Use Efficiency of Industrial Land in China. *J. Clean. Prod.* **2019**, *207*, 1047–1058. [CrossRef]
29. Yuan, X.; Liu, R.; Huang, T. Analyzing Spatial–Temporal Patterns and Driving Mechanisms of Ecological Resilience Using the Driving Force–Pressure–State–Influence–Response and Environment–Economy–Society Model: A Case Study of 280 Cities in China. *Systems* **2024**, *12*, 311. [CrossRef]
30. Kuang, B.; Lu, X.; Zhou, M.; Chen, D. Provincial Cultivated Land Use Efficiency in China: Empirical Analysis Based on the SBM-DEA Model with Carbon Emissions Considered. *Technol. Forecast. Soc. Change* **2020**, *151*, 119874. [CrossRef]
31. Fu, Y.; Zhou, T.; Yao, Y.; Qiu, A.; Wei, F.; Liu, J.; Liu, T. Evaluating Efficiency and Order of Urban Land Use Structure: An Empirical Study of Cities in Jiangsu, China. *J. Clean. Prod.* **2021**, *283*, 124638. [CrossRef]
32. Seto, K.C.; Fragkias, M.; Güneralp, B.; Reilly, M.K. A Meta-Analysis of Global Urban Land Expansion. *PLoS ONE* **2011**, *6*, e23777. [CrossRef]
33. Paulsen, K. Geography, Policy or Market? New Evidence on the Measurement and Causes of Sprawl (and Infill) in US Metropolitan Regions. *Urban Stud.* **2014**, *51*, 2629–2645. [CrossRef]
34. Xia, J.; Yan, Z.; Zhou, W.; Fong, S.K.; Leong, K.C.; Tang, I.M.; Chang, S.W.; Leong, W.K.; Jin, S. Projection of the Zhujiang (Pearl) River Delta's Potential Submerged Area Due to Sea Level Rise during the 21st Century Based on CMIP5 Simulations. *Acta Oceanol. Sin.* **2015**, *34*, 78–84. [CrossRef]
35. Masini, E.; Tomao, A.; Barbati, A.; Corona, P.; Serra, P.; Salvati, L. Urban Growth, Land-Use Efficiency and Local Socioeconomic Context: A Comparative Analysis of 417 Metropolitan Regions in Europe. *Environ. Manag.* **2019**, *63*, 322–337. [CrossRef]
36. Gao, X.; Zhang, A.; Sun, Z. How Regional Economic Integration Influence on Urban Land Use Efficiency? A Case Study of Wuhan Metropolitan Area, China. *Land Use Policy* **2020**, *90*, 104329. [CrossRef]
37. Cao, X.; Liu, Y.; Li, T.; Liao, W. Analysis of Spatial Pattern Evolution and Influencing Factors of Regional Land Use Efficiency in China Based on ESDA-GWR. *Sci. Rep.* **2019**, *9*, 520. [CrossRef]
38. Reynès, F. The Cobb–Douglas Function as a Flexible Function: A New Perspective on Homogeneous Functions through the Lens of Output Elasticities. *Math. Soc. Sci.* **2019**, *97*, 11–17. [CrossRef]
39. Liao, X.; Fang, C.; Shu, T.; Ren, Y. Spatiotemporal Impacts of Urban Structure upon Urban Land-Use Efficiency: Evidence from 280 Cities in China. *Habitat Int.* **2023**, *131*, 102727. [CrossRef]
40. Yan, S.; Peng, J.; Wu, Q. Exploring the Non-Linear Effects of City Size on Urban Industrial Land Use Efficiency: A Spatial Econometric Analysis of Cities in Eastern China. *Land Use Policy* **2020**, *99*, 104944. [CrossRef]
41. Bellassen, V.; Luyssaert, S. Carbon Sequestration: Managing Forests in Uncertain Times. *Nature* **2014**, *506*, 153–155. [CrossRef]
42. Hong, J.; Mao, Y. How Does the Green Efficiency of Urban Land Use Evolve in the Urban Agglomeration of China's Middle Yangtze River? *Nat. Resour. Model.* **2024**, *37*, e12392. [CrossRef]
43. Anselin, L. Local Indicators of Spatial Association—LISA. *Geographical Analysis* **1995**, *27*, 93–115. [CrossRef]
44. Meng, Q.; Pi, H.; Xu, T.; Li, L. Exploring the Relationship between Regional Tourism Development and Land Use Efficiency: A Case Study of Guangxi Zhuang Autonomous Region, China. *PLoS ONE* **2024**, *19*, e0297196. [CrossRef]
45. Lin, Q.; Bai, S.; Qi, R. Cultivated Land Green Use Efficiency and Its Influencing Factors: A Case Study of 39 Cities in the Yangtze River Basin of China. *Sustainability* **2024**, *16*, 29. [CrossRef]
46. Han, D.; Cao, Z. Evaluation and Influential Factors of Urban Land Use Efficiency in Yangtze River Economic Belt. *Land* **2024**, *13*, 671. [CrossRef]

land

MDPI

Article

Spatiotemporal Analysis of the Coupling Relationship Between Urban Infrastructure and Land Utilization in a Shrinking City: A Case Study of Hegang, Northeast China

Siyi Huang [1,*], Yuefeng Lyu [2], Haokun Shi [1] and Cifang Wu [1,*]

[1] Department of Land Resources Management, School of Public Affairs, Zhejiang University, Hangzhou 310058, China; 11622039@zju.edu.cn
[2] School of Public Administration, Zhejiang University of Finance and Economics, Hangzhou 310018, China; lyuyuefeng@zufe.edu.cn
* Correspondence: huangsiyi@zju.edu.cn (S.H.); wucifang@zju.edu.cn (C.W.)

Abstract: Globally, urbanization is accelerating, with China witnessing a significant 40% rise in urbanization rate over the past 4 decades. However, the dynamic changes in the spatial coupling between infrastructure and utilization intensity during the early, middle, and late stages of urbanization are not clear. The trajectory of development and coupling within the urbanization process is crucial for understanding issues such as urban over-saturation and urban shrinkage. Using Hegang in Northeastern China as an example, we utilized high-resolution remote sensing data, examined the construction intensity of urban land use, analyzed the degree of coupling with utilization efficiency, and clarified the dynamic evolution of the binary relationship system between development and coupling. Results show that Hegang's construction intensity has seen a continuous rise from 1992 to 2000, with a 200.06% increase over 28 years, while its coupling with utilization efficiency has experienced a significant drop in the 21st century, suggesting a persistent decline in the utilization of buildings and a notable urban shrinkage phenomenon. Considering development status and coupling degree, we delineate a characteristic urbanization state curve for Hegang, reflecting its progression through stages of "Underdeveloped, Highly coupled," to "Underdeveloped, Weakly coupled", and finally to "Highly developed, Weakly coupled", offering insights into its urban development path. This research not only establishes a foundational data groundwork for future land-use planning in Hegang but also presents a replicable template for urbanization path analysis in other cities, contributing to a broader understanding of urban development dynamics.

Keywords: built-up land; spatial coupling; construction intensity; nighttime light data; Hegang

Citation: Huang, S.; Lyu, Y.; Shi, H.; Wu, C. Spatiotemporal Analysis of the Coupling Relationship Between Urban Infrastructure and Land Utilization in a Shrinking City: A Case Study of Hegang, Northeast China. *Land* **2024**, *13*, 1904. https://doi.org/10.3390/land13111904

Academic Editor: Jūratė Sužiedelytė-Visockienė

Received: 10 October 2024
Revised: 7 November 2024
Accepted: 11 November 2024
Published: 13 November 2024

1. Introduction

Urbanization is one of the most significant human activities, representing a complex process where rural populations migrate to urban areas and land cover transitions from natural to human-dominated landscapes [1], marking a lifestyle change from agricultural to modern urban living [2–4]. Since the year 2000, an estimated 1.57 billion individuals have transitioned from rural to urban lifestyles, comprising 20.2% of Earth's populace by the year 2020. This demographic shift has led to a 56.4% escalation in urban dwellers throughout the 21st century. Concurrently, there has been a 150% expansion in the world's impervious surface expanse, reaching 108,710 square kilometers. However, the rate of land urbanization does not always align with the rate of population urbanization [1], with urban land expansion either outpacing or lagging behind population growth [5,6]. Additionally, the physical infrastructure of a city, as indicated by its building volume, does not always align with the vitality of its socio-economic status. A low ratio of these two variables indicates an overburdened urban system, which can lead to environmental degradation, food security issues, and climate change challenges [7–9]. Conversely, a large ratio suggests urban shrinkage, which involves socio-economic issues like population loss [10], economic stagnation [11], and social depression [11] due to factors such as deindustrialization, aging, and suburbanization [12]. Therefore, assessing the spatial coupling between physical construction space and socio-economic vitality over the past decades of urbanization, understanding the relationship between development and coupling, and deploying appropriate mitigation strategies for the increasingly severe urban shrinkage in recent years are essential.

Since the 1980s, China has undergone reform and opening up, compressing into a few decades what normally would take several times longer for urbanization. However, China appears to have experienced excessive urban land expansion [13]. Since 2000, the urban population has increased by 3.8 million (76%), while the urban built-up land area has increased by nearly 300% [1]. Meanwhile, cities in China's northeastern region, with Hegang being the most severe, have experienced significant population outflow, economic decline, and industrial stagnation in the last decade, leading to the emergence of ghost cities and an increasing risk of urban shrinkage. Thus, it represents a typical area worthy of studying the coupling of physical space with socio-economic elements at various stages of urbanization. According to existing research, many scholars typically employ panel data to calculate urban land-use efficiency within the boundaries of an administrative division by integrating various input–output indicators. Common input indicators encompass the area of built-up land [14], fixed asset investment, labor intensity, and energy consumption [15,16]. The output indicators usually include newly added construction land, urban population [17], Gross Domestic Product (GDP), government revenue, and environmental pollution [18]. The models and methods frequently applied include Data Envelopment Analysis (DEA) [16], the Slacks-Based Measure (SBM) model [15], and Stochastic Frontier Analysis (SFA) [18]. However, panel data are characterized by temporal lag and are constrained by administrative boundaries, which limit the richness of information over time and across geographic areas. These limitations make it difficult to capture long-term annual change information and also complicate the identification of specific spatial details at a fine-grained or building-stock level.

Remote sensing data, known for their high spatial and temporal resolution, have become a prevalent tool in urban land use studies, effectively compensating for the shortcomings of panel data. A multitude of land cover products contribute significantly to the comprehensive analysis of urban growth and the transformation of land use and land cover (LULC) [19,20]. Building height data enhance the examination of urban morphology from a vertical perspective, serving as a crucial dimension for assessing the expansion of urban construction volume [21,22]. The year when impervious surfaces were converted provides this study with a timeline of building construction, which aids in the analysis of temporal changes [23]. Moreover, nighttime light data consistently show a robust positive association with population growth and economic vitality [24] and exhibit a strong link to various urban dynamics, including urban form and growth patterns [25], energy usage [26], and carbon emission levels. Gridded Gross Domestic Product (GDP) data can represent the economic activities within each grid cell. These data offer high-resolution, fine-scale insights into urbanization processes over time, surpassing the spatial and temporal precision achievable with traditional panel data analysis. Furthermore, they provide a perspective on long-term coupling dynamics, which is distinct from the single time-point focus of previous studies. Therefore, this study utilizes long-term, high-resolution remote sensing data to focus on the urbanization process in Hegang city from 1992 to 2020, a city that is facing one of the most severe cases of urban shrinkage in the nation. By clarifying the coupling degree between the intensity of infrastructure construction and the city's socioeconomic vitality within each 1km grid cell, this study employs GDP as an indicator of economic output and qualitatively categorizes each grid within a binary framework of development and coupling. This approach delineates a representative urbanization state curve for Hegang, providing insights into its urban development trajectory. Thereafter, we offer suggestions according to the spatial heterogeneity of construction intensity and coupling degree. The research provides technological guidance for evaluating coupling by fusing multi-source spatial data and lays a foundational data groundwork for future land-use planning in Hegang, offering a replicable template for urbanization path analysis in other cities.

2. Materials and Methods

2.1. Study Area and Data

2.1.1. Study Area

The research was conducted in Hegang, which is located in the northeastern part of Heilongjiang Province, China (Figure 1). The city has six districts and two counties, covering a total area of 14,684 square kilometers. Situated within the "Northeast Sanjiang Plain" formed by the Heilong River, the Songhua River, and the Lesser Khingan Mountains, Hegang borders Russia to the north across the Heilong River. Hegang's economy has historically relied on coal mining and processing, with abundant high-quality coal reserves establishing it as a significant coal production base in China. Historically, the city has been a central energy industrial hub in northeastern Heilongjiang Province. However, the decline of the coal industry has led to significant economic challenges, including population outflow and an urgent need for economic transformation. According to the 2020 census data, the population of Hegang City decreased from 1,058,665 in 2010 to 891,271, a reduction of 167,394 people, which is a 15.81% decrease. Hegang is facing one of the most severe population losses and urban shrinkages in China. How has the coupling of construction intensity and utilization efficiency changed over the last three decades? How has the binary relationship system of development and coupling in Hegang evolved dynamically from 1992 to 2020?

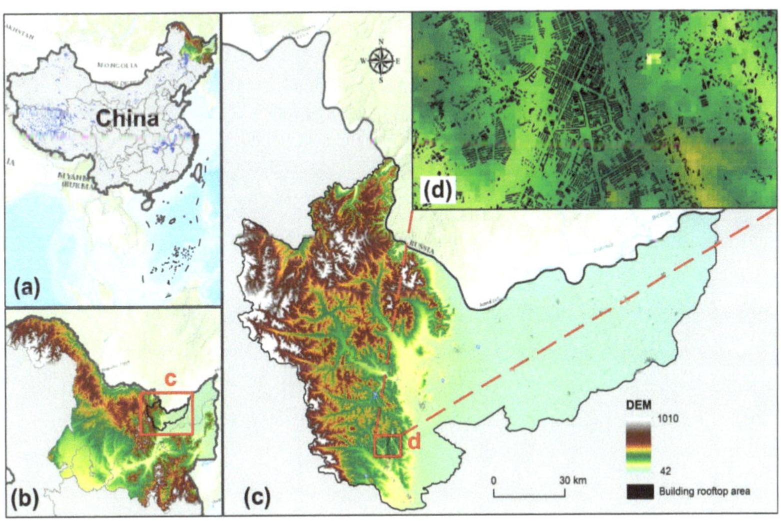

Figure 1. Location of the study area. (**a**) The entire territory of China, (**b**) Heilongjiang Province, (**c**) Hegang City, (**d**) A schematic diagram of rooftops in the center of Hegang City.

In view of the above, this study takes Hegang as an example and tries to analyze the spatio-temporal coupling relationship between its land-use intensity and efficiency, with the hope of providing a reference basis for the construction and development of shrinking cities in China.

2.1.2. Data

The China Building Rooftop Area (CBRA) dataset [27] encompasses a multi-annual collection of rooftop area data with a resolution of 2.5 m, derived from Sentinel-2 satellite imagery spanning from 2016 to 2021. This CBRA dataset is the first to provide full coverage and multi-annual Building Rooftop Area (BRA) data across China. The high resolution of the data allows for a more precise depiction of the density of planar buildings. Furthermore, this study utilizes the GAIA dataset, produced by Gong et al. [23], to attribute construction years to the CBRA, thereby extracting high-precision building horizontal extents for every two years from 1992 to 2020. The building height data, as reported by Wang et al. [22], complement the three-dimensional form of urban buildings from a vertical perspective. By integrating the horizontal CBRA data, the vertical building height data, and the construction year information from GAIA, a comprehensive three-dimensional dataset of urban buildings constructed year by year is delineated.

The Nighttime Light data (NTL), spanning from 1992 to 2021, is sourced from the unified Global Night Time Lights dataset as reported by Li et al. [28]. The DMSP-OLS and NOAA-VIIRS have been capturing light data from the night sky since 1992 and 2012, respectively. However, due to variations in the spatial detail and sensing technology used by these two sources, a calibration process is essential to ensure data compatibility. The resulting dataset offers a seamless, calibrated sequence of nighttime illumination with a resolution of ~1 km. To mitigate the impact of scattered pixels of nighttime illumination in suburban areas, the study area is further refined. By identifying natural urban boundaries produced by Li et al. [29], a spatially contiguous urban space is selected, thereby avoiding interference from scattered impervious surface pixels in agricultural and ecological spaces.

To characterize the level of urban economic development, we utilized GDP raster data with a spatial resolution of 1 km and a temporal span from 1992 to 2019 [30]. Additionally, we employed the 1 km GDP raster data for China created by Yang et al. as the data source for the final year of our study. Terrain data are collected from SRTMDEM 90M resolution raw elevation data, which is part of the Shuttle Radar Topography Mission (SRTM), an international project executed by NASA and the National Geospatial-Intelligence Agency (NGA). The administrative boundary data are collected from http://www.resdc.cn/ (accessed on 10 July 2024). The specific data types and sources are shown in Table 1.

Table 1. Data and sources.

Data	Type	Resolution	Unit	Source
China Building Rooftop Area (CBRA)	Raster	2.5 m	0/1	Liu et al. [27]
GAIA	Raster	30 m	Year	Gong et al. [23]
Building height data	Raster	1 km	m	Wang et al. [22]
Nighttime Light Data (NTL)	Raster	1 km	DN value	Harmonized Global Night Time Lights (1992–2021)—awesome-gee-community-catalog [28]
Urban boundaries	Vector	/	/	Li et al. [29]
GDP	Raster	1 km	Yuan	Chen et al. [30] & https://doi.org/10.6084/m9.figshare.21485682.v1 (accessed on 10 July 2024)
DEM	Raster	90 m	m	http://www.gscloud.cn/ (accessed on 10 July 2024)
Administrative boundary	Vector	/	/	http://www.resdc.cn/ (accessed on 10 July 2024)

2.2. Methodology Framework

This study is structured into four steps, conducting a long-term sequence analysis of land-use changes in Hegang from 1992 to 2020 across four dimensions: construction intensity (measured by Built-up Volume, BUV), its coupling with actual utilization intensity (measured by Lighted Building Index, LBI), the evolution of the binary relationship between coupling and development over the past three decades, and the spatial autocorrelation of both BUV and LBI annually. The detailed technical framework is shown in Figure 2.

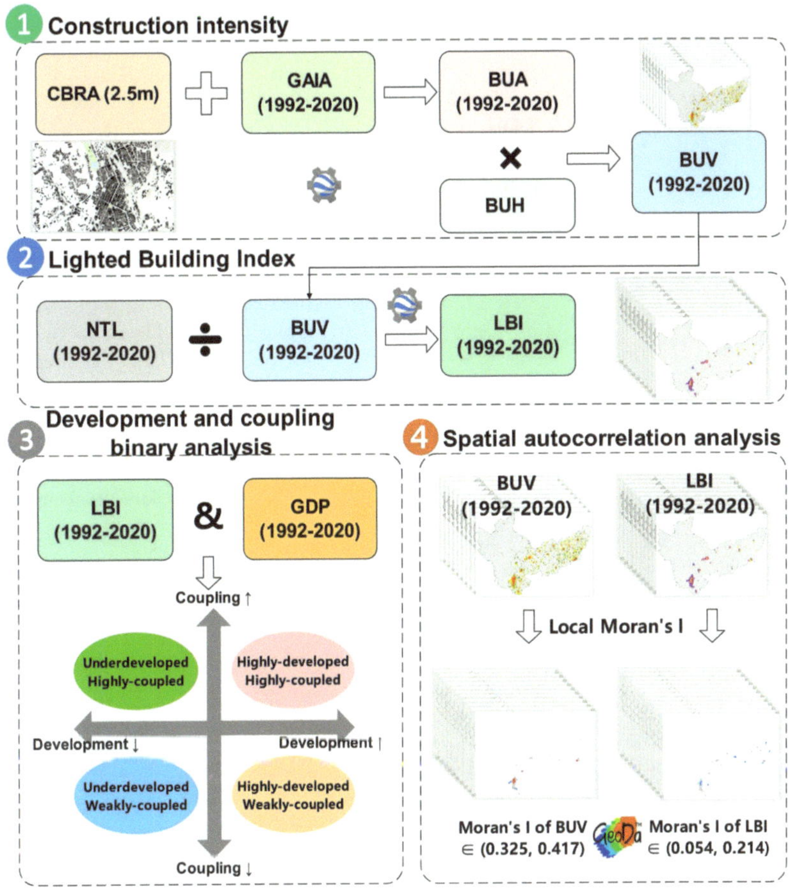

Figure 2. Technical framework.

2.2.1. Construction Intensity of Urban Built-Up Land

We used the 2021 CBRA to represent the existing rooftop areas as of that year. Since GAIA indicates the year of construction for buildings (ranging from 1985 to 2022 on an annual basis), overlaying the CBRA with GAIA allows us to assign the construction date information from GAIA to the corresponding locations in the CBRA. This enables us to obtain rooftop area data for any desired analysis year, resulting in Built-up Area (BUA) for every two-year interval from 1992 to 2020. The intensity of urban construction land use is represented by Built-up Volume (BUV) [31,32]. BUV is calculated as the product of the Built-up Area (BUA) on the horizontal plane and the Built-up Height (BUH) on the vertical plane within a 1 km by 1 km grid. The formula is as follows:

$$BUV = BUA \times BUH, \quad (1)$$

where BUA represents the total area of building rooftops within a 1 km grid for each year, measured in square meters (m^2); BUH represents the average height of buildings within the 1 km grid, measured in meters (m); BUV represents the total volume of buildings within a 1 km grid for each year, measured in cubic meters (m^3). BRA refers to the Constructed Building Rooftop Area with a 2.5 m resolution data set within the 1 km grid for each year. TA is the total area of the 1 km grid. Calculations are performed for every even year between 1992 and 2020, encompassing a total of 15 years of data.

2.2.2. Lighted Building Index

We utilize Nighttime Light data (NTL) to characterize the actual utilization of urban construction land within each 1km grid cell. The Lighted Building Index (LBI) is a metric we have developed in this study to reflect the efficiency of light utilization per unit of construction volume. It is calculated by taking the ratio of Nighttime Lights (NTL) to Built-up Volume (BUV). This study employs LBI to quantify the coupling between construction intensity and actual utilization, attempting to conduct a long-term coupling analysis from 1992 to 2020 (Equation (3)).

$$LBI = \frac{NTL}{BUV}, \quad (2)$$

where LBI represents Lighted Building Index (unit: DN value/m^3); NTL refers to Nighttime Light data, which is typically derived from the fusion of two commonly used remote sensing sources, DMSP-OLS and NPP-VIIRS. BUV represents the total volume of buildings within a 1 km grid for each year, calculated in Section 2.2.1. To mitigate the impact of scattered pixels of nighttime illumination in suburban areas during the calculation of the Landscape Biodiversity Index (LBI), the study area is further refined. By identifying natural urban boundaries, a spatially contiguous urban space is selected, thereby avoiding interference from scattered impervious surface pixels in agricultural and ecological spaces.

2.2.3. Bivariate Quadrant Trajectory Analysis

We conducted an integrated analysis of the binary relationship between the total GDP and the Lighted Building Index (LBI) in Hegang City from 1992 to 2020. The two-dimensional space defined by these two indicators was divided into four quadrants, representing 'Highly developed, Highly coupled', 'Highly developed, Weakly coupled', 'Underdeveloped, Highly coupled', and 'Underdeveloped, Weakly coupled'. By performing pixel-by-pixel time series analysis and spatial time series analysis on the GDP and LBI data for every other year over the 1992–2020 period, spanning 15 phases, we mapped out the developmental trajectories across different years. From these analyses, we summarized the evolution of the relationship between coupling and development.

2.2.4. Spatial Autocorrelation Analysis

Spatial autocorrelation analysis, including Moran's I and Local Moran's I, are utilized to examine the variability in spatial distribution of Built-up Volume (BUV) and the Lighted Building Index (LBI). Moran's I assesses the overall spatial clustering within the data, identifying whether values are concentrated in specific areas or dispersed randomly. Local Moran's I, on the other hand, identifies specific spatial units with high or low values relative to their neighbors, revealing distinct spatial association patterns. The calculation of Moran's I is as follows:

$$I = \frac{n}{\sum_{i=1}^{n}\sum_{j=1}^{n} w_{ij}} \times \frac{\sum_{i=1}^{n}\sum_{j=1}^{n} w_{ij}(x_i - \overline{x})(x_j - \overline{x})}{\sum_{i=1}^{n}(x_i - \overline{x})^2}, \quad (3)$$

where I represents Moran's I, which is calculated based on the number of spatial units n within the study area. The value of the ith spatial unit is represented by x_i, while $\overline{x}$ represents the average value across all units. The matrix of spatial weights is denoted by w_{ij}. The selection of the weight matrix is to serve the purpose of explaining spatial autocorrelation. In this study, we considered the use of the adjacency matrix, which effectively reflects the adjacency relationships between observations in spatial data. Specifically, the adjacency matrix is defined based on the boundary contact between regions. If two regions share a boundary, we set w_{ij} = 1; otherwise, w_{ij} = 0.

The calculation of Local Moran's I is as follows:

$$I_i = \frac{x_i - \overline{x}}{\left(\frac{\sum_{i=1}^{n}(x_i - \overline{x})^2}{n}\right)} \times \sum_{i=1}^{n}\sum_{j=1}^{n} w_{ij}(x_j - \overline{x}), \quad (4)$$

where the Local Moran's I index for the ith spatial unit, denoted as I_i, uses the same parameters as previously defined for the Moran's I calculation.

3. Results

3.1. Spatiotemporal Construction Intensity of Urban Built-Up Land

Integrating the horizontal expansion and vertical height of construction land, the results of Built-up Volume (BUV) are illustrated in Figure 3. The construction intensity in Hegang city has been increasing year by year from 1992 to 2020, characterized by a gradual increase in the number of pixels containing buildings and a year-on-year increase in the construction intensity of existing building pixels. Spatially, the northwest region, dominated by the mountainous area of the Lesser Khingan Range, has rugged terrain and less construction land; the southern main urban area is the region with the highest construction intensity in the city; the eastern region is a grain planting development area, mainly plain, with scattered patches of construction land. Over time, the average BUV in Hegang has risen from 204.5×10^3 m^3 in 1992 to 410.3×10^3 m^3 in 2020, an increase of 200.6% over 28 years. Based on the natural break method, construction land is divided into 10 categories. Taking the BUV distribution in 1992 and 2020 as examples, the top 1 category of BUV values (BUV > 240×10^3 m^3) is mainly concentrated in the central area of Hegang city and the centers of various counties, accounting for 22.2% of the total construction land area in 1992 and 45.5% in 2020; In contrast, the lowest category of BUV values (BUV < 1×10^3 m^3) is mainly sporadically found in rural construction land, accounting for 13.5% of the total construction land area in 1992 and 1.9% in 2020. Construction land with moderate BUV exists in the transitional areas between the aforementioned two, that is, the urban–rural fringe areas on the edge of the city. The construction intensity has surged over the span of 28 years, affecting not only the city center but also the suburban areas and the scattered rural construction sites.

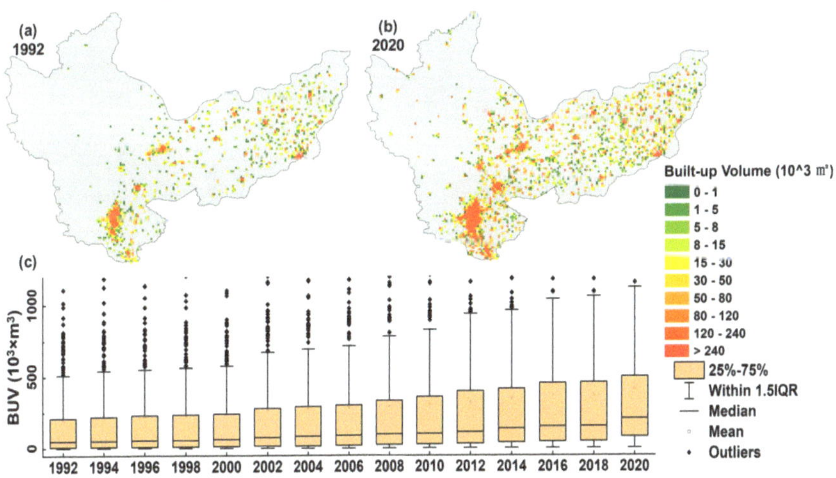

Figure 3. Spatiotemporal distribution of Built-up Volume (BUV) in Hegang city from 1992 to 2020. (**a**,**b**) represent the spatial distribution of BUV for construction land in Hegang for the year 1992 and 2020, respectively; (**c**) BUV statistics of build-up land pixels per year.

3.2. The Coupling Between Construction and Actual Utilization

The Lighted Building Index (LBI) in Hegang city has shown an overall trend of increasing initially and then decreasing (Figure 4), representing a pattern where the coupling degree between urban construction volume and actual utilization first increased and then decreased. Between 1992 and 1998, both the average and median LBI values in Hegang increased significantly. However, from 2000 to 2020, there was a noticeable decline. By 2020, the mean LBI had dropped to 1.68×10^{-3}, which is only 40.5% of its value in 1992

(4.15×10^{-3}), indicating a severe phenomenon of urban shrinkage in Hegang during the 21st century. By 2020 (Figure 4b), the pixels with the lowest LBI values were concentrated in the central area of Hegang city, suggesting that the city center, which should be the most vibrant, is now the most affected by urban shrinkage. This central decline in the LBI is indicative of a larger trend where the areas that were once economic and social hubs are now experiencing a significant reduction in activity and investment. The concentration of low LBI values in the city center could be a result of various factors such as depopulation, economic downturn, or a shift in economic activities to other regions within or outside the city. Moreover, Figure 4c tracks the changes in the LBI over the 28-year period for the top 10 1 km pixels with the highest LBI values in 1992. The top three pixels experienced a dramatic decrease in LBI after the turn of the century. The combined LBI of the top 10 pixels was 1.35 in 1992, with an average value of 0.13; by 2020, this sum and average had reduced to 0.18 and 0.02, respectively. These figures represent only 13.3% and 15.4% of their values in 1992, indicating a more severe decline than the overall mean LBI value for all pixels in Hegang.

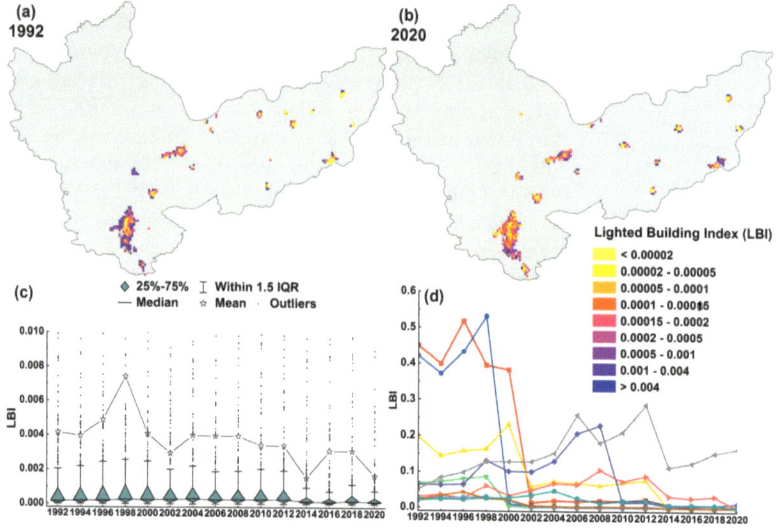

Figure 4. Spatiotemporal distribution of the Lighted Building Index (LBI) in Hegang city from 1992 to 2020. (**a**,**b**) represent the spatial distribution of the LBI for construction land in Hegang for the year 1992 and 2020, respectively; (**c**) LBI statistics of build-up land pixels per year; (**d**) Temporal changes in the LBI from 1992 to 2020 for the top 10 pixels with the highest LBI values in Hegang in 1992.

3.3. *The Evolution of the Relationship Between Coupling and Development*

The average LBI and the total GDP from 1992 to 2020 can reflect the urban land-use coupling degree and the level of economic development, respectively (Figure 5a). Over the period from 1992 to 2020, the coupling degree generally showed a downward trend, while GDP, despite a decline in 2014, has generally been on an upward trajectory. In terms of spatial correlation, Figure 5b shows that pixels with higher GDP in the city center have a lower LBI, whereas pixels with lower GDP in the urban–rural fringe and suburban areas have a higher LBI. This phenomenon may be attributed to excessive construction in the city center due to overconfidence on the part of local governments and fiscal incentives.

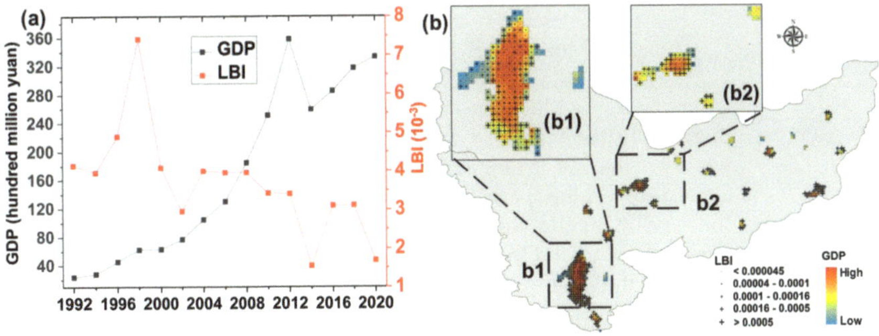

Figure 5. The spatiotemporal differentiation of GDP and LBI in Hegang. (**a**) Time series changes in GDP and LBI from 1992 to 2020; (**b**) The spatial correspondence between GDP and LBI; (**b1**) The center of Hegang City; (**b2**) Sub-center of Hegang City.

We utilized the biennial LBI and GDP of Hegang as the two axes, dividing the space into four quadrants, which represent 'Highly developed, Highly coupled', 'Highly developed, Weakly coupled', 'Underdeveloped, Highly coupled', and 'Underdeveloped, Weakly coupled'. Based on the changes in the two factors for Hegang each year, we can delineate its position within the four quadrants. By connecting these positions over time, we can illustrate the transformation trajectory of Hegang over the past 28 years. Overall, Hegang has experienced three developmental stages from 1992 to 2020: "Underdeveloped, Highly coupled", "Underdeveloped, Weakly coupled", and "Highly developed, Weakly coupled" (Figure 6). Initially, before the 21st century, Hegang was limited by traditional economic constraints, resulting in a low GDP, lower construction intensity, and lower urban building utilization. However, the Lighted Building Index (LBI), an indicator of the utilization per unit of construction volume, was relatively high, suggesting a state of low-level coupling. Thereafter, in the first decade of the 21st century, driven by the broader economic environment in China, Hegang's economy began to improve rapidly. With the incentive of land finance, a large number of new buildings were constructed, but actual utilization did not keep pace with construction intensity, reflected by a decrease in the LBI, indicating a state of low-level decoupling. After 2010, Hegang's economy further improved, with the total GDP exceeding 200 billion yuan. Although GDP declined in 2014, it can be generally recognized that the city entered a high-level stage. However, facing significant population outflow and urban shrinkage, LBI saw a sharp decline, with the 2020 LBI being approximately one-quarter of that in 2010, indicating a state of high-level decoupling. Therefore, in the development process over the past 30 years, Hegang has gone through three stages, with the notable absence of a high-level coupling state.

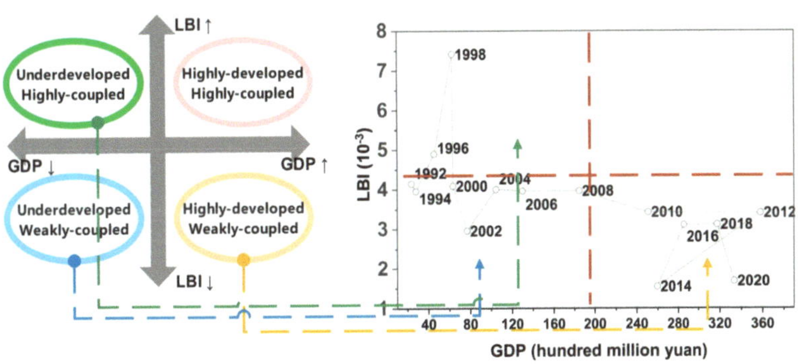

Figure 6. The transformation pathways of Hegang between four coupling–development models from 1992 to 2020.

3.4. Spatial Autocorrelation

Based on Moran's I, BUV and the LBI are positively spatially correlated, respectively, i.e., the higher values of BUV are associated with a greater tendency for pixels to cluster, while the lower LBI values are associated with a greater tendency for pixels to cluster. Specifically, the Moran's I value for BUV from 1992 to 2020 ranges from 0.325 to 0.417, while for the LBI it ranges from 0.054 to 0.214. The results from the Local Moran's I analysis reveal the spatial clustering of Built-up Volume (BUV) and the Lighted Building Index (LBI), highlighting the spatial characteristics of built-up land construction intensity (Figure 7a,c,e,g) and the coupling between building and utilization (Figure 7b,d,f,h). Both BUV and the LBI exhibit spatial agglomeration. The spatial pattern of agglomeration for BUV and the LBI is such that they share the same spatial locations but exhibit opposite values. This implies that pixels which are high-high clusters for BUV are, in contrast, low-low clusters for the LBI. This phenomenon illustrates that areas with high construction intensity, primarily centered around the city center, are paradoxically areas of low coupling, as shown in the low LBI values. This is corroborated by the GDP high-value areas in 2020 (as depicted in Figure 5b), which also coincide with lowLBI-value areas, illustrating a disconnect between the physical space and economic vitality in the city of Hegang from two different perspectives.

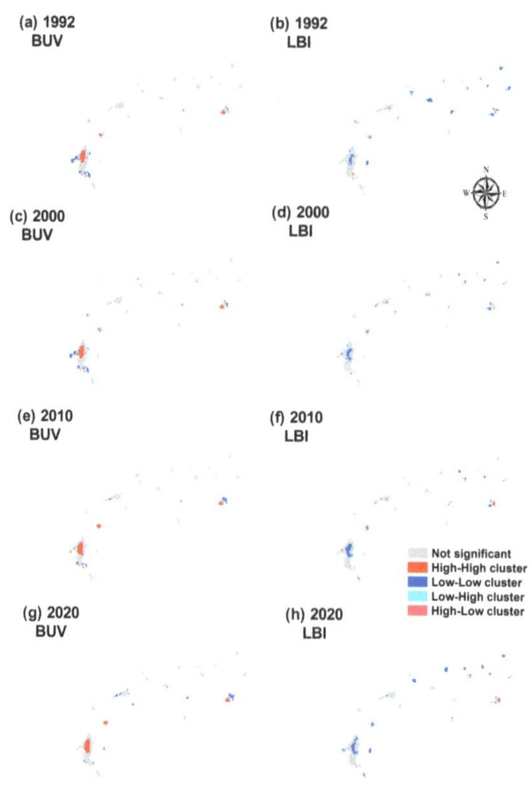

Figure 7. Local Moran's I of BUV and LBI in 1992, 2000, and 2020.

4. Discussion

4.1. Analyzing Longitudinal Urbanization from the Perspective of Spatial Coupling

Relying on long-time, high-resolution remote sensing images, the aim of this study is to analyze the spatial coupling between the physical space of a shrinking city and its actual utilization. While gridded GDP data inevitably introduce errors during their production and distribution process, making them less accurate in reflecting utilization intensity compared to nighttime light data (NTL), NTL, on the other hand, consistently show a robust positive association with population growth and economic vitality [24] and exhibit a strong link to various urban dynamics, including urban form and growth patterns [25], energy usage [26], and carbon emission levels. We employ a 1 km resolution NTL dataset spanning from 1992 to 2020. This dataset is instrumental in analyzing the actual utilization corresponding to the Built-up Volume (BUV) within 1 km grids. Our methodology incorporates a biennial monitoring frequency, an approach that adeptly captures temporal dynamics without redundancy, thus making it scalable to broader regions. Additionally, we have scaled up 2.5 m resolution rooftop data to align with the 1 km grid scale of our NTL data. This upscaling process is critical as it harmonizes with the NTL dataset and minimizes granular errors, ensuring a more accurate representation of BUV. In the calculation of BUV within the 1 km grids, the precision of horizontal building footprints is paramount, significantly influencing the determination of building volume. Recognizing the absence of Chinese regional building contour data in several globally recognized datasets, such as Microsoft's [33] and Google's dataset [34], we have opted for 2.5 m resolution building rooftop data [27]. These higher-resolution data provide a precise delineation of each building's outline, enhancing the precision of construction intensity calculations. By comparing the LBI and GDP within the identical 1 km grid locations, we qualitatively categorized each grid within a binary framework of development and coupling. Utilizing time series data from 1992 to 2020, we have classified the urbanization states of individual grids and the city as a whole over this period. Our findings reveal that Hegang has traversed an urbanization trajectory characterized by stages of "Underdeveloped, Highly coupled", "Underdeveloped, Weakly coupled", and "Highly developed, Weakly coupled", now confronting an escalated risk of urban contraction. Previous studies typically employ panel data to calculate urban land-use efficiency within the boundaries of an administrative division by integrating various input–output indicators. However, panel data are characterized by temporal lag and are constrained by administrative boundaries, which limit the richness of information over time and across geographic areas. This study addresses these challenges by utilizing multi-source long-term and high-resolution remote sensing data, which not only sets a precedent for urbanization studies but also lays a foundational data groundwork for future land-use planning in Hegang, offering a replicable template for urbanization path analysis in other cities.

4.2. Policy Implications

Over the past 28 years, the construction intensity of Hegang city has continuously increased, yet the actual utilization per unit of building space, as indicated by the Lighted Building Index (LBI), has been decreasing throughout this century. Moreover, the spatial heterogeneity demonstrates that areas with high construction intensity, primarily centered around the city center, are paradoxically areas of low coupling, as shown in the low LBI values. This is corroborated by the GDP high-value areas in 2020, which also coincide with low-LBI-value areas, illustrating a disconnect between the physical space and economic vitality in the city of Hegang from two different perspectives. Temporally, the degree of mismatch between the city's physical space and its economic vitality has been escalating, with Hegang facing some of the most severe urban shrinkage issues observed both nationally and globally. The Moran's I value for BUV and the LBI both increased, demonstrating that spatial heterogeneity within the administrative region of Hegang is becoming more severe. Against this backdrop, the Hegang government's continued reliance on traditional incremental planning is likely to exacerbate the disconnection between people and the land in the future. Looking ahead, the local government of Hegang should shift from the traditional land finance-driven extensive growth model to a more refined and sustainable urban development strategy. Like many other resource-oriented cities such as Karamay [35] and Datong [36], Hegang needs to transition into a more compact city. This includes enhancing the optimization of existing building spaces, improving the intensity and efficiency of actual use of construction and focusing on elevating the quality of life for residents to cre a more livable urban environment. The government must prioritize a balance betwe urban planning and land use, curb unregulated expansion, and mitigate the disconnectio between urban spaces and human-land relationships. Through these measures, Hegan can gradually alleviate the pressures of urban contraction and forge a path of sustainable development that meets the demands of the new era. This study provides insights for the development and construction of second-tier cities, as well as for the allocation of national land use indicators.

5. Conclusions

This study utilizes long-term, high-resolution remote sensing data to focus on the urbanization process in Hegang city from 1992 to 2020. By clarifying the coupling degree between the intensity of infrastructure construction (BUV, which has seen a continuous rise with a 200.06% increase over 28 years) and the city's socioeconomic vitality (NTL), we conducted a quantitative analysis of the Lighted Building Index (LBI) across each 1 km grid cell over the past 28 years. Our findings indicate that the LBI initially increased and subsequently decreased, with a significant drop in the 21st century, suggesting a persistent decline in the utilization of buildings and a notable urban shrinkage phenomenon. Additionally, the study uses GDP as an economic output indicator and qualitatively categorizes each grid within a binary framework of development and coupling. This methodology has allowed us to delineate a characteristic urbanization state curve for Hegang, reflecting its progression through stages of "Underdeveloped, Highly coupled", to "Underdeveloped, Weakly coupled", and finally to "Highly developed, Weakly coupled", offering insights into its urban development path. The spatial autocorrelation results reveal that the local Moran's I for BUV and the LBI share the same spatial locations but exhibit opposite values. That is, pixels identified as high-high clusters for BUV are, conversely, low-low clusters for the LBI, indicating that the central urban construction areas are the most severely affected by hollowing out and the emergence of ghost towns. Accordingly, we provide recommendations tailored to the spatial heterogeneity of construction intensity and coupling degree.

This study, while offering novel insights, has some limitations. For example, the Nighttime Light (NTL) data we relied upon are susceptible to the spatial saturation effect, which can overstate the brightness levels in areas with lower nighttime illumination. Moreover, the NTL values, being based on artificial light emissions, may not accurately reflect the economic activity in industrial zones and commercial districts with minimal nighttime activity, potentially leading to an underestimation of their actual utilization. Moreover, building rooftop data are selected as the focus of this research; however, the study does not differentiate between urban and rural rooftops. Consequently, a notable portion of rural built-up land with low BUV is inadvertently included in the analysis. In future work, we will refine our methodology to mitigate the limitations of NTL data, employing additional datasets and analytical techniques to provide a more accurate and comprehensive assessment of urban spatial dynamics. Moreover, we plan to expand our research framework to include a broader geographical scope, aiming to analyze urban coupling degrees across different cities on a regional and national scale. This will enable a comparative study of urban development trends and spatial coupling evolution over the past three decades, offering valuable data for urban planning and policymaking, steering towards sustainable urban growth. Overall, this research offers technical guidance for evaluating coupling by integrating multi-source spatial data, establishing a foundational data groundwork for future land-use planning in Hegang. It also presents a replicable template for urbanization path analysis in other cities, contributing to a broader understanding of urban development dynamics.

Author Contributions: Conceptualization, S.H. and C.W.; methodology, Y.L.; software, S.H. and Y.L.; validation, S.H., Y.L. and H.S.; formal analysis, S.H.; data curation, H.S.; writing—original draft preparation, S.H.; writing—review and editing, Y.L.; visualization, H.S.; supervision, C.W.; project administration, C.W. All authors have read and agreed to the published version of the manuscript.

Funding: This research was funded by The National Natural Science Foundation of China, Grant No. 42401120 and Zhejiang Provincial Natural Science Foundation of China, Grant No. Q24D010026.

Data Availability Statement: Data will be made available on request.

Conflicts of Interest: The authors declare no conflict of interest.

References

1. Ruan, L.; He, T.; Xiao, W.; Chen, W.; Lu, D.; Liu, S. Measuring the Coupling of Built-up Land Intensity and Use Efficiency: An Example of the Yangtze River Delta Urban Agglomeration. *Sustain. Cities Soc.* **2022**, *87*, 104224. [CrossRef]
2. Buhaug, H.; Urdal, H. An Urbanization Bomb? Population Growth and Social Disorder in Cities. *Glob. Environ. Change* **2013**, *23*, 1–10. [CrossRef]
3. Dadashpoor, H.; Azizi, P.; Moghadasi, M. Land Use Change, Urbanization, and Change in Landscape Pattern in a Metropolitan Area. *Sci. Total Environ.* **2019**, *655*, 707–719. [CrossRef] [PubMed]
4. Hatuka, T.; Zur, H.; Mendoza, J.A. The Urban Digital Lifestyle: An Analytical Framework for Placing Digital Practices in a Spatial Context and for Developing Applicable Policy. *Cities* **2021**, *111*, 102978. [CrossRef]
5. Hennig, E.I.; Schwick, C.; Soukup, T.; Orlitová, E.; Kienast, F.; Jaeger, J.A.G. Multi-Scale Analysis of Urban Sprawl in Europe: Towards a European de-Sprawling Strategy. *Land Use Policy* **2015**, *49*, 483–498. [CrossRef]
6. Seto, K.C.; Fragkias, M.; Güneralp, B.; Reilly, M.K. A Meta-Analysis of Global Urban Land Expansion. *PLoS ONE* **2011**, *6*, e23777. [CrossRef]
7. Bai, X.; McPhearson, T.; Cleugh, H.; Nagendra, H.; Tong, X.; Zhu, T.; Zhu, Y.-G. Linking Urbanization and the Environment: Conceptual and Empirical Advances. *Annu. Rev. Environ. Resour.* **2017**, *42*, 215–240. [CrossRef]
8. Deng, X.; Huang, J.; Rozelle, S.; Zhang, J.; Li, Z. Impact of Urbanization on Cultivated Land Changes in China. *Land Use Policy* **2015**, *45*, 1–7. [CrossRef]
9. Sun, R.; Lü, Y.; Yang, X.; Chen, L. Understanding the Variability of Urban Heat Islands from Local Background Climate and Urbanization. *J. Clean. Prod.* **2019**, *208*, 743–752. [CrossRef]
10. Sutradhar, U.; Spearing, L.; Derrible, S. Depopulation and Associated Challenges for US Cities by 2100. *Nat. Cities* **2024**, *1*, 51–61. [CrossRef]
11. Jiang, Z.; Zhai, W.; Meng, X.; Long, Y. Identifying Shrinking Cities with NPP-VIIRS Nightlight Data in China. *J. Urban Plan. Dev.* **2020**, *146*, 04020034. [CrossRef]
12. Niva, V.; Horton, A.; Virkki, V.; Heino, M.; Kosonen, M.; Kallio, M.; Kinnunen, P.; Abel, G.J.; Muttarak, R.; Taka, M.; et al. World's Human Migration Patterns in 2000–2019 Unveiled by High-Resolution Data. *Nat. Hum. Behav.* **2023**, *7*, 2023–2037. [CrossRef] [PubMed]
13. Li, Y.; Pu, S.; Zhao, Y.; Yao, J.; Qian, K. Refractive-Index-Matched Coupling Generated by Magnetic Fluid around Cladding Removed Multimode Optical Fiber. *Opt. Fiber Technol.* **2020**, *59*, 102326. [CrossRef]
14. Masini, E.; Tomao, A.; Barbati, A.; Corona, P.; Serra, P.; Salvati, L. Urban Growth, Land-Use Efficiency and Local Socioeconomic Context: A Comparative Analysis of 417 Metropolitan Regions in Europe. *Environ. Manag.* **2019**, *63*, 322–337. [CrossRef]
15. Jiang, H. Spatial–Temporal Differences of Industrial Land Use Efficiency and Its Influencing Factors for China's Central Region: Analyzed by SBM Model. *Environ. Technol. Innov.* **2021**, *22*, 101489. [CrossRef]
16. Zhu, X.; Zhang, P.; Wei, Y.; Li, Y.; Zhao, H. Measuring the Efficiency and Driving Factors of Urban Land Use Based on the DEA Method and the PLS-SEM Model—A Case Study of 35 Large and Medium-Sized Cities in China. *Sustain. Cities Soc.* **2019**, *50*, 101646. [CrossRef]
17. Du, Y.; Li, B.; Chen, M. Surges Induced in Building Electrical Systems during a Lightning Strike. *Electr. Power Syst. Res.* **2016**, *139*, 68–74. [CrossRef]
18. Liu, S.; Xiao, W.; Li, L.; Ye, Y.; Song, X. Urban Land Use Efficiency and Improvement Potential in China: A Stochastic Frontier Analysis. *Land Use Policy* **2020**, *99*, 105046. [CrossRef]
19. Kussul, N.; Lavreniuk, M.; Skakun, S.; Shelestov, A. Deep Learning Classification of Land Cover and Crop Types Using Remote Sensing Data. *IEEE Geosci. Remote Sens. Lett.* **2017**, *14*, 778–782. [CrossRef]
20. Xu, M.; Yang, Y.; Deng, Q.-W.; Shen, J.-T.; Liu, W.-F.; Yang, W.-J.; Liu, K.-X. Microarray Profiling and Functional Identification of LncRNA in Mice Intestinal Mucosa Following Intestinal Ischemia/Reperfusion. *J. Surg. Res.* **2021**, *258*, 389–404. [CrossRef]
21. Esch, T.; Zeidler, J.; Palacios-Lopez, D.; Marconcini, M.; Roth, A.; Mönks, M.; Leutner, B.; Brzoska, E.; Metz-Marconcini, A.; Bachofer, F. Towards a Large-Scale 3D Modeling of the Built Environment—Joint Analysis of TanDEM-X, Sentinel-2 and Open Street Map Data. *Remote Sens.* **2020**, *12*, 2391. [CrossRef]
22. Wang, K.; He, T.; Xiao, W.; Yang, R. Projections of Future Spatiotemporal Urban 3D Expansion in China under Shared Socioeconomic Pathways. *Landsc. Urban Plan.* **2024**, *247*, 105043. [CrossRef]
23. Gong, P.; Li, X.; Wang, J.; Bai, Y.; Chen, B.; Hu, T.; Liu, X.; Xu, B.; Yang, J.; Zhang, W.; et al. Annual Maps of Global Artificial Impervious Area (GAIA) between 1985 and 2018. *Remote Sens. Environ.* **2020**, *236*, 111510. [CrossRef]
24. Chen, Z.; Yu, B.; Ta, N.; Shi, K.; Yang, C.; Wang, C.; Zhao, X.; Deng, S.; Wu, J. Delineating Seasonal Relationships Between Suomi NPP-VIIRS Nighttime Light and Human Activity Across Shanghai, China. *IEEE J. Sel. Top. Appl. Earth Obs. Remote Sens.* **2019**, *12*, 4275–4283. [CrossRef]

25. Zhang, Q.; Fang, K.; Chen, J.; Liu, H.; Liu, P. The Role of Sectoral Coverage in Emission Abatement Costs: Evidence from Marginal Cost Savings. *Environ. Res. Lett.* **2022**, *17*, 045002. [CrossRef]
26. Jasiński, T. Modeling Electricity Consumption Using Nighttime Light Images and Artificial Neural Networks. *Energy* **2019**, *179*, 831–842. [CrossRef]
27. Liu, Z.; Tang, H.; Feng, L.; Lyu, S. China Building Rooftop Area: The First Multi-Annual (2016–2021) and High-Resolution (2.5 m) Building Rooftop Area Dataset in China Derived with Super-Resolution Segmentation from Sentinel-2 Imagery. *Earth Syst. Sci. Data* **2023**, *15*, 3547–3572. [CrossRef]
28. Li, X.; Zhou, Y.; Zhao, M.; Zhao, X. Harmonization of DMSP and VIIRS Nighttime Light Data from 1992–2020 at the Global Scale. *Sci. Data* **2020**, *7*, 168. [CrossRef]
29. Li, X.; Gong, P.; Zhou, Y.; Wang, J.; Bai, Y.; Chen, B.; Hu, T.; Xiao, Y.; Xu, B.; Yang, J.; et al. Mapping Global Urban Boundaries from the Global Artificial Impervious Area (GAIA) Data. *Environ. Res. Lett.* **2020**, *15*, 094044. [CrossRef]
30. Chen, J.; Gao, M.; Cheng, S.; Hou, W.; Song, M.; Liu, X.; Liu, Y. Global 1 Km $\times$ 1 Km Gridded Revised Real Gross Domestic Product and Electricity Consumption during 1992–2019 Based on Calibrated Nighttime Light Data. *Sci. Data* **2022**, *9*, 202. [CrossRef]
31. Dutta, D.; Rahman, A.; Paul, S.K.; Kundu, A. Estimating Urban Growth in Peri-Urban Areas and Its Interrelationships with Built-up Density Using Earth Observation Datasets. *Ann. Reg. Sci.* **2020**, *65*, 67–82. [CrossRef]
32. Geiß, C.; Leichtle, T.; Wurm, M.; Pelizari, P.A.; Standfuß, I.; Zhu, X.X.; So, E.; Siedentop, S.; Esch, T.; Taubenböck, H. Large-Area Characterization of Urban Morphology—Mapping of Built-Up Height and Density Using TanDEM-X and Sentinel-2 Data. *IEEE J. Sel. Top. Appl. Earth Obs. Remote Sens.* **2019**, *12*, 2912–2927. [CrossRef]
33. GlobalMLBuildingFootprints. Available online: https://github.com/microsoft/GlobalMLBuildingFootprints (accessed on 10 August 2024).
34. Sirko, W.; Kashubin, S.; Ritter, M.; Annkah, A.; Bouchareb, Y.S.E.; Dauphin, Y.; Keysers, D.; Neumann, M.; Cisse, M.; Quinn, J. Continental-Scale Building Detection from High Resolution Satellite Imagery. *arXiv* **2021**, arXiv:2107.12283.
35. Peng, Y.; Fujiyama, M.; Ishida, T. Spatial Evolution of an Oil City: A Case Study of Karamay, Northwest China. *Land* **2022**, *11*, 2011. [CrossRef]
36. Xue, M.; Zhang, X.; Sun, X.; Sun, T.; Yang, Y. Expansion and Evolution of a Typical Resource-Based Mining City in Transition Using the Google Earth Engine: A Case Study of Datong, China. *Remote Sens.* **2021**, *13*, 4045. [CrossRef]

MDPI AG
Grosspeteranlage 5
4052 Basel
Switzerland
Tel.: +41 61 683 77 34

Land Editorial Office
E-mail: land@mdpi.com
www.mdpi.com/journal/land

www.ingramcontent.com/pod-product-compliance
Lightning Source LLC
Chambersburg PA
CBHW041403210926
51850CB00013B/3167
* 9 7 8 3 7 2 5 8 3 7 8 5 4 *